PHYSICS AND POLITICS IN REVOLUTIONARY RUSSIA

Studies of the Harriman Institute

CALIFORNIA STUDIES IN THE HISTORY OF SCIENCE

J. L. HEILBRON, Editor

The Galileo Affair: A Documentary History, edited and translated by Maurice A. Finocchiaro

The New World, 1939–1946 (A History of the United States Atomic Energy Commission, volume 1) by Richard G. Hewlett and Oscar E. Anderson, Jr.

Atomic Shield, 1947–1952 (A History of the United States Atomic Energy Commission, volume 2) by Richard G. Hewlett and Francis Duncan

Atoms for Peace and War, 1953–1961: Eisenhower and the Atomic Energy Commission (A History of the United States Atomic Energy Commission, volume 3) by Richard D. Hewlett and Jack M. Holl

Lawrence and His Laboratory: A History of the Lawrence Berkeley Laboratory, Volume 1, by J. L. Heilbron and Robert W. Seidel

Scientific Growth: Essays on the Social Organization and Ethos of Science, by Joseph Ben-David, edited by Gad Freudenthal

Physics and Politics in Revolutionary Russia, by Paul R. Josephson

STUDIES OF THE HARRIMAN INSTITUTE
Columbia University

Founded as the Russian Institute in 1946, the W. Averell Harriman Institute for Advanced Study of the Soviet Union is the oldest research institution of its kind in the United States. The book series *Studies of the Harriman Institute*, begun in 1953, helps bring to a wider audience some of the work conducted under its auspices by professors, degree candidates and visiting fellows. The faculty of the Institute, without necessarily agreeing with the conclusions reached in these books, believes their publication will contribute to both scholarship and a greater public understanding of the Soviet Union. A list of selected *Studies* appears at the back of the book.

PHYSICS AND POLITICS IN REVOLUTIONARY RUSSIA

Paul R. Josephson

University of California Press
Berkeley   Los Angeles   Oxford

Permission to reprint has been granted by the following publishers:

"Science Policy in the Soviet Union, 1917–1927," *Minerva*, vol. 26, no. 3 (Autumn 1988), 342–369.

"Physics and Soviet-Western Relations in the 1920s and 1930s." *Physics Today*, vol. 41, no. 9 (September 1988), 54–61.

"Physics, Stalinist Politics of Science and Cultural Revolution," *Soviet Studies*, vol. 11, no. 2 (April 1988), 245–265.

"Early Years of Soviet Nuclear Physics," *Bulletin of the Atomic Scientists*, vol. 43, no. 10 (December 1987), 36–39. From the *Bulletin of the Atomic Scientists*. Copyright (c) 1987 by the Educational Foundation for Nuclear Science, 6042 South Kimbark, Chicago, IL, 60637, USA.

University of California Press
Berkeley and Los Angeles, California

University of California Press
Oxford, England

Library of Congress Cataloging-in-Publication Data

Josephson, Paul R.
  Physics and politics in revolutionary Russia / Paul R. Josephson.
    p. cm.—(California studies in the history of science)
  Includes bibliographical references and index.
    ISBN 0-520-07482-3 (alk. paper)
    1. Physics—Soviet Union—History—20th
  century. 2. Physics—Research—Soviet Union—
  History. 3. Physics—Political aspects—Soviet Union—
  History. 4. Science and state—Soviet Union—
  History—20th century. 5. Science—Political aspects—
  Soviet Union—History. I. Title. II. Series.
    QC9.S65J67 1991
  306.4'5—dc20   91-16584
                                                            CIP

Printed in the United States of America

1 2 3 4 5 6 7 8 9

The paper used in this publication meets the minimum requirements of American National Standard for Information Sciences—*Permanence of Paper for Printed Library Materials, ANSI Z39.48-1984* ♾

To Cathy and Zack

Contents

Illustrations

APPENDIX B: THE LENINGRAD PHYSICO-
TECHNICAL INSTITUTE

APPENDIX C: PUBLICATION

Acronyms and Abbreviations

Dom Uchenykh	The Scholars' Club
DFTI	Dnepropetrovsk Physico-Technical Institute
FMF, Fizmekhfak	Fiziko-mekhanicheskii fakul'tet
Glavnauka	Main Scientific Administration of Narkompros
GOI	Gosudarstvennyi Opticheskii Institute
Gosplan	State Planning Administration
GFTRI	Gosudarstvennyi Fiziko-Tekhnicheskii Rentgenologicheskii Institut
GRRI	Gosudarstvennyi Rentgenologicheskii i Radiologicheskii Institut
IEB	International Education Board
KEPS	Kommissiia po Izucheniiu Estestvennykh Proizvoditel'nykh Sil
LEFI	Leningrad Electrophysical Institute
LFTI	Leningrad Physico-Technical Institute
LFTL	Leningrad Physico-Technical Laboratory
LIKhF	Leningrad Institute of Chemical Physics
Narkommash	Narodnyi Komissariat Mashinostroeniia
Narkompros	Narodnyi Komissariat Prosveshcheniia
Narkomtiazhprom	Narodnyi Komissariat Tiazheloi Promyshlennosti
NTO	Nauchno-tekhnicheskii otdel Vesenkha
Rabfak	Workers' faculty (special school)
RAF	Russkaia Assotsiatsiia Fizikov
RFKhO	Russkoe Fiziko-khimicheskoe Obshchestvo
SARRPN	Svobodnaia Assotsiatsia dlia Razvitiia i Rasprostraneniia Polozhitel'nykh Nauk
SFTI	Siberian Physico-Technical Institute
Sovnarkom	Council of Peoples' Commissars
TsEKUBU	Tsentral'naia Komissiia po Uluchsheniiu Byta Uchenykh

UkFTI   Ukrainian Physico-Technical Institute
Vesenkha   Vysshii Sovet Narodnogo Khoziaistvo (Supreme Economic Council)
Vydvizhenie   "Advancement" of workers during Cultural Revolution

Plates

Plates

Acknowledgments

I have had the good fortune to have the support of a large number of individuals and institutions in the production of this book at various stages along the way. Loren Graham provided lucid criticism and generous guidance. Cathy Frierson offered astute advice on the substance of the matter and helped turn cumbersome prose into readable passages. George Hoberg patiently explained cost-benefit analysis or played softball with me as appropriate. Peter Buck was extremely important in providing ideas and input in the early stages of my graduate education and this project.

Erwin Heibert and Donald Blackmer served on the dissertation committee which predated this book, and helped me to understand physics (Heibert) and politics (Blackmer) better. John Heilbron, Mark Adams, and Spencer Weart commented on earlier drafts and offered a number of suggestions which I hope have improved the final product. I would also like to thank Roberto Clemente, Roe Smith, Thane Gustafson, Mark Kuchment, and Vlad Toumanoff for providing inspiration, advice, beer, or a combination of the three. It has also been a pleasure to work with the University of California Press. Elizabeth Knoll helped me to complete the manuscript with judicious advice in matters of style and substance. Shirley Warren saw to it that a polished manuscript made it to the typesetter without any delay.

The Program for Science, Technology, and Society at MIT provided a home for my earliest work in the history of science and the conception of this project. Carl Kaysen, the former director of the Program, encouraged or cajoled as appropriate. The administration of Sarah Lawrence College underwrote this book by allowing me to explore the history and politics of science in a supportive environment. A number of my colleagues at Sarah Lawrence College have also given encouragement. They include Ray Seidelman, Frank Roosevelt, Lydia Kesich, and Frank Randall, as well as others in the division of social sciences. They ensured that I adopted a multidisciplinary approach to my teaching and hence to this study.

My colleagues in physics and history of science in the USSR have seen this manuscript improve over time, providing cogent criticisms and sharing their ideas. Viktor Iakovlevich Frenkel' has been a true friend in this regard. Since we first began to correspond in 1978 we have carried on a long discussion about the history of Soviet physics. Dr. Frenkel' and his family have graciously opened their home to me during many long Leningrad winter nights. Academician V. M. Tuchkevich, formerly director of the Leningrad Physico-Technical Institute, ensured that my first stay in Leningrad was productive. Academician Zhores Ivanovich Alferov, director of the Leningrad Physico-Technical Institute and head of the Leningrad Scientific Center, somehow found the time to be a friend and gracious host to me, as well as the source of informative anecdotes. I am grateful to Zhores Ivanovich and the administration of the Institute. The photographs in this volume have been made available by the Leningrad Physico-Technical Institute.

My thanks also go to scholars at the Institute of History of Science and Technology of the Academy of Sciences in Moscow, and its Sector of the History of Physics. Vladimir Vizgin, Boris Iavelov, and Gennadii Gorelik read parts of my work and offered suggestions for improvements. Vladimir Kirsanov, Natalia Vdovchenko and Ol'ga Kuznetsova made me feel welcome among the community of historians of science in Moscow.

Academicians E. L. Feinberg and V. L. Ginzburg at the Physics Institute of the Academy of Science and L. P. Pitaevskii at the Institute of Physical Problems also helped make my research proceed more smoothly, and I benefited from their counsel.

A number of different organizations supported the research leading to this book. The American Institute of Physics provided grants-in-aid. The International Research and Exchanges Board helped me to the Soviet Union on two occasions (1984–85 and 1985–86), the second time with funding from the Fulbright-Hays Doctoral Dissertation Program. I participated in the U.S.-U.S.S.R. Academy of Sciences Exchange Program in the Fall of 1989 through the kindness of the Office of Soviet and East European Affairs of the National Academy of Sciences.

The directors of the Archive of the Academy of Sciences in Moscow and its Leningrad Division, the Archive of Moscow State University, Archive of the Physics Institute of the Academy of Sciences, Archive of the Leningrad Physico-Technical Institute, Archive of the Institute of Physical Problems, and the Rockefeller Foundation Archive all willingly facilitated my research. The staff of the Manuscript Division of the Saltykov-Shchedrin Public Library in Leningrad and the library of the Institute of Scientific Information on the Social Sciences (INION) also facilitated my research.

These individuals and institutions were helpful in such a way that I am already looking forward to my next project. I hope the reader feels the same way.

Sarah Lawrence College
Bronxville, NY

Introduction

This book is about science and politics. It grows out of a tradition which argues that it is impossible to separate scientific research from its social, cultural, and political context. It describes the development of the physics discipline in Soviet Russia between 1900 and World War II, focusing on the relationship between physicists and the state, and the changes in science policy brought about by political revolution—the downfall of the Tsarist regime, the Bolshevik seizure of power in 1917, and the institution of Stalinist policies toward science in the 1930s.

The Soviet Union is among world leaders in solid-state and semiconductor, space, nuclear, elementary particle, and theoretical physics. The work of I. E. Tamm, A. D. Sakharov, and L. A. Artsimovich in controlled thermonuclear synthesis led to the development of the tokamak fusion reactor; Soviet elementary particle physicists have constructed some of the largest particle accelerators in the world and contributed to our understandings of the ultimate constituents of matter; in low-temperature physics and superconductivity such scholars as L. D. Landau and P. L. Kapitsa have made pioneering discoveries; N. G. Basov and A. M. Prokhorov received the Nobel Prize in 1964 for their work in quantum electrodynamics and the creation of the laser; and A. F. Ioffe created a world-class institute for the study of the solid state and semiconductors.

The Soviet Union today possesses one-fifth of the world's physicists and more scientific workers than any other country. In spite of these achievements and the sheer size of the Soviet research apparatus, the physics establishment labors under externally imposed constraints on the productivity and efficiency of research: excessive interference of Party and governmental organizations in the administration and funding of research, overemphasis on applied research at the expense of fundamental science, poorly managed

1

resources and overlap of personnel and projects, and, until recently, constant ideological supervision. The result is that Soviet science and technology lag behind Western science even in areas designated as priorities, and often fail to keep a lead in such areas as tokamaks where Soviet scholars have been pioneers.

The roots of many of these problems are found in the politics of science before World War II. This book describes how scientists and policymakers tried to reach accommodation over the issue of how best to organize, support, and administer fundamental physics research; how physicists strived to maintain a wide degree of latitude in the direction of research programs; and how the state and Communist party gradually placed the research-and-development (R and D) apparatus under the firm control of economic and political organs, resorting ultimately to coercion and terror in the late 1930s.

This investigation weighs the influence of social and cultural factors on the politics of science. It describes the history of the Leningrad physics community: the extent to which physicists shared certain common beliefs by virtue of specialized training; their expertise and its utility for the state; and their relationship to other groups and classes in a time of rapid social and economic upheaval. What was the position of such scientific specialists in Tsarist Russia? To what extent did the state embrace science and technology and its practitioners as a path to modernization? How did the role of the so-called Tsarist or bourgeois expert change after the October Revolution and the creation of a state founded on Marxian notions of workers' control of the means of production? In a word, this book evaluates the extent to which the aspirations of a scientific intelligentsia meshed with those of the state in light of the broader cultural context. The establishment of a series of large-scale government-financed institutes of fundamental and applied science provided the forum in which physicists and the government reached accommodation.

Since the turn of the twentieth century, science and technology have become increasingly expensive and equipment intensive, and the orientation of researchers has become increasingly collective. At the same time, governments have recognized the importance of supporting research and development for the purposes of national security, economic growth, and the health and welfare of their citizens. The government-funded central research institute is a modern phenomenon, and a case can be made that we can best understand the politics of twentieth-century physics through institutional history. This is certainly true of the Soviet Union, where institutes of the Academy of Sciences are the locus of fundamental scientific research.

Hence, an institutional history lies just beneath the surface of political, social, and cultural history in this book. The following chapters explore how research institutes were created in revolutionary Russia and the problems physicists faced in securing support, material, and equipment, in establishing foreign contacts, and in training promising young cadres. It examines the extent to which scientists and the government shared interests and goals, and where they came into conflict. Through the window of the Leningrad Physico-Technical Institute, it evaluates the impact of political, cultural, and scientific revolution on fundamental physics research.

The true source of power and influence of scientists vis-à-vis the state rested not in professional societies but within the walls of research institutes. The commensurability of the research programs of physicists and the economic interests of the state permitted them to secure financial stability for research without immediate applicability, and to embark on such promising new fields as nuclear physics. The structure of the institutes largely protected physicists and their research programs from encroachment by the state or by overzealous young Communist workers who saw them as a haven for bourgeois "remnants." On the eve of World War II, after four decades of tumultuous change within their discipline, Soviet physicists were among world leaders in nuclear, theoretical, and solid-state physics in a series of institutes created only since the Revolution.

The Leningrad Physico-Technical Institute (LFTI) played a central role in the development of Soviet physics, by supporting the establishment of a professional organization, the Russian Association of Physicists; by mediating disputes between the Bolshevik regime and physicists; in its pathbreaking role in establishing new areas of research; in the training of a future generation of physicists; and in resurrecting international contacts after the Revolution. From the start, within its walls and around its seminar tables the leading physicists of the USSR gathered. Fifty-two Academicians and twenty-two corresponding members of the Academy of Sciences have worked at LFTI at one time or another, including its founder and the dean of Soviet physicists, A. F. Ioffe; P. L. Kapitsa, founder of the Academy's Institute for Physical Problems in Moscow in 1937; I. V. Kurchatov, father of the Soviet atomic bomb project; A. P. Aleksandrov, president of the Academy of Sciences for thirteen years; and N. N. Semenov, director of the Institute of Chemical Physics and later vice-president of the Academy.

LFTI physicists dominated publication in Soviet Russia, publishing between one-quarter and three-fifths of all physics articles in the major Soviet journals every year between 1919 and 1939 (see app. C). Fifteen other scientific research institutes were formed from personnel and equipment split off from LFTI, including facilities in Tomsk, Kharkov, Dnepropetrovsk, Sverdlovsk, Tashkent, and indirectly the Kurchatov Institute of Atomic Energy. Tens of other institutes have been formed on the pattern of LFTI, including many of the institutes of Science City (Akademgorodok) in Novosibirsk. Soviet scholars refer to the Leningrad Physico-Technical Institute as the "cradle" of Soviet physics, and there is no reason to dispute this evaluation.

In the pages that follow, I tell the story of the Leningrad physics community: its efforts to gain recognition and support from the Tsarist government and the reasons for its limited successes; the impact of the abdication of the tsar on the scientific community and the inability of the Provisional government to support physicists at the level they required for research and professional activities; the absence of coherent science policy in the first years of the Soviet regime, and how physicists took advantage of the situation to establish a series of large-scale research institutes; and the revolutionary changes in the physics research enterprise brought about by the introduction of Stalinist science policies.

Between the turn of the century and 1940, the poorly funded, understaffed infant science of physics became the leading science in Soviet Russia, in terms of numbers of institutes, quality of scientists, and importance to government economic programs. Before 1917 there were no more than one hundred physicists and a handful of research laboratories. By 1940 there were, according to some estimates, over one thousand physicists in the USSR, and tens of well-equipped research institutes. The professional aspirations of physicists, thwarted throughout the Tsarist era, met with success in the early 1920s when they finally established a viable national association (Russkaia assotsiatsia fizikov, the Russian Association of Physicists) which represented their interests to the state.

The physics enterprise initially grew in fits and starts after the Revolution because of the political uncertainties of civil war, international isolation, and unsuccessful Bolshevik economic development programs. Scientists encountered the life-threatening physical and psychological obstacles of famine, cold, shortages of basic necessities, and the lack of the simplest equipment or publications. But the calming effect of the NEP (New Economic Policy) on the economy, the reestablishment of international relations, and the growing strength and stability of the Party facilitated the sustained growth of Soviet physics, the return of research and publication to normal, and the development of an international reputation. These were the golden years of Soviet physics when the Russian Association of Physicists met regularly in larger and larger congresses, frequently with foreign scholars in attendance, to discuss contemporary developments in physics as well as relations with the government.

Toward the end of the 1920s, budgetary constraints began to give way to constraints of another kind: pressure for research of an applied nature, often at the expense of fundamental science; centralization of science policy within bureaucracies whose sole purpose was to increase industrial production as rapidly as possible; and the introduction of five-year plans, all to put research in service of "socialist reconstruction." This marks the introduction of Stalinist policies toward science and the creation of an R and D apparatus in whose roots are the strengths and weaknesses of Soviet science and technology to this day. Physicists on the whole were in tune with these pressures, and in a number of cases not only endorsed an applied-scientific research agenda but called for its expansion. They participated in "socialist reconstruction": national programs in electrification, telecommunications and radio, machine building, heat engineering, construction, and metallurgy. And they participated in a number of policymaking bodies in the formulation of five-year plans for physics research.

At the end of the 1920s another assault on the autonomy of

Soviet physicists accompanied the introduction of Stalinist policies of science from above: cultural revolution from below. Physicists failed to recognize that their professional aspirations were inherently political, or that they would arouse the opposition of the Communist party and the increasingly militant young cadres who entered it in the late 1920s and 1930s. During cultural revolution, Soviet physicists, most of whom had been trained under the Tsarist system of education, came to be seen as "bourgeois remnants" whose interests invariably conflicted with those of the working class. They were the object of class war, were attacked for being aloof from Soviet economic development programs and politics, and faced constant pressure to cede control of their discipline to the Party and its representatives, based on notions of proletarian science.

The final aspect of political history to be discussed in this book concerns the scientific revolution in physics. In the Soviet case, the revolution in science brought about by relativity theory and quantum mechanics went far beyond the abandonment of prior conceptions of space, time, matter-energy, and causality required by the reworking of Newtonian or classical representations of physical phenomena into electromagnetic ones. This was because the notion of a qualitatively different "proletarian science" gained currency in many circles.

According to its adherents, proletarian science would arise in the socialist Soviet Union under the watchful eye of the Communist party. It differed from "capitalist" science since it served the interests of the masses and the government, not the private capitalist and the few; it was not exploitative; and because of the organic union of theory and practice, it led directly to applications, whereas bourgeois science encouraged "ivory tower reasoning." Even in terms of methodology and epistemology, proletarian science was progressive while bourgeois science was subservient to idealist tendencies and politically reactionary.

Initially, epistemological issues failed to ensnare physicists. Soviet physicists participated actively in the development of quantum mechanics and relativity theory, the new physics, as it was called, and were not unduly constrained by the epistemological problems raised by physical relativism, indeterminacy, and complementarity. But in Stalin's Russia, physicists and Marxist philosophers eventually became embroiled in a debate over the validity of new physical ideas within the dialectical materialist framework. While physicists were successful in avoiding some of the ideological interference which was so damaging to biology and genetics under T. D. Lysenko, a number of them were censured and removed from their posts, while others were shot or perished in forced labor camps.

In the 1930s, while the Leningrad Physico-Technical Institute grew and expanded in terms of cadres and research programs, its position as the major physics institute in Soviet Russia began to wane for several reasons. First, such new institutes as S. I. Vavilov's Physical Institute of the Academy of Sciences and P. L. Kapitsa's Institute of Physical Problems, both of which were in Moscow, began to rival its influence. More important, LFTI came under attack from both within and outside of the physics community. Such theoretical physicists as L. D. Landau criticized Ioffe for his alleged overattention to applied science. They also found fault with his "empire building" in the creation of a network of physico-technical institutes in Kharkov, Tomsk, Sverdlovsk, and Dnepropetrovsk, and with his imperious style of leadership.

Stalinist ideologues and Party officials shared this criticism of "empire building" and pressured the institute to conduct more research of an applied nature in line with the dictates of the five-year plans and the needs of heavy industry. During this period physicists lost much of their academic autonomy to the forces of cultural revolution, and faced pressure to bring more Communists within the walls of their institutes and subject research programs to Party scrutiny. Scientists at LFTI were also attacked at this time for philosophical idealism, failure to uphold the tenets of dialectical materialism, and for following the lead of Western science rather than establishing the leading position of Soviet physics in the world. The criticism culminated in a special session of the Academy of Sciences in March 1936.

The Great Terror of 1936 to 1938 during which millions upon millions perished is the final chapter in this story. The Terror led to the decimation of the Leningrad physics community and in particular its young theoreticians. Coupled with Leningrad's decline as a scientific and cultural center, the transfer of the Academy of Sciences to Moscow in 1934, the purge of the Leningrad Party apparatus, and the devastation of Leningrad's population and physical plant by German armies in World War II, the March 1936 session marked the end of LFTI's preeminence in the history of Soviet physics.

In many respects, this story is like that of physicists in any Western country. Germany, France, England, and the United States also saw such developments as the rapid changes in science under the influence of relativity theory and quantum mechanics, the creation of government-supported research institutes, and the displacement of chemistry by physics as the leading science. But the Soviet experience was unique for a number of reasons. International and domestic isolation, persistent government interference, and politicized philosophical debates led to obstacles nowhere else encountered, as this social, cultural, and political history of the Leningrad physics community will demonstrate. In sum, the history of Soviet physics provides a unique opportunity to examine the influence of external factors on the development of modern science.

1

The Politics of Tsarist Physics

Tsarist Russia was the last of the major powers to recognize the importance of science and technology for economic development, national security, and the health and welfare of its citizens. The reforms of the October Manifesto, brought about by the Revolution of 1905, which created a limited constitutional monarchy, did little to promote policies that supported the scientific enterprise. The governments of Germany, England, France, and the United States had recently begun to fund scientists on a larger scale, although only the experience of World War II would finally convince modern states of the importance of increasing annual budgetary allocations for research and development. But in Tsarist Russia, as the case of physics demonstrates, conservative statesmen actively inhibited the development of a national program for the support of science and technology, advancing shortsighted and often contradictory policies. They could not fathom the need for large-scale, well-funded, and centralized research institutes in modern society.

The elder generation of Russian physicists did little to dispel ingrained notions of the secondary importance of science and technology for modernization. Given the material backwardness of existing laboratories, and convinced of the need to follow European research programs rather than embarking on new ones, they adopted a pragmatic but narrow view of the importance of government financial support. They sought out funding for larger laboratories, new equipment, and several physics journals. But they failed to see that modern science required a new institutional setting, or that their physics association might advance their professional interests before government and society. Until new leadership developed within the community, physicists avoided confronting the government about the inadequacies of its position.

On the eve of the Revolution, however, a new generation of physicists, trained in both theoretical and experimental physics, took control of the discipline. They nurtured a vision of big science, organized on an increasingly collective and expensive scale, and proclaimed the need for indigenous research and development for reasons of personal pride, national patriotism, and international security. They recognized the need to create scientific institutes which would draw together researchers from a number of different subdisciplines whose efforts might complement theirs. Abram Feodorovich Ioffe led this new generation of physicists. His visions attracted promising young scholars to his laboratory. And after the fall of the Tsarist regime, he convinced the new Soviet government to support the creation of a massive new physics institute, the Leningrad Physico-Technical Institute, now known as the "cradle" of Soviet physics. This chapter looks at Tsarist policies toward physics research, its organizational setting on the eve of the Revolution, and the replacement of the elder generation by younger, more professionally oriented, and more farsighted physicists.

The problems caused by the absence of a national policy for physics notwithstanding, between 1900 and 1917 the physics enterprise began to grow. Both in Moscow and in Petersburg, under the leadership of such physicists as P. N. Lebedev, D. S. Rozhdestvenskii, and A. F. Ioffe, "schools" of research, the kernel of Soviet physics institutes, began to form. Drawing on the intellectual leadership, vision, and organizational abilities of their mentors, the "schools" offered a vital, if poorly funded, forum for physicists to discuss recent developments in physics. They embarked on research in atomic and molecular physics independent of the European tradition. Relying upon on an increasingly active professional association, they learned how to articulate their disciplinary and social concerns in the face of financial and organizational problems caused by the incompetence of the Tsarist government.

THE ORGANIZATION OF
PREREVOLUTIONARY PHYSICS

In 1900, physics in Russia, as in Europe, had only begun to develop as a discipline. It was poorly funded, often restricted in its focus of study to the physics of observation and measurement in such traditional organizations as the Imperial Academy of Sciences, and limited to a few major cities and research centers. The hesitancy of the Tsarist system to support physicists' research and professional activities forced them to turn increasingly to scientific societies and philanthropies. Especially after many of the more progressive faculty of Moscow University were purged in 1911, Russian scientists turned away from the Imperial government to other sources of funding. When called upon to join the Tsarist regime in the war effort against Germany, they nonetheless served willingly. They participated on joint government-academic advisory boards, and they developed new wartime industries in explosives, chemical and gas weapons, and instrument building.

Because of the political conservatism of the Academy of Sciences and its emphasis on study in the humanities, the university was the locus of physics research in the Russian Empire. It was in the university that an experimental and theoretical tradition developed, where P. N. Lebedev conducted his famous experiment on light pressure, A. A. Eikhenval'd performed a series of tests on magnetism, and N. A. Umov wrote a series of articles on early quantum theory. The physics of the Academy was largely limited to improvements in observational techniques and measurements, hence its major areas of excellence: geophysics and astronomy. But in terms of budget, the Academy commanded greater support from the government.

Founded by Peter the Great in 1725, the Academy of Sciences was a conservative body, dominated by humanists and traditional, older scholars. Only twenty-eight of seventy full members elected to the Academy of Sciences between 1890 and 1917 represented natural scientific disciplines, and there were only seven Academicians in physics or mathematics as of 1916. Members were often elected on the basis of allegiance to political ideals rather than scientific achievements. Two notorious examples of this tendency were the rejections of D. I. Mendeleev, creator of the periodic table, and the physicist A. G. Stoletov for membership in the Academy. In 1880, a group of Academicians led by the Petersburg mathematician P. L. Chebyshev and the organic chemist A. M. Butlerov nominated Mendeleev to become a full member of the Academy of Sciences. The uproar which accompanied the defeat of his nomination showed how pervasive was the view among scientists and society that the Academy was not a national scientific institution but dominated by foreign interests. The rejection of

Stoletov was particularly striking because the Academy elected in his place Prince B. B. Golitsyn, whose dissertation, *Issledovanie po matematicheskoi fiziki*, Stoletov himself had rejected in 1893, causing Golitsyn to turn to seismology. Until the election of A. P. Karpin-skii in 1917, the president of the Academy was a political appointee and rarely had a strong scientific background. This not only served to dampen any attempts to develop autonomy but indicates the effort of the government "to create an 'academic science,' science untouched by the contaminating influence of ideological conflict and social unrest." The result was inadequate support for complex research, and funding for projects which rarely reflected national scientific interest.[1]

The Academy supported physics well, however, in two areas: astronomy and seismology. After beginning seismological and spectrographic research, Prince Golitsyn sought government funding to transform the Academy's physical laboratory into the equal of any European one in terms of instrumentation. Between 1894 and 1914 he acquired over 300 new instruments to add to the existing 980, including a large spectrograph and an interference refractometer. In 1895 the budget of this laboratory was doubled to 2,000 rubles. While improving material conditions within the laboratory, Golitsyn was still unable to rival foreign physics institutions in terms of personnel and research, and therefore turned to seismographic work. By 1912 he had secured a yearly budget of 6,000 rubles, as well as a second assistant. To complement the work of the physical laboratory, in 1901 he established a seismic commission, which at first received over 10,000 rubles per year, and in 1910 received almost 75,000 rubles for equipment for seismic stations and a yearly budget of 46,912.[2] The Tsarist government also heavily supported the Academy's Main Physical Observatory. To the detriment of other areas of physics, the physical laboratory and observatory remained the only major Academy-sponsored physics research centers.

In spite of uneven support of the sciences in the period 1884–1917, three major changes occurred in the Academy of Sciences. First, the Academy, which had always been dominated by foreign, primarily German membership, "at long last became a national institution in both human composition and cultural orientation." Second, it began to grow in terms of institutional apparatus, laboratories, and personnel, and its ties with other scientific centers such as universities expanded. Last, although it required a national emergency—war with Wilhelmian Germany—the Academy became involved in applied science and other problems of national significance, with KEPS being the best example of this trend.[3]

Many of Russia's leading scientists, V. I. Vernadskii, A. N. Krylov, N. S. Kurnakov, A. P. Karpinskii, and I. P. Borodin among them, founded the Commission for the Study of the Productive Forces (Kommissiia po izucheniiu estestvennykh proizvoditel'nykh sil, or KEPS) in 1915 to study the natural resources of Russia, which at that time remained virtually unexplored. These resources assumed great importance in the face of the cutoff from the West of mineral ores and of material and equipment for the chemical, machine building, electrotechnical, and other infant industries during World War I. The commission studied the great mineral, geographical, and geological wealth of Russia, including the Kursk Magnetic Anomaly, and set out to identify and catalog flora and fauna.

Reflecting the growing belief among natural scientists that modern science required systematic government support, KEPS also addressed the issue of the need for a national body to coordinate the activity of the scientific forces of Russia. Several scholars suggested that the Tsarist government create a series of central scientific research institutes, perhaps modeled on the institutes of the Kaiser Wilhelm Gesellschaft.[4] But such efforts of science and government as KEPS to join forces were cut short by World War I and the Revolution. While on several occasions it had attempted to develop policies for the empire's scientific activity, the Academy of Sciences simply was not geared to be a national, coordinating body.

Universities, while more autonomous and somewhat better equipped, labored under subordination to the Ministry of Education and such conservative Tsarist ministers of education as D. A. Tolstoi (in the 1870s), I. D. Demianov (1882–1897), and L. A. Kasso (just before the war and Revolution). Still, universities are more important to the history of Russian physics than the Academy by virtue of their more extensive laboratories (although modest and rather primitive by European standards) and larger number of active scientists. While the Tsarist bureaucracy instituted quotas to restrict the numbers of Jewish and middle- and lower-class students, universities turned out to be a major avenue for social mobility in any event. Jews chose in particular physics and mathematics, which, as newer disciplines, were less likely to be closed to them by quotas or dominated by anti-Semitic faculty. The universities thus served as the training ground for the entire first generation of Russian physicists, many of whom were Jewish.

The growth of the educational system accompanied the modernization and expansion of Russian society and economy in general, and the number of university students grew in spite of quotas. However, the number of students with advanced degrees did not increase at the same time: Between 1904 and 1913, institutions of higher education awarded only 345 masters degrees and 491 doctor of science degrees. Of these graduates, only one-sixth were physicists or mathematicians and two-thirds were medical doctors.[5] Advanced students turned to Europe to continue graduate work. In addition, it was extremely difficult for poor students to matriculate, in spite of the presence of legal aid societies. There were many fewer scholarships available in Russia than in the U.S., Great Britain, or France.[6]

For the universities the period 1884–1914 was one of financial uncertainty and increasing government interference. During the 1890s, in response to unrest and violence, the autocracy became more repressive and threatening to the student body and professoriat. Following the 1905 Revolution, the autonomy of the universities was restored, in part because of the political and social reforms granted by the October Manifesto, a document giving Russia a limited constitutional monarchy. But the tension between the state and scholars over academic autonomy returned with a vengeance to the universities in 1910, after a speech to the Third Duma in which V. M. Purishkevich, a leader of the reactionary and anti-Semitic Black Hundreds group, called for the minister of education, L. A. Kasso, to purge the universities of undesirable elements. This act precipitated student unrest; many students were arrested, and the provisional regulations of the October Manifesto of 1905 dealing with education were abrogated. The rector of the university and his two chief assistants resigned, and the government responded by dismissing all professors who had been members of the university council. Within a month, by February

1911, almost one-third of the faculty had resigned or been dismissed, including many of Russia's leading physicists—P. N. Lebedev, N. A. Umov, V. I. Vernadskii, and others.

Throughout 1912 and 1913 conditions for teaching and research deteriorated at higher-educational institutions. The Kasso affair and the outbreak of World War I forced many physicists to look toward private capital and professional societies to continue their research. Umov, an experimentalist and theoretician with contributions in early quantum theory, divorced himself completely from university activities, ceased his lectures, and refused to show his face in the Moscow University physics laboratory, of which he was director. He confined himself to research at the private laboratory of V. F. Luginin and worked with such independent organizations as the Ledentsov Society for the Advancement of the Experimental Sciences and Their Practical Application (as president), the Society for the Study and Dissemination of the Physical Sciences (as chairman), and the Society for a Scientific Institute in Moscow (as a member).[7]

Scientific societies took on major importance in the face of mounting obstacles to normal physical research. Over 114 new organizations were formed between 1900 and 1917, of which 80 were independent scientific societies and 30 were intergovernmental bureaucratic committees and commissions.[8] Of importance to physics were the Ledentsov Society, the Lebedev Moscow Physical Society, the Moscow Society for a Scientific Institute, and the Russian Physico-Chemical Society.

The Kh. S. Ledentsov Society for the Advancement of the Experimental Sciences and Their Practical Application played a major role in supporting physicists after the Kasso affair. The society began operation in 1909, with N. A. Umov as the president of the society and editor of its organ, *Vremennik obshchestva im. Ledentsova*. From interest earned on its endowment, it awarded grants totaling 60,000 to 80,000 rubles annually to such scholars as the physiologist I. P. Pavlov, the physicist P. N. Lebedev, and the biogeophysicist V. I. Vernadskii in the "leading theoretical and practical" sciences, based on something akin to peer review. Awards ranged from 35 rubles for individuals to 1,500 rubles for laboratories, and covered start-up costs, patent awards, and an occasional publication subvention for such organizations as the Russian Physico-

Chemical Society.[9] And, when Moscow physicists departed the university en masse in 1911, the Ledentsov Society enabled many of them to continue their research in conjunction with other organizations, such as the Moscow Society for a Scientific Institute and the Moscow Physical Society.[10]

P. N. Lebedev, N. A. Umov, A. A. Eikhenval'd, and P. P. Lazarev founded the Moscow Physical Society (MFO) on March 16, 1911. Lebedev (1866–1912), who studied in Strasbourg with A. Kundt and became a full professor upon returning to Moscow, was a brilliant experimentalist. His work on light pressure demonstrated the interaction of light and matter and underlined the fundamental nature of Maxwell's theory by showing that the pressure of light corresponds with the magnitude predicted by Maxwell's equations.[11] Lebedev's professional accomplishments were as important as his scientific ones. He created the first Russian physics "school," whose members included P. P. Lazarev (biophysics), A. K. Timiriazev (kinetic theory of gases), and V. K. Arkad'ev (magnetism). When forced to leave Moscow University in 1911 because of the Kasso affair, Lebedev turned to private foundations for his work, and had begun to receive support to establish a scientific research institute through the MFO, but died in March 1912 without seeing it come to fruition.

The Kasso affair was a blow to the development of physics in Russia and in particular to the Lebedev School. But university scholars, rather than give up the ghost, united in the Moscow Physical Society to keep their community alive. The society arranged for popular scientific lectures, occasionally with other societies; it took part in the planning and calling of conferences of mathematicians and physicists throughout Russia; its members published in Russian and foreign journals. The MFO had a membership of over one hundred, which included the leading representatives of Russian physics, primarily from Moscow.[12] In the war years the society discussed issues of immediate concern in the fight against Germany; for example, V. K. Arkad'ev and A. A. Eikhenval'd gave a presentation entitled "On Gas Warfare." Until the early 1920s, the MFO played a role in the development of Soviet physics, holding weekly discussions by such scholars as the crystallographer Iu. V. Vul'f and the future president of the Academy of Sciences, S. I. Vavilov.[13] The Russian physics community also attempted to establish a "Lebedev Physics Laboratory" in Shaniavskii University. But like so many other hopes of Russian physicists before the Revolution, this plan failed to materialize.[14] Until well into World War I, the Tsarist government displayed little interest in physics research, and private societies were unable to raise enough funds to take up the slack.

Independent industrial research laboratories gained importance in physics research and development (R and D) in the Tsarist empire only during World War I. While Russia's iron, mining, textile, and chemical industries supported R and D activities, on the whole this research was an afterthought or repetition of foreign developments, especially since much of Russia's industry was controlled by foreign investors. And while the Tsarist educational system trained engineers of some skill and ingenuity, these scholars lacked adequately equipped laboratories in which to work. By the eve of the war, a number of industrial laboratories existed throughout the empire: between 1914 and 1916 several large factories— Putilovskii, Obukhovskii, and Okhtinskii—had established R and D facilities. But industrial R and D lagged far behind that of Germany and the United States in its scope and funding.[15]

The belief that science and technology would spur Russia's modernization and economic growth dates to the reign of Peter the Great. It became central in government policies at the turn of the century, with the establishment of higher technical schools and the attraction of foreign capital for industrialization under the leadership of minister of finance Sergei Witte.[16] But the problem for engineers and scientists remained that in its final years the government mistrusted independent scientific activity, and failed to support research in the way that governments in Germany, England, or the United States did.

When the war revealed the extent of Russia's economic backwardness and vulnerability to isolation and embargo, at long last the Tsarist bureaucracy recognized the need to create a national plan for the production of war matériel, synthetic fuels, and poison gas. The government and its private citizens, industrialists and scientists, joined together in a national research effort centered at industrial laboratories. One result was the establishment of a military-industrial committee (*voenno-promyshlennyi kommitet*). The committee represented the government's realization that scientific research served the national interest. But the war only exacerbated such problems as bureaucratic interference and fear of scientific initiative, which already belabored scientific research in the Academy and universities.

The chemist V. N. Ipatieff described his frustration as a member of the military-industrial committee, whose attempts to build a modern Russian war machine were hampered by unrealistic production targets, incompetent bureaucrats, and the backwardness of Russia's industry and science in comparison to the West.[17] The few steps toward cooperation between the Tsarist state and scientists in the committee, brought about only by the life-and-death situation of war, confirmed that successful cooperation between society and state was indeed possible. But the state went no further. There was no effort to create broad-profile scientific institutes beyond the short-lived and ineffective Central Scientific-Technical Laboratory of the Ministry of Defense.[18]

The war created significant problems for the nascent physics discipline. First, it grew isolated from the international scientific community. Second, many young scholars perished at the front, while others died in Petrograd and Moscow as cold and famine began to take their toll. Third, the war effort was so costly that the government curtailed its already modest support of physics research and organizations. It is not surprising that the majority of Russian physicists welcomed the abdication of the tsar in March 1917, and the establishment of the Provisional government with its promise of greater support for the scientific enterprise.

Unlike Tsarist officials, scientists learned about the promise of a national science policy through the hastily established military-industrial committee. By war's end they had begun to push for a national approach to the organization of the scientific enterprise—state support for scientific journals, a network of research laboratories, scholarships and grants, and technological innovation and invention. However, as the physicist V. A. Mikhel'son and others lamented, Russia's lack of a scientific worldview, compared with Europe and the United States, and its cultural backwardness, handicapped these efforts.[19]

THE PHYSICS DISCIPLINE IN
TSARIST RUSSIA

In addition to organizational and financial handicaps that were reinforced by the Tsarist system, Russian physicists faced other significant obstacles in their attempts to develop a modern discipline. These included the relatively small size and isolation of their community, the competing demands of teaching and research, which stifled originality and creativity, and the need for curriculum reform.

One of the major impediments to the creation of larger, better-equipped physics facilities was the isolation of physicists throughout the empire. Prerevolutionary Russian physics evolved almost exclusively within universities in a handful of cities: Moscow, Petersburg, Kharkov, Kiev, Kazan, Tartu (Yurev), and Odessa. Other than the Academy of Sciences, the only government institution conducting physics research was the Palace of Weights and Measures, directed by D. I. Mendeleev, and following his death, by N. G. Egorov. This led A. G. Stoletov (1839–1896), who studied with Kirchhoff in Heidelberg, taught at Moscow University, and was an early popularizer of Maxwell's and Hertz's work, to spend the last years of his career struggling to overcome the "inadequate material conditions" of Tsarist research facilities.[20]

There were less than one hundred physicists in the entire empire, of whom no more than fifteen held doctorates and for most of whom scientific activity was a sideline to pedagogical work. Those ten to fifteen physicists who lived outside of the two capitals, Moscow and Petersburg, were largely isolated from their colleagues.[21] In terms of such categories as regular annual physics laboratory budgets, salaries and fees of professors, posts and students by institute, and so on, Russia lagged behind the West.[22]

Until late in the last century, physics instruction and facilities in Petersburg, too, were on a primitive level. There was little if any laboratory experience to accompany the dry lectures which professors delivered. E. Kh. Lents (1804–1865), who helped to found the physics section of the Russian Physico-Chemical Society, preferred to work in the Academy of Sciences laboratory and forbade students to cross its threshhold until they completed their studies.

F. F. Petrushevskii (1828–1904), who took over the physics department at Petersburg University in the fall of 1864, showed more interest in students and the university laboratory. When he inherited it, the entire physics department occupied two rooms in a small building. One room had no natural lighting and windows only in the hallway. There were over 20,000 rubles' worth of instruments, including barometers, various apparatuses of tube construction, and relatively sensitive scales, but these were used for lecture demonstrations and "nothing to be proud of."[23] The rooms were so filled with instruments, in fact, that there was no space for experiments.

Conditions began to improve with the arrival of I. I. Borgman, N. A. Gezekhus, and N. G. Egorov as laboratory assistants in the 1870s. Petrushevksii curtailed his pedagogical activity, turned entirely to the development of the physics laboratory, and acquired more space in a new building, a small library, increased funding, and new equipment including a dynamo and a second, more powerful Otto internal combustion engine to conduct experiments with a reliable electrical supply. The number of students grew from 7 in 1867 to 60 in the mid-1870s, and expanded to 100 to 150 per year by the beginning of the 1880s. This increase indicates the growth of interest in physics, the general quality of instruction under Borgman, Gezekhus, and Egorov, and expanding university enrollments as a result of the liberalized admission standards which held sway until the law of 1884.

Throughout his last years Petrushevskii sought out additional funds for the physics department. In 1884, as chairman of the Committee for Improvement of Educational Institutions, he requested 100,000 rubles from the Ministry of Education; he received the allotment only in 1896 when the ministry discovered a surplus of 1.5 million rubles assigned in 1890 for construction at the university. In 1897 Petrushevskii and Borgman submitted plans for a new laboratory, which finally opened in 1900.[24] A series of other problems continued to slow the development of Petersburg physics, however.

One of these problems concerned the necessity of curriculum reform. Under Petrushevskii, physics instruction at the university, and in fact in all higher schools of Petersburg, was founded on the physics and methods of measurement as the basis of exact knowledge. The first year of instruction was devoted to the description of measuring instruments, and only in the second year did students cover laws of heat, electricity, magnetism, optics, and acoustics. Experimental practica became an obligatory aspect of study only in 1884.

The university-wide system of master's examinations which required students to pass higher-level mathematics courses also prevented students from completing physics degrees. For several decades, with the exception of P. S. Ehrenfest, not one Petersburg physicist passed the exams. Such mathematicians as A. A. Markov and V. A. Steklov in the physico-mathematical department did not differentiate between scholars with an interest in mathematics and those physicists for whom mathematics was an analytical tool. Only with the arrival in Petersburg of several young physicists— P. S. Ehrenfest, A. F. Ioffe, D. S. Rozhdestvenskii, and D. A. Rozhanskii—and the cooperation of elder statesmen I. I. Borgman and O. D. Khvol'son did Russian physicists break through this "mathematical barrier" and introduce a special mathematics program for physicists in 1911.[25]

Finally, there was only one chair in theoretical and mathematical physics for the entire Russian school system until the Russian Revolution.[26] While this situation does not differ greatly from that in the United States and Great Britain, which had three and two chairs in theoretical physics respectively, two other factors made the situation in Russia more significant. The first was the overly mathematical emphasis of the curriculum. The second was the formal organization of the physics department, which was combined with mathematics and astronomy and was not as yet seen as an independent entity.

The quality of university instruction suffered under these circumstances. A heavy teaching load hampered efforts to stay current in recent events in physics.[27] In spite of the presence of O. D. Khvol'son and I. I. Borgman on its faculty, the level of scientific work at the Physics Institute of Petersburg University was not first-rate. Ioffe maintains that "the brilliant but primarily phenomenological" lectures of O. D. Khvol'son "did not inspire any impulse toward scientific creativity."[28] This was true in spite of his remarkable five-volume *Kurs fiziki*, which was written and revised over a twenty-year period (with the first volume appearing in 1897). The

Kurs was, in fact, symptomatic of Khvol'son's belief that it was better for Russian physicists to follow tradition, and to help to order it. In what has become an apocryphal story, Ioffe remembers that his proposal to work on some new, unsolved problems produced Khvol'son's response, "As if it is possible to think up something new in physics? For that it is necessary to be [a] J. J. Thomson."[29]

In the same way, I. I. Borgman (1849–1914), known among students for his progressive views in spite of having been a teacher of Nicholas II, was an excellent organizer and lecturer, but often failed to foster originality in research. Borgman closely followed the work of Faraday and Maxwell and focused primarily on electromagnetism, electrical charges in gases, and radioactivity, but preferred organizational and professional activities to research and teaching. From 1875 until 1900, he was the editor of the first Russian physics journal and with Petrushevskii established the Physics Institute at Petersburg University. He was its director from 1901 until his death, and became rector of the university in 1905, although he resigned in 1911 in solidarity with the Moscow physicists.

Borgman is responsible for helping to develop a sense of community among Russian physicists upon which Ioffe and Rozhdestvenskii could build. One way he did this was through the publication of occasional collections of papers in the series *Novye idei v fiziki* (New ideas in physics), which he edited between 1911 and 1913. The six editions of *Novye idei* sought to familiarize "readers with differing contemporary views on the most important questions of physics. Every edition will be devoted to the explanation of one or two questions and will include articles in which the authors examine these questions from different points of view."[30] For example, the first volume focused on the structure of matter and included articles written by J. J. Thomson, Jean Perrin, and Borgman himself, with the latter rejecting energetics and arguing that we "cannot as yet answer if it is possible to have the existence of atoms of positive electricity [protons] separate from atoms."[31] Volume 2 focused on the ether with articles by Philip Lenard, Thomson, Norman Campbell, and Max Planck; volume 3 covered relativity theory; volume 4 examined the physics of light, as did volume 5, which was dedicated to P. N. Lebedev and included his famous paper on light pressure; and, finally, in volume 6, articles by Khvol'son, Planck, Wien, and H. Callendar examined the nature of heat.

While Borgman was successful on one level in building professional awareness among his colleagues, he could in no way single-handedly address the problems facing Petersburg physicists on the eve of World War I. Like their Moscow counterparts, the old-school Petersburg scholars—Khvol'son and Borgman—closely followed the European tradition, failed to recognize the need for curriculum reform, and did little to improve the quality of university instruction. The fact that a scholar of Lebedev's scientific reputation and achievement should encounter so many difficulties in conducting research convinced even the most conservative physicist that the physics section of the Russian Physico-Chemical Society would have to become more active in developing national professional consciousness in the face of a government increasingly reticent to consider their disciplinary needs. Only this would ensure the creation of modern laboratories, the establishment of independent directions of research, and the success of nascent research "schools."

THE RISE OF PROFESSIONALISM:
THE PHYSICS SECTION OF THE RFKhO

The physics community of Tsarist Russia grew increasingly politicized in the first two decades of the twentieth century. This was brought about by four interrelated factors: the politicization of society as a whole after the 1905 Revolution; the growing domination of the discipline by younger scholars who tended to be more liberal in their social and scientific views than their predecessors; the revolution in physics brought about by quantum and relativity theory; and the general growth of the discipline. The physics section (*fizicheskoe otdelenie*) of the Russian Physico-Chemical Society (Russkoe fiziko-khimicheskoe obshchestvo, or RFKhO) served as the forum for the articulation of physicists' interests.

Before the Russian Revolution the physics section was the only organization to represent the professional aspirations of physicists and provide a place for discussion of contemporary research topics. It was national in membership, although the majority of its members and its central office were located in Petersburg. It promoted the organization of independent, well-funded physics institutes, support for journals to publish their works and publicize their interests, and the maintenance of international scientific contacts.

As early as the 1860s, scientists in Petersburg began to meet regularly at the house of one or another physicist to discuss the creation of a national association. E. Kh. Lents and F. F. Petrushevskii founded the Physical Society in 1872, and joined with the Chemical Society in 1878 on the initiative of D. I. Mendeleev to create the RFKhO. Its goals were to ensure regular scientific communication and to coordinate and publicize the research activities of its members, since, as the specialist in mechanics N. A. Gezekhus lamented, Russian physics was "in complete infancy and at a full stand-still." It had neither tradition nor research "schools," neither student practica nor adequate equipment without which it was nearly impossible to train good experimentalists. Finally, "there was no organ, nor meetings which would give the possibility to exchange thoughts or become excited to action; there was not even the mention of these things then."[32] But by the turn of the century some of Gezekhus's concerns had been answered.

After fits and starts, the RFKhO began to meet regularly, corresponded with other scientific societies and institutions, allocated moneys to scientific endeavors in spite of limited funds, and established a professional library with book and journal exchanges with scientific organizations throughout Russia and the world. The society solved yet another problem: before the early 1870s there was no physics journal in Russia, whereas in Western Europe and the United States several were published, and a number had existed for many decades. In 1873, the *ZhRFKhO* was first published, and in 1907 it was split into two parts, physical and chemical. The physics journal was published until 1931 when it was reconstituted as *Zhurnal eksperimental'noi i teoreticheskoi fiziki* (or *ZETF*, which is published to this day). The society grew throughout the prewar years, reflecting in part its own successes and in part the general expansion in the number of trained experts in a society undergoing rapid economic growth and industrialization.

The membership of the RFKhO grew consistently between 1890 and 1917 (see app. A), with the number of chemists keeping pace with that of physicists, reflecting chemistry's ascendance as the major science in Tsarist Russia—as in the rest of the world at that time—as well as physics' slowness to amalgamate as a profession. Only after the Revolution did physics begin to dominate Russian science in terms of numbers of scientists and institutions and level of government support. This is because on issues of practical importance, organization, institutional structure, and other matters "that touch upon the development and refinement of a professional scientific discipline, physics achieved maturity relatively late when compared with the professional status of several other natural sciences."[33] This is especially true in Tsarist Russia.

More important than increases in absolute numbers for the rise of modern physics in Tsarist Russia was the replacement of the older generation of physicists by young scholars in universities and the RFKhO, beginning in 1905 (the date of the Einsteinian and first Russian revolutions). The period immediately preceding the February Revolution was one of intense intellectual and organizational activity. Three factors in particular demonstrate physicists' increasing self-awareness and confidence: the day-to-day activity of the physics section; their response to World War I; and physicists' reaction to the Kasso affair.

After the 1905 Revolution, between thirty and sixty members of the RFKhO physics section gathered regularly to discuss recent research and address a number of practical problems: the need for domestic production of physical instruments; involvement in the war effort; and budgetary problems. In fact, the Tsarist regime's inadequate support for physics plagued the section throughout its history. The physics section ran a deficit on the order of two to three thousand rubles per year from 1914 to 1917 (see app. A, table 1). Subscription fees and membership dues produced roughly three thousand rubles income annually, but small subsidies from the Ministry of Education and Petrograd University could not make up the difference in operating and publication expenses. In the fall of 1915, the dues and the journal subscription rate increased. However, a decline in subscriptions from 257 in 1913 to 212 in 1915 offset these gains, as did a deficit of nearly two thousand rubles which had been carried over from 1914. The section asked the Ledentsov Society for three thousand rubles and the Ministry of Education for another two thousand.[34] But in 1916 the Ministry of Education ceased its subsidy to the physics section altogether, a move surprising in view of the help the Tsarist regime expected from physicists in World War I.[35]

Russian physicists were active in the war effort, both through the Moscow Physical Society and through the RFKhO physics section. In September 1915 the council of the section discussed how its members might assist the government in the fight against Germany. In response to a request from a committee for military-technical assistance, the council decided that the physics section itself could do little in the production of any equipment on a large scale, but it would study specific problems, produce analyses, and consider special requests as they arose. The physicists entered into direct communication with the military-industrial committee, with A. F. Ioffe and D. S. Rozhdestvenskii serving as representatives. In October 1915, the Commission for the Study of the Productive Forces (KEPS) also invited a representative of the physics section to participate in its activities; D. S. Rozhdestvenskii was selected.[36]

As the war progressed, the physics section had difficulties of other than a financial nature, reflecting the vagaries of dealing with the Tsarist bureaucracy. First among these was the problem of publications, especially after physicists lost government support for their journal, *ZhRFKhO*. In addition, the war brought about the isolation of Russian physicists. Through painstaking effort and at great expense, the section had developed journal and book exchanges with scientific societies in other countries and throughout Russia. In 1914 the society received over eighty-three journals of which thirty-nine were foreign, including the *Proceedings of the Royal Society, Philosophical Magazine, Physikalische Zeitschrift*, and one American publication, *Astrophysical Journal*. By the end of the next year, however, only fifty-nine journals were received, of which ten were new; but only nineteen came from abroad, and the thirty-two German journals had been dropped. Such physicists as A. F. Ioffe were deeply concerned about the war with Germany. In a talk entitled "Achievements in Physics in 1914," given on February 10, 1915, Ioffe expressed his concern that war would bring about the ruin of science and culture, let alone interrupt the flow of scientific instruments from such corporations as Siemens to Russia.[37] All in all, the war brought about painful and prolonged scientific isolation: a lack of scientific correspondence or journal exchange with the West, the absence of foreign research travel, and the inability to purchase instruments and equipment. When one recalls that most Russian physicists had studied in Germany, it is clear that the rupture with Germany was particularly devastating to Russian physics. This setback would be overcome only in 1922.[38]

The Kasso affair radicalized the physics section, which in 1910–1911 responded in solidarity with the Moscow physicists. At first only the younger, more progressive members of the section, A. F. Ioffe, P. S. Ehrenfest, A. A. Dobiash, A. A. Friedmann, and N. A. Gezekhus, pressed the council of the section to protest Kasso's heavy-handed destruction of Moscow University and express its support of the departed professoriat. At a March 1911 meeting, the council, under the chairmanship of N. G. Egorov, refused to act, arguing that such an action would not correspond with the organization's charter.[39] As Sominskii argues, this decision was surprising because the Moscow University events affected a large number of their colleagues, forcing them to leave their laboratories and work.[40] The stormy debate which followed indicated that a majority of members stood against the council's intransigence. Egorov then resigned, and the physics section expressed solidarity with the Moscow physicists, voting almost three-to-one in favor of a resolution condemning the government's actions. From this point on, the council never failed to denounce the arbitrary behavior of their government.[41] In December 1916, when the Tsarist bureaucracy dismissed the sympathetic and moderate minister of education, Count P. N. Ignat'ev, the physicists sent the count a telegram expressing sincere sympathy and deep regret at his unexpected removal from office. The physics section also criticized other governments, condemning the Paris Academy in January 1911 when it failed to elect Marie Curie to full membership in spite of her Nobel Prize–winning research, simply because she was a woman.[42]

Notwithstanding the disruptions of the Kasso affair, the war, budgetary constraints, and scientific isolation, on the eve of the Revolution the physics section was a growing and active organization, reflecting the vitality of its youngest members and their knowledge of contemporary physics. These scholars—Ioffe, Rozhdestvenskii, Ehrenfest, and others—also provided the impetus for the development of prerevolutionary Russian physics. The center of their activities was the Petersburg physics seminar, where a group of young scientists with a theoretical orientation provided the leadership necessary to create a physics "school" in Tsarist Russia.

THE PETERSBURG PHYSICS SEMINAR AND
THE FORMATION OF THE IOFFE SCHOOL

In 1907 a circle of young physicists including A. F. Ioffe, D. S. Rozhdestvenskii, and P. S. Ehrenfest began to gather unofficially in Petersburg to discuss recent developments in physics. These physicists debated the implications of relativity and quantum theory. They discussed the increasingly complex relationship between theoretical and experimental physics and the role of mathematics in twentieth-century physics. And they set out to rectify the inadequacies of Tsarist physics. Because of the vision of its leaders, the circle grew to have two major goals. The first was the creation of a central scientific institute to undertake a world-level research program. The second was the development of a training ground for physicists in Russia, so that the best young talent would not have to go abroad for graduate study. These goals were realized after the Revolution in the founding of the Leningrad Physico-Technical Institute and the physico-mechanical department of Petrograd Polytechnic Institute, directly across the street from each other.

A. F. Ioffe was the "glue" for the Petersburg physics circle, which served as the basis for his research "school," the Russian functional equivalents being *shkoly, seminary,* and *kruzhki.* He provided the circle with intellectual leadership, identified new areas for study, and ensured the publication of results. He attracted a steady stream of well-qualified, motivated students. He fostered creativity and initiative by proposing that advanced students conduct their own research under the watchful eye of a critical lab director as part of a close-knit research group. His experience in Germany where he visited the Kaiser Wilhelm research centers and in Russia where he worked in the Putilovskii industrial laboratory convinced him of the need to organize modern science on a large scale. This required both new forms of organization and a new relationship with the government to secure financial support for the creation of a central research institute.[43]

Abram Feodorovich Ioffe was born in 1880 in Romny, a small provincial town in the Poltava region of the Ukraine, into a typical middle-class Jewish family.[44] His grandfather had been in handicrafts before turning to bookkeeping. His father was a midlevel bank official. His mother was a housekeeper. He was the firstborn in a family of two brothers and three sisters. After finishing *Realschule* in 1897 he entered St. Petersburg Technological Institute. At the time it was the only higher-educational institution (or *vuz*), other than the universities, where it was possible to study physics. Ioffe found teaching too "formal," since it forced students to know, not to understand. He was interested in the physics of the ether and the propagation of light, and the physics of smell, but had little opportunity to discuss his scientific interests with professors since they were "bureaucrats and very narrow technical specialists." Ioffe was critical of the poor instruction, the absence of experimental work, and the stress placed on the physics of measurement.

Early on, Ioffe gained insight into the relationship between industrial production and physics research. He spent his first summer practicum in a laboratory in the Putilovskii metal works, a huge factory involved in building boilers and military equipment, and one of the first to have electric lights, but known today for its history of labor unrest during the Revolution. He graduated from the technological institute in 1902, and during the summer, as was common practice for Russian engineers, he gained practical experience by working for the state on a construction project, in this case two railroad bridges near Kharkov. In that year, on the advice of N. A. Gezekhus, chairman of the physics department of the technological institute, he traveled to the Physics Institute of Munich University to work with W. K. Roentgen, realizing he could go no further in Russia with his present studies.

The three years that Ioffe spent in Munich had a significant impact on his methods of study, his subjects of research, and his personal style as a scientific administrator. He first joined approximately twenty other students in fulfilling a practicum of some one hundred problems, a task normally accomplished in two months. Physics in Munich at this time had a strongly classical experimental tradition (it would change drastically to a theoretical orientation with the arrival of Arnold Sommerfeld and his assistant, Peter

Debye, and the appointment of Max von Laue in the years 1906–1908). Roentgen, on the other hand, was a rather conservative educator and scientist, in spite of his sensational discovery of X rays in 1895. At first he did not pay special attention to Ioffe, even though the latter completed the practicum in one month. Roentgen then asked Ioffe to compare two methods for the measurement of dielectrical loss in insulation based on the application of high-frequency electrical oscillations, recently suggested by the German physicist Paul Drude. Upon Ioffe's successful completion of this work, Roentgen gave him full independence in the institute, offering only "scrupulous" criticism of his method of measurements and the quantitative data produced thereby. Many Soviet physicists have pointed to Ioffe's experience in Munich under Roentgen's guidance, and specifically to Roentgen's methods of criticism, as a major influence on Ioffe's style of leadership as it developed at LFTI: Ioffe was a conservative but able administrator, capable of fostering independent lines of investigation and at the same time placing more emphasis on experimental than on theoretical work.[45]

Ultimately Ioffe turned to research suggested by Roentgen for his doctoral degree. He studied the elastic aftereffects in quartz crystals, work which led him to a number of fundamental discoveries related to the mechanical, electrical, and photoelectrical properties of crystals. This work grew into the study of the internal photoeffect in X-rayed crystals, the spectra of absorption of light in them, and the transformation of F-centers into U-centers and back. In Roentgen's laboratory Ioffe also investigated the influence of electrical current through ionic crystals at a wide range of temperatures and the phenomenon of high-voltage polarization in similar crystals (in which current flows not by electrons but by ions). Ioffe continued to collaborate with Roentgen and to consult with him throughout the years, but most of their joint efforts were not published and in fact were burned on Roentgen's orders after his death in 1923.[46]

Ioffe returned to Russia in 1906 with a doctorate in hand from Munich University. But his Jewish background and the overly bureaucratic Tsarist system of education kept him from finding a position as more than a laboratory assistant in the physics department of St. Petersburg Polytechnical Institute, founded in 1903 at the suggestion of Count Witte. This was a fateful move, since it was at the polytechnical institute that Ioffe established the physics circle, which served as the kernel for LFTI. Ioffe also taught classes at the Commercial Real School throughout 1907. He soon organized a laboratory group in which I. S. Shcheglinev, A. I. Tudorovskii, and F. A. Miller, all of whom later worked in LFTI, took part. Ioffe also began to seek better ties with Moscow physicists. In general, these years saw the first steps toward the development of the Ioffe school.[47]

Ioffe gradually filled his laboratory with instruments and equipment, and successfully attracted his first students and co-workers, primarily from the university. In 1912, P. L. Kapitsa and P. I. Lukirskii, later prominent Soviet physicists, entered the polytechnical institute as graduate students; in the following year Ioffe became docent at the university and gained an additional following, including N. N. Semenov, K. F. Nesturkh, and Ia. I. Frenkel'. This group became the core of the Petrograd physics seminar beginning in 1916 and ultimately of the LFTI. Ioffe recognized the importance of training young scholars for independent research; before this time, capable students who wished to undertake independent physics activity had been urged by their professors to conduct research conceived and initiated abroad.

In 1913 Ioffe turned down a position at Kharkov University, partly on the advice of Paul Ehrenfest, to become Extraordinary Professor at the polytechnical institute. He declined in order to stay in Petersburg, despite the fact that such a position would have provided him with more time to publish and proximity to D. A. Rozhanskii, another budding young physicist. But Kharkov University had the disadvantage of being under the jurisdiction of the Ministry of Education, and was weaker in terms of students and facilities, whereas the better-equipped polytechnical institute was under the more liberal Ministry of Finance.

Ioffe's professional and research accomplishments marked him as a leader of Russian physics. In 1912 he defended his *magister* dissertation and in 1915 his doctoral dissertation on the elastic aftereffect in quartz crystals. He was elected deputy chairman of the physics section in 1913; he joined the editorial board of *ZhRFKhO* in December 1914; and in 1915 he became chairman of the physics section and vice-president of the RFKhO. In the nine years since his return to Russia, Ioffe had succeeded in establishing himself as a major figure within the physics community, both through his own research efforts and by organizing a network of promising young physicists. It remained for Ioffe and young physicists like him—D. S. Rozhdestvenskii and P. S. Ehrenfest in particular—to overcome historical weaknesses in the Russian physics community.

By 1916 Ioffe conducted a weekly seminar that was reputed to be one of the most progressive centers of physics in the empire and that attracted the best young talent to discuss papers on contemporary physics problems. The roots of this seminar are found in a students' physics circle whose motive force was P. S. Ehrenfest. Ehrenfest, who lived and worked in Petersburg from 1907 to 1912, spurred the development of theoretical physics in Russia.[48] In this way, theory and experiment came together in modern Russian physics.

Paul Ehrenfest (1880–1933), the well-known theoretician who worked in the areas of thermodynamics, statistical mechanics, nuclear physics, and relativity and quantum theory, was a native of Vienna and received his doctorate from its university in 1904 under Ludwig Boltzmann, having spent time in Göttingen working with Felix Klein along the way. He met his future wife, Tat'iana Alekseevna Afanas'eva, in Göttingen. Tat'iana Alekseevna was born in Kiev in 1879, and moved to Petersburg to live with her uncle because of the illness of her father. She took classes through the Petersburg Higher Women's Courses where Khvol'son and Borgman taught, and went to Germany to continue her studies with Klein. After Ehrenfest finished his dissertation, the couple married and moved to Petersburg in 1907 for family, not professional, reasons as agreed upon earlier. This move caused Ehrenfest little dismay since he had met Ioffe in Munich in a cafe in 1905 and looked forward to working with his friend again in Russia. Ehrenfest remained in Russia until 1912, leaving because of the Kasso affair.[49] He joined Hendrik Antoon Lorentz in Leiden, but his role as a go-between for Russian and European physics was equally important after World War I, and lasted until 1933.

A number of contemporaries recalled the importance of Ehrenfest and Ioffe in providing leadership for a students' circle away from the formal, traditional, and somewhat stultifying approach to physics at the university. This circle met at roughly two-week intervals and pointedly excluded the elder statesmen of physics, Borgman and Khvol'son. Physicists who attended the circle normally included such teachers from Petrograd's higher-educational institutions as Baumgart, Dobiash, L. D. Isakov, and M. A. Levitskaia, and several university physics students: V. R. Bursian, G. G. Veikhart, V. V. Doinikova, Iu. A. Krutkov, and V. M. Chulanovskii, as well as the mathematicians S. N. Bernstein, Ia. D. Tamarkin, and A. A. Friedmann. The circle focused on contemporary issues. At a 1910 meeting called by Dobiash, Rozhdestvenskii, and Baumgart, the discussion centered on the practical problem of the need for an ultra-microscope at the university, and the question of the dependence of the resistance of metals on their magnetic field as observed in the Hull effect.[50] As I. V. Obreimov wrote,

> The fact of the matter is that the young physicists of Petersburg organized a closed physics circle. They did not invite the physics professors of the University, I. I. Borgman and O. D. Khvol'son, in view of their hostile relationship to the new physics of Einstein, Planck, to the theory of relativity, and personally to P. S. Ehrenfest who was the organizer and soul of the circle. The circle gathered on Sundays from ten o'clock until noon in someone's room, or in secret from Borgman and Khvol'son in a room at [the Physics Institute]. In the fall of 1912, in October, G. G. Veikhart introduced me to the circle. On this occasion it met in Baumgart's room. Ioffe reported briefly on the beginnings of his work to measure the charge of the electron. In fact, the Viennese physicist Ehrenhaft had discovered a subelectron. The American physicist Millikan established the charge of the electron, or as can be said, "had weighed an electron," which adhered to an oil drop. Ehrenhaft reproduced Millikan's experiment, but not on one isolated drop but in drops in the form of a cloud.[51]

After Ehrenfest moved to the chair in theoretical physics at Leiden, Ioffe inherited direction of the seminar. Semenov, later director of the Institute of Chemical Physics of the Academy of Sciences and a Nobel Prize winner, was a first-year student at the university when he first met Ioffe in 1913. While he read such university texts as Khvol'son's *Kurs*, Semenov found the lectures boring, and stopped attending classes. He then heard that Ioffe was giving an advanced course in physics at the polytechnical institute, which he began to attend. Semenov credits Ioffe with inspiring him to study such new phenomena of twentieth-century physics as quantum theory. After completing his second year of university physics studies, Semenov jumped at the opportunity to continue in Ioffe's physics seminar with such classmates as P. L. Kapitsa, Ia. I. Frenkel', P. I. Lukirskii, and Ia. G. Dorfman. Semenov welcomed the long discussions at the physics seminar:

> [We] were in the center of the general achievements of contemporary science itself. During these discussions Ioffe, and to a small extent we, too, began to build suppositions about the near future of these or other scientific regions, coming out with new points of view on various questions, and referred to concrete experiments of the present period. Ioffe taught us rigor and definitiveness in judgments, boldness of thought, flights of fancy, and passion for science.[52]

Ia. G. Dorfman, later known for his work in magnetism, as deputy director of the Ural Physico-Technical Institute in the 1930s, and as a historian in the Institute of History of Science and Technology, recalled Ioffe's bold introductory course in physics. Ioffe's lectures on the contemporary view of matter—atomic and molecular structure and Brownian motion—created in Dorfman and his classmates so deep an impression that they "first recognized physics as a vital and majestic science."[53] Other elements that attracted young physicists to Ioffe included his reputation as a liberal, his carefully prepared lectures which he read freely as if improvising and his use of models to clarify explanations. Most exciting for Dorfman, however, was the six o'clock Thursday evening seminar which Kapitsa, Frenkel', Semenov, K. F. Nesturkh, M. V. Kirpicheva, and Dorfman attended. Here Dorfman observed Ioffe's ability to summarize talks precisely. He would turn the group's attention to some article, point out its insufficiencies and problems, and suggest solutions to the problem. "In these discussions," Dorfman writes, "all participants of the seminar took part with equal rights. [Ioffe] never pressured us, nor utilized his authority; he patiently heard out all objections and comments. A friendly, cordial, thoughtful atmosphere always reigned at the seminar, and thereby the true scientific collective as formed whose inspiration and soul was A. F. Ioffe."[54]

Ia. I. Frenkel' also credits Ioffe with providing the spirit and leadership for the Petersburg physics seminar, and for creating an environment in which he found encouragement to pursue his theoretical interests. The mathematics school at the university

under A. A. Markov and V. A. Steklov was outstanding; under
N. A. Bulgakov, who replaced Borgman at the physics institute of
the university, Frenkel' could indeed study mathematical but not
theoretical physics. Bulgakov was a good calculator, capable at carrying out complex analyses, but his approach was unquestionably
closer to mathematics than to physics. In September of 1916,
Frenkel' took advantage of the opportunity to talk with Ioffe and
accepted his invitation to participate in the physics circle. Two
times per week he trekked to the polytechnical institute. The seminar attracted young physicists "who were interested in the newest
developments in physics and desired to work in this area. . . .
A. F. Ioffe determined the choice of material to be discussed, and his
stimulating discussions had a deep impression on all young participants, including myself."[55] Soon after Bulgakov, Rozhdestvenskii,
and Khvol'son decided to accept Frenkel' into graduate studies at
the university, Ioffe asked him to give a talk at the polytechnical
institute, and later to address the RFKhO physics section on
November 8, 1916, on his senior (*diplomnaia*) thesis. This talk, in
fact, generated lively discussion, and Frenkel' was elected to membership in the section on the spot. Ioffe also encouraged Frenkel' to
submit his work, which was based on a simple idea of the application of the Rutherford-Bohr planetary model of the atom to solid
and liquid states, to *ZhRFKhO*, where it was published.[56]

The physics seminars in Petersburg lacked well-equipped laboratories, although conditions continued to improve until the war.
But what they lacked in equipment they more than compensated
for by providing a group of young scholars with able leadership at
the cutting edge of physics. Here Ioffe's personal research agenda
in molecular and atomic physics played a major role. Ioffe concentrated his efforts in two main areas of physics, which together
would serve as the initial research program for his laboratory at
LFTI, and both of which were pressing contemporary problems: the
elementary photoeffect and the physics of solid bodies, in particular dielectrics and crystals.

Ioffe's interest in the electrical and mechanical properties of
crystals dates to his first days in Roentgen's laboratory in Munich
when he investigated the piezoelectricity of quartz crystals, which
the Curie brothers, Jacques and Pierre, had discovered in 1885.[57]
The Curies experimented with crystals found in nature, which

made it difficult to study the process of piezoelectricity because of heterogeneities of structure and impurities. At first unaware of the influence of heterogeneities on current, physicists concluded that the passage of electricity through crystals gives rise to unexplained phenomena, which came to be referred to as "the dielectrical anomaly" in the literature.

Roentgen became interested in this "anomaly" and suggested that Ioffe study the problem of elastic aftereffects in crystals, specifically the piezoelectric effect in crystalline quartz. But Ioffe changed the phrasing of the problem and looked at such phenomena as piezoelectricity which occurred in the solid state during the action of mechanical forces.[58] In his study of the electrical conductivity of rock salt crystals, Ioffe also demonstrated that during irradiation by X rays they become tinted and sensitive to visible light; it is precisely under the action of light that electrical conductivity increases many times. Thus, Ioffe had discovered the phenomenom of photoconductivity in X-rayed crystals. This finding excited his interest in the mechanism of deformation of crystals, to which he would return after Laue, Friedrich, and Knipping developed an X-ray method for studying crystalline deformation. Ioffe turned to detailed X-ray analysis of crystals, and first reported the work of Laue to Russian scholars in 1912.[59] Working with M. V. Kirpicheva in the period 1914–1916, Ioffe conducted a series of important experiments to study the mechanism of the passing of current through dielectrics, and thus helped significantly to clarify the "dielectrical anomaly." Ioffe also examined the influence of temperature and irradiation by ultraviolet and X rays on electroconductivity and showed that impurities play a great role in the physics of the current. Ioffe and Kirpicheva then turned toward the production of "electrically pure" crystals of ammonium, because of the great influence of impurities on conductivity.[60]

As a result of research on both natural and artificial crystals, Ioffe established that current passing through quartz crystals follows Ohm's law, that electroconductivity of the crystal is a fully defined magnitude, and that the "anomaly" is connected with the formation of volume charges within the crystal during the application of current. He concluded that conductivity arises as a result of thermal dissociation of the crystal lattice. Ioffe and Kirpicheva also showed that the electrical conductivity of the majority of ionic crystals carries an electrolytic character and is conditioned in part by the presence of these impurities and in part by the thermal dissociation of the lattice itself. In 1915 Ioffe defended his doctoral dissertation on the elastic and electrical property of crystals based on these research efforts. In a 1916 article with Kirpicheva on pure crystals, Ioffe first presented results of research on ionic conductivity which led to proof of the motion of ions through interstitial spaces (the electrolytic mechanism of conductivity of ionic crystals). Later, within LFTI, the Ioffe group—N. N. Davidenkov, M. V. Klassen-Nekliudova, I. V. Obreimov, and others—undertook detailed analysis of the electrical and mechanical properties of crystals.

Ioffe also undertook research on the quantum theory of light and the recently published theory of Einstein on the photoeffect, which advanced the hypothesis that light had a corpuscular nature. By 1912 the well-known work of R. Millikan and W. Wright, which verified Einstein's work in the area of visible and ultraviolet light, had been published, and Ioffe deemed no further work necessary.[61] Ioffe then sought to verify experimentally the statistical character of the photoeffect, studying the elementary photoelectric effect—that is, the flight of individual electrons from metallic foil, suspended in an Ehrenhaft condensor. Ioffe described the apparatus he applied for this goal in his *Elementarnyi fotoelektricheskii effekt* (1913), for which he received the prestigious S. A. Ivanov Prize, awarded by the Academy of Sciences every four years for original work in Russian in the natural sciences.[62] Ioffe presented his work on the magnetic field of cathode rays and the elementary photoeffect in his master's thesis.

All in all, it can be seen that Ioffe provided the ingredients necessary to create a physics "school" in Tsarist Russia: leadership, scientific style, an outlet for publication, *ZhRFKhO*, and a research program on the cutting edge of solid-state, molecular, and atomic physics, which generated interest among other physicists who were subsequently drawn to his fold. Ioffe commanded authority among his colleagues and accordingly received recognition in the RFKhO, which elected him chairman of the physics section by 1914. Finally, his personality—he was shy and polite, yet decisive and critical, fair and egalitarian, as revealed in the memoir literature—contributed to the development of this school. Every student had the opportunity to be an equal partner in the physics circle. What Ioffe lacked—funding to develop the program and put it on a world-class level—would be secured only after the Revolution.

CONCLUSIONS

When the Bolsheviks seized power in October 1917, the first generation of Soviet physicists had already entered adulthood. At the time of the Revolution they were, respectively, A. F. Ioffe, age thirty-seven; D. S. Rozhdestvenskii, age forty-one; Ia. I. Frenkel' and P. L. Kapitsa, age twenty-three. These men had entered the discipline of physics through the Tsarist system of education and research, which laid the foundation for later Soviet achievements. Science would be critical in the construction of the socialist Soviet Union, and physicists were very much a part of the revolutionary society. Their ability to take a leading role in Soviet science was not only a product of the Soviet experience, however, but also of their success in building upon the achievements of prerevolutionary Russian physics.

Between 1900 and 1917 physicists overcame many of the constraints on the development of physics imposed by the Tsarist system. These years saw the consolidation of scientific forces. There were better-equipped laboratories, though in absolute numbers there were but a handful of physics centers in 1917. The number of physicists reached somewhat less than one hundred. Most important, physics schools developed, first under Lebedev and then under Ioffe and Ehrenfest. These scholars had no strong national tradition to oppose or uphold; they established research in areas with great promise; and they were prepared to accept new developments such as relativity theory and quantum theory without undue philosophical concerns.

The Tsarist regime proved incapable of supporting physics at the level necessary for its continued growth. In part because of innate conservatism and in part because of an inability to see the implications of industrialization in Russia or the role of trained specialists in the process, the regime provided piecemeal funding for physics research. The bureaucracy opposed the academic freedom and autonomy sought by scientists. Its arbitrary behavior in the universities led ultimately to the loss of one-third of all professors at Moscow University. The Tsarist regime supported the Academy of Sciences with looser pursestrings, but controlled its membership and thereby its research program. The result was the emphasis on the humanities, and support for a physics almost entirely devoted to observatories and seismographic stations.

This led to a weak discipline, limited to a few major cities and research centers, poorly supported in terms of budget, and reliant on Europe for intellectual leadership and training. Even the growth and development of industrial laboratories and scientific research institutions lagged in Russia. Only the demands of war forced the Tsarist regime to cooperate with industrialists and scientists, and then the war effort only highlighted the inadequacy of Russian laboratories in comparison with Western ones.

However, because of the efforts of an increasingly active professional organization and capable leadership centered in Petersburg, and because of the presence of a young but growing physics community, the locus of future research schools, with mature knowledge of developments in world physics, Russian physicists stood ready to take advantage of the promise of greater support from the state under the Bolshevik regime.

2

The Russian Revolution and the
Search for a National Science Policy[1]

Unlike their predecessors, the Tsarist regime and the Provisional government, the Soviets recognized the importance of science and technology in matters of "government construction," a catchall phrase, prominent in early Soviet political literature, that signified measures supportive of the increased administrative power and stability of the new government and economy. But civil war, economic decay, and disagreement among decision makers over which policy was appropriate for science prevented the adoption of a unified approach. In spite of rhetoric in support of science and technology and a call for a national science policy, officials provided funding on an irregular schedule, often making promises which they failed to keep. Combined with cold and famine in Petrograd and Moscow, this financial uncertainty placed an unbearable burden on Russia's scientists.

Scientists had welcomed the February Revolution. They had assumed that the Provisional government, as the kind of government they desired, would willingly accept their advice and provide the leadership necessary for the full flowering of Russian society. The promise of academic freedom and support offered by the liberal government had never materialized, however, owing to the fact that the Provisional government was short-lived and more concerned with life-and-death matters—world war, anarchy in the countryside, and the threat of insurrection from the right and the left—than with science and technology.

The February Revolution released scientists from traditional institutional, political, and social affiliations. Within weeks of the abdication of the tsar they had mobilized resources to establish a series of such national organizations as the Free Association for the Development and Dissemination of the Positive Sciences, based on utilitarian notions of the value of science for Russian society and culture. They sought to provide guidance for the government in matters of social and economic policy, believing that democracy would succeed only with the participation of the intelligentsia.

When the October Revolution had passed, scientists intended to avoid political involvement or to play an active role in serving the new regime. They saw the Bolshevik government as a pretender. But by the middle of 1918, scholars would have accepted almost any regime that offered the hope of domestic stability and financial support. After years of war and civil war, the academic community's ever more precarious financial, physical, and psychological condition forced it into an uneasy alliance with the government.

During the first years of their rule, while supporting the successful establishment of a series of major research institutes, Bolshevik policies and inaction actually made conditions worse for the scholar. His life was at risk and he was internationally isolated. He was starved literally for food, figuratively for scientific literature and camaraderie. This required the establishment of the Central Commission for the Improvement of the Living Conditions of Scholars, first established by the intelligentsia and then adopted by the government as a means to give the scientist, artist, and writer emergency rations, salary, and community. The scholars' lot had come full circle. They had come from conditions of great risk under the tsar, to hope under the Provisional government, and significant risk again after the October Revolution.

This chapter examines scientists' professional activities in response to the February and October revolutions, and their attempts to foster the development of a scientific worldview in Russia; the hardships that scientists endured during the years of revolution and civil war, which required rapprochement between the forces of science and government, and emergency action on the part of the Bolshevik regime; and the first tentative steps toward the development of science policy from the point of view of both the government and the scientist.

SCIENCE AND THE RUSSIAN REVOLUTION

Scientists greeted the Revolution with a firm conviction that the new government would support the scientific enterprise. They set out to reorganize their discipline, seeking financial commitment from the government, establishing new research institutes, and joining together in new societies and professional associations. Scientists in Petrograd alone created eleven societies in 1917 and 1918, and another thirty-six by 1925; only half of the societies existing before the Revolution, however, survived the first eight years of revolution.[2] At first, such societies were broad-based, popular organizations open to all. Scientists who joined them often shared a positivistic and utilitarian perspective. Members believed that knowledge was the source of all power, and that science was value-free. They believed that on the basis of scientific thought a truly democratic and free Russia could be created, and assumed that the government would embrace these ideals.

The most significant example of a broad-based, socially and culturally oriented organization was the Free Association for the Development and Dissemination of the Positive Sciences (Svobodnaia assotsiatsiia dlia razvitiia i rasprostraneniia polozhitel'nykh nauk, or SARRPN). The Free Association was founded on March 28, 1917, in a meeting at the Women's Medical Institute on the initiative of Academicians V. I. Vernadskii, I. I. Borodin, A. N. Krylov, V. A. Steklov, and I. P. Pavlov; Maxim Gorky, the Bolshevik engineer L. B. Krasin, the physicists N. G. Egorov and D. S. Rozhdestvenskii; and others. Many of the natural scientists had been active in KEPS, and pursued the same goal of a national science policy through SARRPN. In all, ninety-six "representatives of the exact sciences" attended the meeting. At the suggestion of Pavlov, the participants voted to name the aggregate of scientific-educational institutions represented by those in attendance the "Institute in Memory of 27 February 1917." The association would apply science to all of Russia's historical cultural problems: poverty, ignorance, backwardness, and unfulfilled economic promise.[3]

The public response to the Free Association was so overwhelming that it held three open sessions in April and May in Petrograd and Moscow at the Bolshoi Theater with such leading Russian political figures as P. N. Miliukov, A. F. Kerenskii, and N. N. Sukhanov, a member of the executive committee of the Petrograd Soviet, in attendance.[4] The Free Association played to larger and larger public audiences throughout the spring. Throughout the summer, its executive committee met irregularly to discuss questions of organization, finance, and activities.[5]

The leading membership of SARRPN included many of Gorky's associates from *Letopis'*, a journal he established in December 1915 as a vehicle to tap the power of "positive," exact science for the social and political purposes of raising Russian culture to a "Western level." The contributors to *Letopis'* included Ivan Bunin, Sergei Esenin, Vladimir Maiakovskii, Aleksandra Kollontai, Aleksandr Shaliapin, and the biologist and agronomist K. A. Timiriazev, who became the science editor for *Letopis'*. The attitude toward science as a panacea for Russia's cultural backwardness, so prominent in the writings of the contributors to *Letopis'*, carried over into the creation of SARRPN. The membership of the Free Association sought to convince the liberal Provisional government that "science"—Western culture based upon the principles of the experimental method of natural science—was the key to Russia's future.

There are two major features of the "science" espoused by SARRPN. One was the belief that applied research and technology would spur Russia's modernization and economic growth. The second major feature of the science of SARRPN was its positivistic and technocratic bent, as exemplified in the views of Gorky and the mathematician V. A. Steklov. In his addresses to the Free Association forums, Steklov, an academician, argued that Russia had two enemies, "Germany and ignorance." Regarding the latter, the members of the association shared the belief that Russia was culturally backward, that it needed to embrace Western, "positive" science to overcome this backwardness, and that its members' advice should be accepted fully by the government. The association was established as a freely and broadly constituted organization to manifest the creative genius of the nation in the natural sciences, for the happiness and prosperity of the people. This could be done only through "the development and perfection of the exact sciences and their methods." SARRPN's members also desired to apply scientific research, discoveries, and inventions to industry and "the needs of life in general." A third task, broadly connected to the first two, was to bring science to the masses, "primarily in the form of finished courses of a theoretical and practical character, and first of all in workers' groups." Finally, Steklov pointed out that the association hoped to establish in honor of the February Revolution an "institute of the positive science," with well-equipped research divisions, laboratories, libraries, museums, and studies.[6]

Through association with such practicing scientists as K. A. Timiriazev and the physiologist I. P. Pavlov, Gorky soon came to view science as a system of natural laws that could be mastered by empirical investigation. In as much as Western science exerted a positive influence on all social phenomena the embrace of science would lead inevitably to the creation of democratic institutions in all societies. Gorky's most direct elaboration of these ideas came in a lecture, "Science and Democracy," given in Moscow and Petrograd in 1917 at organizational meetings of the Free Association. Gorky came to see "science" as the positive science of Auguste Comte, developed on the "soil of exact observation, directed by the iron logic of mathematics." Gorky spoke of an international, universal science as the path to unity, freedom, beauty, and democracy. Russians in particular, he argued, needed to be "inoculated with respect for and love of reason, and feel its universal force." The problem was that "Russian history had knit for our people a dense network of conditions that had instilled . . . in the masses a suspicious, even hostile relationship to the creative force of reason and the great conquests of science."[7] The nobility had brought West European culture to Russia, but the masses, under the influence of the church, monarchy, and the pressures of day-to-day life, remained largely ignorant and inclined to the metaphysical and mystical. Science, under democracy, on the other hand, would destroy these disbeliefs which were "rooted in the Russian *narod* [people]."[8]

Toward these ends, Gorky envisaged the creation of a rather Baconian "city of science, a series of temples where each scholar is a priest who is free to serve his god," a city of well-equipped laboratories, museums, libraries, and an atmosphere of freedom to encourage creativity and "love for reason."[9] He acknowledged that his thoughts might seem utopian but offered the technological achievements of airplanes, submarines, the wireless telegraph, and the discovery of radioactivity as evidence that his utopia—a society built upon "the laws of reason and experimental science—could indeed be realized."[10]

Gorky adopted an almost technocratic view of science. He was not technocratic in the sense wholeheartedly embracing Taylorist, Fordist, and "Americanist" doctrines that promised productivity, expertise, and optimalization for the purposes of escaping class confrontation and social division.[11] For Gorky technocracy meant rule by scientific and technical elites, with science seen as a realm of knowledge and truth, politics as one of power, action, and corruption. He did not explicitly address the notion of technocracy or call for its introduction, but the general thrust of his writings and thoughts indicates that he envisaged a prominent role for the scientific intelligentsia in the government of Soviet Russia.

Steklov and Gorky hoped to attract "men of science" and especially biologists to the "Institute of February 27" to help combat disease, raise food production, and to overcome the generally low level of Russian life. This institute—with well-equipped research laboratories and freedom to work without government pressure—somewhat resembled Solomon's House in Francis Bacon's *New Atlantis*. The association worked throughout 1917 and 1918 to acquire a building for its staff and for the special institute, but budgetary problems continued to get in the way. It was frustrating to the members of the association that private donations such as those from Carnegie or Hopkins in the United States were no longer an option.[12]

The Free Association also sought to bring science to the masses. Beginning in spring 1918, it sponsored popular scientific lectures on botany, zoology, physics, chemistry, and general scientific subjects by the geologist and Academician A. E. Fersman, the physicist K. K. Baumgart, and others, often attracting large audiences. SARRPN's lecture bureau of G. A. Tikhov, Rozhdestvenskii, and Gorky organized twenty-two public lectures in 1918. This led to the creation of the Joint Association of Scientific and Higher-Education Institutions to acquaint the public with recent scientific advances through popular talks. The Joint Association also sponsored "field trips" to southern universities with the help of 237,000 rubles from The Commissariat of Enlightenment (Narodnyi komissariat prosveshcheniia, or Narkompros). By October, in addition to Petrograd

University, seven scientific research institutes and *vuzy* (institutions of higher education) had joined the Joint Association.[13]

Having failed to gain financial backing from the Provisional government, the membership attempted to work with the Bolshevik regime to secure support. Steklov was convinced that the Bolsheviks would support the development of the "positive sciences." Narkompros's V. T. Ter-Oganesov viewed the association as a powerful concentration of the best scientific forces of the capital, acting like a "parliament of scientific opinion" to help the government in all kinds of matters.[14] But on several occasions, while authorizing money for the "Institute of February 27" and other activities of SARRPN, the government failed to come through on its promises.[15] The long-term friendship and association of Gorky with A. V. Lunacharskii, the commissar of enlightenment, helped the Free Association to secure 1.8 million rubles from Narkompros. The Bolshevik government in turn pressed SARRPN for contributions to the "new construction of knowledge."[16] However, scientists did not accept Soviet power as willingly as they had the liberal government, and even avoided recognition of it until financial circumstances required it. There was, in fact, an environment of suspicion and mistrust, if not hostility toward the Bolshevik regime, which limited the influence and impact of SARRPN.

It was almost inevitable that Gorky and Lenin would clash over the activities and attitudes of SARRPN, as Gorky grew increasingly critical of Bolshevik policies. Gorky still believed, as he had during his close association with Lunacharskii and Bogdanov a decade earlier, that Russia, being eighty percent peasant, was not ready for socialism. He repeatedly called upon the intelligentsia to take action to rescue the country, to "cast aside political differences and unite in the common cause of saving the masses."[17] Having seized power, however, the Bolsheviks were unlikely to relinquish it to scientists and technologists—a group that in general preferred some sort of capitalist democracy to socialism. In response to repeated criticism in Gorky's *Novaia zhizn'*, the Bolsheviks shut down the newspaper several times, and eventually suppressed it on Lenin's orders.

Perhaps because of its wide range of activities, SARRPN did not achieve its goals. The members had quite unrealistic ideas about the role of science in society. They thought that simply appealing to "Western culture" or "science" would convince any government to adopt its methods for building democracy throughout Russia. Similarly, they believed that scientific activity was apolitical, not recognizing that their demands to have their program adopted required political activity. Gorky himself naively argued that science and politics could and must be kept in separate spheres. "When science invades or is forcibly drawn into the bloody filth of politics," he wrote, "it is not only the purity and freedom of science that suffers, but also the finest ideas and hopes of mankind, and the reasoning power of the entire world is destroyed."[18]

The Provisional government had not survived long enough to heed the advice of the members of the Free Association. After the Bolsheviks seized control, the Free Association faced a government which grew increasingly suspicious of its seeming political aspirations and social programs, especially during a period of civil war and War Communism (the period 1918–1992, characterized by strict state control of the economy, mobilization of labor, and forced requisition of food supplies from the countryside). In the midst of increasing administrative disorder throughout 1917, SARRPN's programmatic offerings and requests for support fell on deaf ears. Broadly based scientific societies proved inadequate as vehicles to express scientists' demands or meet the needs of government; scientists therefore soon turned to more narrow professional organizations, created along disciplinary lines (e.g., physicists, chemists, geologists), to articulate their interests and gain support to resume their research.

While political uncertainties prevented Gorky and other members of SARRPN from seeing their scientific utopia come to pass, the international isolation, shortages of materials and supplies, and famine and cold brought about by war with Germany, revolution, and civil war ultimately frustrated attempts to organize and harness science to achieve Russia's democratic potential. When such disruptions began to cause scientists to endure extreme physical hardships, Gorky again stepped forward, this time to establish the Central Commission for the Improvement of the Living Conditions of Scholars. As in Weimar Germany, where such institutions as the Notgemeinschaft der deutschen Wissenschaft "were created to help support German science and scholarship,"[19] the same individuals who had once pressed the government to utilize the intelligentsia to achieve Russia's great promise now had to organize to save scholars, through TsEKUBU, from personal and professional ruin.

TSEKUBU

In 1922, the Central Commission for the Improvement of the Living Conditions of Scholars (Tsentral'naia komissiia po uluchsheniiu byta uchenykh, or TsEKUBU), a governmental and scientific organization, was created to lessen the burden on Russian scientists, artists, and writers of increasingly harsh conditions—famine, cold, isolation, and confiscation of housing and books—through the so-called scholar's ration, and financial, health, cultural, and housing assistance, including monthly gold ruble allowances based on seniority and reputation. The intelligentsia also had access to dormitories, the Scholars' Club, and five specially established sanitaria.[20] The founding and activity of an ad hoc organization to assist scientists reveals the extent to which science was at risk in the early chaotic years of the new Soviet state, and how Gorky's technocratic and utopian views of science could not have been further from reality.

Conditions of scientific life had deteriorated during the first years of Soviet power as cold and shortages of heating oil and food restricted scientific activity. Schools closed. Many professors emigrated or moved southward. Many died, including the historians A. S. Lappo-Danilevskii and A. A. Shakhmatov and the economist M. I. Tugan-Baranovskii. Because of the paucity of information, it is hard to judge if scientists as a group were worse affected than the population at large; death lists are incomplete and often do not give the date or city of death. But the extent of dislocation, suffering, and death can be seen in Petrograd, whose population dropped 1.2 million between 1910 and 1920. Between 1918 and 1920, deaths exceeded births by 115,429, due mostly to evacuation but also to disease and starvation. Mortality reached 70 per 1,000 in 1919 as 65,347 died, over 7,000 of them by starvation.[21] The Academy of Sciences, located in Petrograd, experienced similar losses.[22] A 1921 list of the 174 scientists who died between 1918 and 1921 includes 10 Academicians and such noteworthy physicists and mathematicians as N. A. Gezekhus, N. E. Zhukovskii, L. S. Kolovrat-Chervinskii, A. M. Liapunov (by suicide), M. A. Rykachev, S. Ia. Tereshin, and D. K. Chernov. In 1922 another 100 were added to the list, of whom a number had died in earlier years, and many had died of typhus or been killed during the Civil War, including the physicists A. R. Kolli, K. N. Egorov, A. I. Efimov, I. I. Kosonogov, and A. V. Ignat'ev, and three more Academicians.[23]

Gorky turned his efforts to saving Soviet Russia's scientific talent from the physical hardships of the Revolution. After his election to the presidium of the executive committee of the Petrograd Soviet of Workers' and Peasants' Deputies, he sought to enlist "bourgeois specialists" in the reconstruction of Russia's productive forces (although he recognized that many were hostile to the current rulers), and to bring science and technology to the masses.[24] But scholars were unable to assist the government; they were ill-prepared, by virtue of their lifestyles, to struggle with workers for pieces of bread. Gorky therefore constituted an unofficial committee to alleviate the situation, gathering around himself a large number of volunteers and philanthropists. Gorky's group ran afoul of F. E. Dzerzhinskii and the secret police, or Cheka, who believed that all semiofficial activities should be approved and organized through the Council of Peoples' Commissars (Sovnarkom). However, when Pavlov petitioned Lenin for permission to go abroad to continue his scientific activity, and Academician A. A. Shakhmatov died, Lenin finally recognized the need to constitute an official organization in support of scholars.

Lenin reproached the Petrograd Party executive committee and its chairman, Zinoviev, for not doing enough to help the academic world deal with the physical hardships of Petrograd. He turned to Gorky for advice; during the next months, Gorky traveled to Moscow frequently for discussions with Lenin and his close associate, V. D. Bonch-Bruevich, about the activity of Gorky's committee.[25]

Commissar of Enlightenment A. V. Lunacharskii charged his deputy, the historian M. N. Pokrovskii, "with the goal of improving the conditions of scholars." On December 13, 1919, Narkompros began to centralize efforts to improve supply and procurement of goods. A committee was organized which included Timiriazev, Gorky and his unofficial group, and the chairman of

Narkomprod (the commissariat of supply), who was ordered by Bonch-Bruevich to establish emergency transportation of produce to Petrograd.[26] Finally, on December 23, the Sovnarkom resolved

> 1) to grant increased allowances to specialists in those scientific areas which are essential for the solution of problems referred to earlier; 2) to free these specialists from any kind of service (labor, military, etc.) not having relation to their scientific work; 3) to create for the scientific work of these specialists living conditions which guarantee . . . conveniences.[27]

Narkompros and the Scientific-Technical Department of the Supreme Economic Council (Nauchno-tekhnicheskii otdel vysshego soveta narodnogo khoziaistva, or NTO Vesenkha) were charged with compiling a list of specialists, and quickly assembled the names of 550 scholars, including 50 writers, to receive special rations with specified norms for flour, cereal, grain, sugar, meat, fish, sunflower oil, salt, tea, and tobacco.[28] Over the next five years TsEKUBU's list of scholars grew until it included nearly all of Russia's scientific and literary intelligentsia!

The new organization immediately ran into problems of governance, power, and inadequate access to resources. Largely decentralized, it distributed rations through local KUBUs which had to share their power with local authorities. Throughout 1920 and 1921, problems over rations and distribution of foodstuffs continued. In April 1920, Gorky wrote the Sovnarkom pleading for additional rations.[29] Later that year he wrote again, protesting the attempt of local governmental organs to confiscate ("requisition") provisions allotted by Narkomprod for PetroKUBU. The provisions were needed since the scholars gave six percent of their rations to workers at the Scholars' Club, to their laboratory assistants, and to library workers in other institutions, as well as sharing them with their families.[30] In February 1921, the Sovnarkom again attempted to cut rations and transfer the functions of local KUBUs to Narkomprod. But negative response from such larger committees as PetroKUBU and from individual scientists largely neutralized these efforts. Finally, Sovnarkom decrees of December 6, 1921, and January 16, 1922, created a central institution for support of scholars, with adequate resources and authority, uniting PetroKUBU, MoscowKUBU, and the provincial committees into TsEKUBU. TsEKU-

BU grew rapidly; its administrative staff numbered one hundred by the end of 1922.[31]

TsEKUBU provided financial, moral, medical, and "dietary" support to scholars, primarily through the so-called scholar's ration (*uchënyi paëk*). TsEKUBU established norms of rations of butter, salt, oats, flour, and, when available, meat, fish, and flour. In 1923, scholars' relatives became eligible for rations. TsEKUBU also provided monthly gold ruble allowances (based on seniority and reputation), five specially established sanitaria, and access to dormitories and the Scholars' Club.

In creating the scholar's ration, TsEKUBU encountered several interesting problems. Who would receive support? What forms would it take? And how would workers respond to governmental directives giving "bourgeois specialists" seemingly greater rights than their own?[32] An expert commission established in 1922 began to collect *ankety*, or questionnaires, on scholars, and, based on scholarly reputation in all disciplines rather than on the more narrow concerns of government economic and political interests, established a rating system for the *uchënyi paëk*.[33] In forty-two meetings in 1922 the commission rated some 7,500 scholars in five categories ranging from world-class to novice; most scholars lived either in Moscow or Petrograd, where the problems of cold and food supply were the greatest.[34] In addition, TsEKUBU lists in the countryside included a number of priests; rarely was a scientist, let alone a person with a "Marxist point of view," to be found in the countryside, except in the lowest ratings.[35] Not surprisingly, the government pressured TsEKUBU from the start to be more flexible in its ratings, and to take into consideration the "governmental significance" of the scholars, favoring applied scientists over those from the arts and literature. The government was also concerned that the ratings excluded several leading scholars—for example, the philosopher A. M. Deborin—who represented the newest, "most progressive" areas of science, such as Marxist thought.[36]

The commission processed *ankety* very slowly, leaving large numbers of scientists in a dire situation. And even with rations, scientists often had to rely on others and pool their resources. The physicist N. N. Semenov had three students, including the future Academician Iu. B. Khariton, staying with him on the *uchënyi paëk.*[37] The quantum mechanician V. A. Fock received a special ration through his close association with D. S. Rozhdestvenskii's Atomic Commission when it became clear in 1918–1919 that Fock was losing his strength from lack of food.[38] By 1924 through Gorky's efforts TsEKUBU had begun to function smoothly. It expanded rapidly throughout the New Economic Policy, serving almost twenty-one thousand scholars at its peak (see app. A, table 2).[39]

TsEKUBU also served a vital function for scholars in terms of financial support, health care, clothing, and professional contacts. In addition to rations, a scientific worker received gold rubles based on his ratings as a scholar. These awards significantly improved scholars' ability to purchase basic necessities and scientific equipment and supplies, since the scientific worker had "found himself with a salary of rubles that dwindled rapidly to less than the five-hundredth part of their original value."[40] The government awarded annually between 646,000 and 1,746,000 gold rubles under the TsEKUBU program from 1921 to 1927.[41] In addition, under TsEKUBU's supervision, sanitaria were built to help the elderly and infirm.[42]

With the assistance of PetroKUBU and MoscowKUBU, twenty local, provincial KUBUs worked "toward the elimination of arbitrary action against scholars in locales."[43] This meant redistribution of apartment space equitably among specialists and workers, reversing years of arbitrary and random confiscation of property. In October 1920, Gorky wrote VKUBU (the All-Union Committee for the Improvement of Living Conditions of Scholars, another forerunner of TsEKUBU) about problems of confiscation and shortages of laboratory, library, and living space. He noted that often when scientists undertook fieldwork, they returned to find their books, letters, and furniture missing. He asked the Petrograd executive committee to protect scholars from the "masses [*vseleniia*]."[44] The state recognized that the scholar required space to live in and to work in, and a library for his books. His family, too, now began to receive special consideration. In August 1924 the All-Union Central Executive Committee and the Sovnarkom placed limits beyond which space could not be taken from scholars to help solve the living space crisis. A Sovnarkom circular in 1925 helped resolve many of the remaining conflicts, but it was the common industrial worker who was shortchanged in these attempts to provide better living conditions for the scholar.[45]

Although it hardly resembled Solomon's House, the Scholars' Club (Dom uchenykh), established in 1921 under Gorky's leadership, was a vital center for maintenance of professional contacts through meetings, conferences, excursions, and presentations on recent research which Russia's scientists had miraculously managed to complete. After establishing the scholars' ration, Gorky directed his efforts toward these clubs which the visiting writer H. G. Wells described as "salvage establishments." There, Wells wrote, workers "not only draw their food rations, but they can get baths and barber, tailoring, cobbling and the like conveniences."[46] Another important function served by the Scholars' Club was the access it provided to current Western literature. Russia had been cut off from Western scientific contacts and publications since 1914, and domestic publication returned to a normal level only in 1924.[47] Dom uchenykh was the central point for receipt of foreign literature and talks on recent scientific research. For example, in its first years, L. S. Berg, F. Iu. Levinson-Lessing, E. V. Tarle, S. F. Platonov, S. P. Kostychev, and I. S. Osadchii spoke there on one or another area of natural science. Physicists P. P. Lazarev, O. D. Khvol'son (on atomic structure), Iu. A. Krutkov (on general relativity), and A. F. Ioffe (problems in contemporary physics) also gave talks.

It was here that H. G. Wells met Academicians S. F. Ol'denburg, A. P. Karpinskii, and I. P. Pavlov, all looking "careworn and unprosperous." The scholars pressed Wells about recent scientific progress outside of Russia, and impressed upon him the importance of sending them scientific publications. They made a list of all the books and publications that they needed, which Wells brought to the Royal Society in London with the hopes that the three or four thousand pounds required for their purchase and transport might be forthcoming.[48] Gorky later thanked Wells for the thirteen volumes received at the Scholars' Club, each of which was reviewed by a different member and read publicly at one of the Saturday meetings.[49] Ultimately, the Club subscribed to well over one hundred foreign journals.[50]

Many of the privileges accorded scholars through TsEKUBU and Dom uchenykh were taken largely against the wishes of the Workers' Opposition and their supporters in the Communist party. The Workers' Opposition faction sought to ensure the worker's preeminence both within the Party hierarchy and in the administration of industry.[51] In the latter case this meant that the worker should take precedence over the bourgeois specialist, manager, engineer. But the Opposition, which included one of Gorky's former collaborators on *Letopis'*, Aleksandra Kollontai, believed that the Party had abandoned the ideal of the workers' state, turning too much power over to the bourgeois specialist of the Tsarist era, and abandoning collective administration in favor of one-man management. It was anathema to see scientific workers—individuals who worked with their minds, not with their hands—receiving more food and material goods than the manual worker, whose lot was no better. Nevertheless, Bolshevik leaders recognized that without the assistance of those specialists, they might never realize "socialist construction." They therefore approved a program "to move toward the path of socialist construction" with TsEKUBU as "the necessary organ of government assistance," with the help of Narkompros, Vesenkha, and other bureaucracies.[52]

Through its programs, TsEKUBU successfully rescued a large number of Russian scholars from the displacement of the Revolution. Even the American Relief Administration, a major effort to relieve the famine, supported Soviet scientists through TsEKUBU. In 1924 six physicists at the Leningrad Physcio-Technical Institute, M. A. Levitskaia, Ia. G. Dorfman, V. K. Frederiks, P. I. Lukirskii, and F. A. Miller, received ARA rations in this way.[53] While the Revolution saw the creation of tens of new scientific and artistic-literary societies, however, such broad-based and loosely defined groups as SARRPN and TsEKUBU failed to make inroads in terms of generating support for research and training activities.

The reasons for this failure are not hard to find. First, broad-based associations appeal to higher ideals, not practical needs. In the decaying urban centers of Petrograd and Moscow, scientists might be interested in discussing the true mission of Russian culture and the role that scientists would play, but war, civil war, famine, and cold generated more immediate personal concerns. The leading members of the Free Association also had other things on their minds: the resumption of research, the publication of results, the reestablishment of ties with their Western colleagues.

The large-scale associations founded on idealistic conceptions of the relationship between science and government soon gave way to societies created along narrower, disciplinary lines, such as the Russian Association of Physicists, and to newly founded institutes where the true research activities of science took place.

EARLY SOVIET SCIENCE POLICY

When the Bolsheviks seized power in October 1917, they quickly, and somewhat haphazardly began to nationalize industry, to promulgate agricultural programs, and to establish a legal and administrative apparatus. How soon did science enter into these problems of "government construction"? What was Bolshevik policy toward science and technology? How did that policy determine the direction of scientific activities? And in what way did they differ from the policies of the Provisional government?

The failure of the Provisional government to support the scientific enterprise is not surprising. The February Revolution had a disruptive impact on universities and scientific research institutes, from which it was hard to recover. Few universities reopened in the spring of 1917, and the increasing politicization of the student body accelerated "the complete dislocation of academic life."[54] Intense economic and financial depression and broad social displacement, combined with the incessant interference of the soviets, in particular the Petrograd Soviet of Workers' and Peasants' Deputies, prevented the government from gaining popular confidence, making good on the advice of the intelligentsia, or developing programs for its support.[55]

The Provisional government took several steps to support science and education. Professor Manuilov, the new minister of education, who had been fired by the notorious L. A. Kasso as rector of Moscow University in 1911, recalled all students expelled for political reasons. He then dismissed all faculty appointed by Kasso, allowing those released in 1911 to resume their duties. Universities also gained autonomy from local administration and were given "a full measure of self-government."[56]

Nonetheless, a crisis imperiled Russian science and education. The Revolution and the prolongation of the war exacerbated the processes of economic dislocation and skyrocketing inflation, putting strains on already tight resources and making it nearly impossible for the weakened machinery of administration to adjudicate the competing demands of an ever-increasing number of organizations on the government. The Provisional government can be blamed for being too passive in the face of these problems. But widespread urban decay, anarchy in the countryside, and political uncertainties generated by the "dual power" of the government on the one hand and the soviets of workers' and peasants' deputies on the other diverted attention from science until after the Bolshevik coup.[57]

There was also no systematic science policy in Soviet Russia until around 1926. A number of departments, institutions, and individuals advanced programs and opinions about "science"; they ranged from the extreme left where the Proletarian Culture, or Proletkult, movement presented its program for "proletarian science," to the Main Scientific Administration (Glavnoe upravlenie nauchnymi, nauchno-khudozhestvennymi, muzeinymi i po okhrane prirody uchrezhdeniiami, or Glavnauka) of Narkompros with its support of fundamental research and academic autonomy, and to the Scientific-Technical Department of Vesenkha, which emphasized applied research in the interests of the state.

Scientists took advantage of the divergence of opinions and the absence of a unitary agency to expand scientific activities far beyond the limitations of the Tsarist regime; they sought out and received financial and administrative support from among the competing ministries and other institutions in a time of severe financial constraints. While such bureaucracies as Narkomzem and Narkomzdrav—the commissariats of agriculture and health—had scientific research institutes under their respective jurisdictions, the majority of institutes of the fundamental sciences—chemistry, biology, and physics—were under the control of Glavnauka and the Scientific-Technical Department of Vesenkha.

Until the formation of the Scientific-Technical Department and Glavnauka, the government worked through the Academy of Sciences to communicate with scientists. The Academy of Sciences did not play an important role in fundamental research in the early history of Soviet science, although six physics institutes were formed under its jurisdiction before 1922. The Communist party's subjugation of the Academy during the First Five-Year Plan, the transfer of the Academy to Moscow, and the creation of the Physics Institute of the Academy (FIAN) under S. I. Vavilov in 1934 led to the establishment of large-scale, government-funded physics research institutes within the Academy. Over the next five years, such institutes as LFTI were also transferred from the control of other bureaucracies into the Academy. Since World War II the Academy has been the site of the most important fundamental research undertaken within the USSR.

The Academy was, however, important as a channel for communication between the Bolsheviks and scientists in the first days of the Revolution. It is also a good case study of the relationship between the state and science, and how scientists and government officials learn to cooperate or are coopted to meet the other's needs. The Academy maintained its distance from the political events of 1917, insisting upon its autonomy "before any government"[58] and rejecting Narkompros's overtures until financial constraints required that it agree in the spring of 1918 to assist the new regime in "government construction." Until that time, its members preferred to conduct the relations between scientists and government through such organizations as KEPS and SARRPN. Finally, at the request of commissar of enlightenment A. V. Lunacharskii and other high party officials, a special committee of the Academy met to consider a government request that it look into the range of activities it could perform.

At first this special committee voted to inform the government that it would answer each query depending upon the scientific content of the request and the resources available to the Academy. At a general meeting in February 1918, however, the Academy voted to support Bolshevik endeavors but emphasized that it saw itself as a purely scientific institution rather than as an economic one dedicated to applied or technological research. Later, at the urging of Narkompros, and after the receipt of over 2.2 million rubles for publication and other activities, the Academy agreed to participate in the efforts of the government to rebuild the economy, primarily through KEPS.[59] By the summer of 1918, several other Academy and government organizations had been established to examine the relationship between science and government and to consider how to bring scientists to participate in governmental activities. These included Vesenkha's Committee on the

Attraction of Specialists, its Scientific-Technical Department, and several predecessors of Glavnauka. Glavnauka and the Scientific-Technical Department competed for influence over matters of science and technology policy until the late 1920s, each striving to become a sort of commissariat of science, and each hoping to increase the number of scientific research institutes and workers under its jurisdiction, largely because of philosophical disagreements over the proper approach to take toward scientists and their enterprise.

Those who thought that scientists best knew how to run scientific research institutes, that they were the ones who could best formulate research programs, identify fruitful problems and subjects for research, and control publication and contacts with Western scholars, tended also to believe that scientific advances would find their way into the economy quickly, of their own accord, and without interference from the government. Accordingly, such persons would be inclined to believe that decentralized administration and increased autonomy for the scientist in selection of research programs and operation of research institutes were paramount. Those individuals who believed that the scientist under the Soviet regime—who was a product of the Tsarist era—would support ivory tower programs stressing knowledge for knowledge's sake tended also to believe that leaving the administration of institutes in the hands of such a person would be no guarantee of a sense of responsibility to the government and of little use to Bolshevik leaders trying to rebuild a war-torn economy, educate the illiterate peasantry, and develop a socialist society with the help of applied science and technology. The first position came to be associated with Glavnauka and the second in varying degrees with the Proletarian Culture movement and the Scientific-Technical Department.

The Scientific-Technical Department
of Vesenkha

Throughout the 1920s—indeed until the very present—there was pressure in the Soviet Union to establish a supreme body, a sort of commissariat of science and technology, for the coordination and administration of scientific research. Beginning with the Scientific-

Technical Department of Vesenkha, and continuing with the Department of Science of Gosplan (the State Planning Commission) in the 1930s, the State Committee for the Introduction of New Technology into the National Economy of the USSR (Gosudarstvennyi komitet vnedreniia novoi tekhniki v narodnoe khoziaistvo SSSR, or Gostekhnika, founded in 1947), and the current State Committee for Science and Technology (Gosudarstvennyi komitet po nauke i tekhnike or GKNT, founded in 1965), Soviet policymakers have sought to create an administrative solution that would increase the productivity and quality of national scientific performance. Such a body would assign problems for research proposed by various commissariats to research institutes and would watch over the research done within individual institutions; it would strive to eliminate duplication in scientific activity; it would consider, if not control, all budgets for R and D. The centralization of chemistry and physics research within the Commissariat of Heavy Industry in 1931 was perhaps the result of this pressure. The demand for decentralization of research, on the other hand, has been equally powerful, with Soviet scientists seeking to control the research programs and budgetary allocations of their institutes, with a minimum of government scrutiny. This opposition to centralization of R and D, together with the haphazard creation of research institutes and the burgeoning Soviet government bureaucracy throughout the early and mid-1920s, led to the creation of several separate bureaucracies, each entrusted with the administration of science and technology.

Soviet leaders stressed quite early the importance of improving the administration and coordination of scientific research. By the beginning of 1918, "there was talk of an independent body" with those responsibilities.[60] N. P. Gorbunov, apparently at the behest of Lenin, was charged with the establishment of a commissariat of science and technology. From the start, however, Narkompros opposed the creation of this body.[61] Narkompros sought to establish a Russian association of science to unite all scientific establishments under its jurisdiction to ensure its ascendancy in these matters. Both Lenin and Gorbunov disapproved of this proposal.[62] But the opposition of Narkompros to Gorbunov's plan, together with that of the Academy, led instead to the creation of a body responsible primarily for applied science, the Scientific-Technical Department. Established by a decree of the Sovnarkom on August 16, 1918, the department was to have broad functions: organizing scientific research throughout industry; facilitating contact between institutions and scholars in the Soviet Union and abroad; assisting in the creation of new institutions; and coordinating scientists' activities on a national scale; however, it never became the independent commissariat that Gorbunov envisaged.[63]

In a letter to Lenin in November 1918, Gorbunov expressed his frustration in working with Russia's scientists, his failure to get results in the organization of science, and his need for Lenin's moral support. He asked, Was his work important or necessary for the republic? He would continue to try to attract scientists to "government construction," but spoke of how hard it was to get them to overcome the inertia that had for decades frozen them to the same spot. There had been some progress, with such representatives of the old professoriat as N. G. Egorov, director of the Palace of Weights and Measures, coming over to the Soviet side. And, referring apparently to the newly established Atomic Commission funded by Vesenkha, he pointed to atomic physics as a science touched by the organs of Soviet power and an area where the fire of creativity had begun to burn. But these examples were too few. The reorganization of science was a task of gigantic proportions. According to Gorbunov there were not enough Communists working on science, and he fretted that nothing would come of his efforts.[64] Although Lenin's response is not extant, it is clear from the next letter Gorbunov wrote to him that Lenin provided the necessary encouragement to get him to continue with his previous energy.[65]

From the start, centralizing and decentralizing tendencies competed in the department. At its first meeting on September 10, 1918, Gorbunov, an adherent of the principle of "accountability," pushed for the creation of the Scientific-Technical Department as a centralized body designed to ensure fruitful ties between science and industry, and to overcome problems in the administration and supply of science in the Soviet Union, including shortages of literature, chemicals, and apparatus.[66] Ultimately, Vesenkha created a central scientific collegium whose members were drawn from the Academy of Sciences, KEPS, the Moscow Scientific Institute, and other organizations, and whose purpose was to "lead, unify, and coordinate" the research performed within its institutions in support of government economic programs. By 1923 the central apparatus of the department, which grew rapidly to develop the administrative capability to impose accountability, consisted of a secretariat and administrative-finance, economic, and revision sections, ten central scientific-technical councils, which represented such different regions of science and technology as physics, chemistry, and the paper, food, and metallurgical industries; several standing committees; libraries and publishing houses; a patent office; five industrial enterprises; and a growing number of scientific institutes and laboratories. Excluding the ten scientifictechnical councils, the department had 1,708 employees of whom 49 worked on the central staff, 1,036 in institutes and laboratories, and 515 in the enterprises.[67]

Effectively, centralized administration and strict accountability did not follow from this organization for four major reasons. First, independent NTOs (scientific-technical departments) grew up in several cities—Moscow, Petrograd, Kiev, and Kharkov—and the central organ found it difficult to administer the activity of all institutions within its purview, let alone to evaluate their budgetary requests, and in fact encouraged them early on to be independent, under the supposedly watchful eye of the collegium of the Scientific-Technical Department. The Moscow and Petrograd NTOs (MONTO and PONTO), for example, were organized independently of each other and had a great deal of autonomy, with respect to the central body, over the funding and coordination of research; each had subsections covering the major region of science.[68] Although in some sense MONTO and PONTO overlapped in terms of scientific interest, they did not do so in terms of territory or the research institutes under their jurisdiction. Furthermore, scientists, so-called bourgeois specialists, occupied most of the positions of responsibility on the ten councils of the national body, ensuring their domination of the science policy process. The duplication which Gorbunov had sought to avoid thus already existed and persisted throughout the early history of the department.[69]

There were similar problems with the effort to centralize the administration in the Scientific Department of Narkompros, the forerunner of Glavnauka. On one occasion, in 1919, the department took its Petrograd division to task for not having informed

Table 1.   Growth of Scientific Research Institutes of NTO VSNKh,
1923–1930

Year	Number of Research Institutes	VSNKh Budget for (in rubles)
1923–24	13	2,100,000
1924–25	11	3,100,000
1925–26	15	11,900,000
1926–27	32	17,800,000
1927–28	34	32,500,000
1928–29	41	58,000,000
1929–30	50	108,600,000

Sources: B. I. Kozlov, *Organizatsiia i razvitie otraslevykh nauchno-issledovatel'skikh institutov Leningrada, 1917–1977* (Leningrad, 1979), pp. 31–33, 41; *Bibliograficheskaia ukazatel' trudov NTO VSNKh, ego institutov i laboratorii* (Moscow, 1926); F. E. Dzerzhinskii et al., *Nauchnye dostizheniia*, p. 34; *Trudy nauchno-tekhnicheskogo upravleniia, ego institutov i laboratorii* (Moscow, 1928); M. Ia. Lapirov-Skoblo, "Puti rosta i itogi NIRy promyshlennost' za 12 let," in *Sotsialisticheskaia rekonstruktsiia i NIR* (Moscow, 1930), p. 31; and Iu. N. Flakserman, *Promyshlennost' i nauchno-tekhnicheskie instituty* (Moscow, 1925), pp. 20–160. The latter work provides a detailed analysis of the activities of eleven different institutes under the jurisdiction of the Scientific-Technical Department.

the center of funding it had approved for the first meeting of the Russian Association of Physicists.[70] In the mid-1920s Glavnauka fixed rules for the calling of congresses, and required organizational bureaus of those congresses to report to it.[71]

Second, the rapid increase in the number of scientific research institutes naturally led to decentralization of decision making in the central Scientific-Technical Department. The number of scientific research institutes in Russia, including experimental stations, publishing houses, libraries, and full-fledged centers of research activity, grew from 21 to 85, and perhaps to as many as 105 between 1918 and 1928. Leningrad was a major center of scientific activity; there were 13 scientific research institutes under the Scientific-Technical Department in Leningrad alone in 1925. Since this excludes the Academy of Sciences, the number no doubt is still higher. The growth of the Scientific-Technical Department was especially rapid in the late 1920s (see table 1).

Third, the secure financial position of the Scientific-Technical Department and its ability to support research institutes in a number of different ways—*spetssredstva*, contracts, and consultancy agreements—contributed to the broad discretion the institutes had in developing their own research programs and in maintaining autonomy vis-à-vis the government. *Spetssredstva*, a special government fund for the payment of unforeseen expenses and to increase the salaries of workers, attracted scientists and their institutions to come under the umbrella of the Scientific-Technical Department; this also led to difficulties in the staffing of institutions under Narkompros.

From 1921 through 1923, Ioffe's institute encountered problems in maintaining staff size, or getting permission from Glavnauka to hire new workers; in the Soviet system some higher instance always had to approve hiring and firing. Ioffe therefore sought out *sverkhshtatnye*, additional staff which could be hired above the allowable number with funds from other government sources, production contracts, or *spetssredstva*. He was adept at securing funds from outside of Glavnauka to exceed the limits placed on LFTI staff. By 1926 LFTI had thirty-four scientific personnel, sixty-two technical personnel including fifty-three *sverkhshtatnye*, and twenty-three administrative personnel including ten *sverkhshtatnye*.[72] (See app. B, table 1, for staff of the Leningrad Physico-Technical Institute excluding *sverkhstatnye*.)

The Scientific-Technical Department also had funds for "consultants" and contracts. The Leningrad Physico-Technical Institute came under the jurisdiction of Glavnauka. But A. F. Ioffe was able to get funding from such other bureaucracies as the Scientific-Technical Department through all three means. He consulted on his specialty, which was "X-ray analysis of the solid state," bringing needed income to his institute. Revenue from contractual services appears to have been far more important. Although it has not been possible to calculate just how much income from contracts scientific research institutes received, it is clear that most of them relied on contracts for supplementary income. The Scientific-Technical Department's subdivision on applied physics under the chairmanship of Iu. V. Vul'f focused its resources on questions of experimental and theoretical physics, especially concerning the production of precision instruments, contracting with the State Optical Institute for the construction of a glass instrument factory and with Ioffe's institute to produce vacuum tubes and instruments, among others. In fact, many of the chief administrations for each branch of industry, or *glavk*, of Vesenkha could contract directly with scientific research institutes, without the approval of the department and with their own funds; the chief administration of the electrotechnical industry, Glavelektro, contracted with Ioffe's institute.[73]

In one final important way—through the structure and function of their academic councils, which were their chief administrative bodies—scientific institutes managed to maintain relative autonomy in the face of efforts to centralize authority under the Scientific-Technical Department. As the case of LFTI demonstrates, the academic council of the institute, and not Vesenkha's NTO, was the single most important body in the determination and coordination of research programs. The minutes of academic council meetings at the institute reveal that physicists devoted attention to questions of physics research, not government relations (see app. B, table 3).[74]

Physicists used formal reporting requirements to communicate to the government audience that investigations completed or in progress had an applied component which was commensurate with the needs of Soviet power. In a report written in the summer of 1922, V. R. Bursian detailed the expansion of LFTI's four-part program of research and defended its seeming distance from the concerns of "government construction." He noted that physicists not only undertook research of a scientific nature but also used it in "fulfilling practical technological tasks."[75] Bursian stressed the importance of planning in uniting problems of pure science with those of everyday life.

Scientific research institutes did not have complete control over their destiny, however. The Scientific-Technical Department pressed the institutes to account for all aspects of their activities, from internal administration to international scientific contacts and publication. But here, too, the economic recovery of the NEP facilitated expansion of research activities, and relative autonomy prevailed until the late 1920s.

How successful was the Scientific-Technical Department in promoting programs which ensured that scientific research remained close to the "productive process"? Leading administrators of Vesenkha sought to organize research institutes within the leading branches of industry—electrotechnology and applied physics, the metal, chemical, paper, and textile industries, fuel and heat engineering technology—and to keep them under the supervision of the scientific councils of the Scientific-Technical Department. Discussions of the proper role of the department via-à-vis these scientific research institutes often took the form of criticism of Glavnauka's "ivory tower" approach to science. P. A. Bogdanov, chairman of Vesenkha RSFSR, noted that rather than having a well-thought-out and coordinated science policy, capitalist Russia had independent industrial laboratories, scientific institutes, and schools of individual scholars. Glavnauka had adopted these organizational forms, preserving prerevolutionary attitudes about the importance of theoretical science and ignoring the importance of applied science and production. The department would solve this problem, bringing science closer to the demands of production through its research institutes for purposes of the workers' government.[76] Factory laboratories, which might, for instance, have played a significant role in uniting science and production, had a limited impact on R and D in early Soviet Russia until the late 1920s.[77]

But it was difficult for the Scientific-Technical Department to steer between the demands of the research institutes under its jurisdiction and the demands of industry. Scientists dominated the scientific-technical councils, and often sympathized with the interests of the institute scientists in undertaking fundamental research at the expense of applied research. Feliks Dzerzhinskii, head of the Cheka—the secret police (later the GPU)—and chairman of Vesenkha, advanced a program which supported the relative autonomy of institutes and the decentralization of their management while encouraging research that would further industrialization. He aimed at a balance between direction by the Scientific-Technical Department and individual initiative and independence for the scientific institute; he warned against making scientific research too dependent on the promotion of industry. He believed that research and industrial interests should be recognized as different from each other, that scientific research institutes should have broadly defined purposes, not the narrow interests of an industrial trust.[78] In so doing, Dzerzhinskii fought centralization of R and D and won the confidence of engineers and scientists.[79] It was after the death of Dzerzhinskii, when V. V.

Kuibyshev became director of Vesenkha, that centralizing tendencies appeared in the administration of the scientific research institutes by the Scientific-Technical Department.[80]

GLAVNAUKA

In contrast to NTO, Glavnauka, the Main Scientific Administration of Narkompros, experienced major financial difficulties in the first years of its operation. It was charged with providing the basic operating expenses of most of the institutes of fundamental science—chemistry, biology, and physics, including LFTI—in the early postrevolutionary period. Institutes were given enough to continue to operate, if not to expand. This led some physicists to complain that their organizational and research activities had been hindered by the penuriousness of Glavnauka.[81]

One of the reasons for the hesitancy with which Glavnauka moved to fund the scientific research institutes under its jurisdiction was an almost Proletkultist attitude toward science shared by many of its officials. The proponents of proletarian culture, largely political activitists to the left of the Bolshevik party, had only an occasional scientist among their number; many had a mistrustful attitude toward "bourgeois specialists" and science in general. Proletarian science would be based on the principle of "collective" scientific activity, and on planning to avoid the duplication endemic to bourgeois science. The new Bolshevik organizations would supervise, regulate, control, and finance science to protect it against the vestiges of bourgeois society. Many scientists, on the other hand, feared that the Proletkultist position would diminish, if not entirely eliminate, the autonomy of their science. While Proletkult organizations had little influence on the actual conduct of research, they had great importance in setting the direction of science policy within Narkompros in the first years of Soviet power. The Proletkult movement was strongest in such cities as Petrograd and Moscow where a majority of scientific research institutes were located; many of the Proletkultists also worked within such government organizations as Glavnauka. They were thus able to influence the direction of national policy toward science. Such bureaucrats as A. V. Lunacharskii, who helped found Proletkult after the Revolution but was inclined to give former Tsarist scientists and educators leeway in the management of their institutes, therefore also paid attention to the notion of "proletarian science."

While the organizations of Proletkult did not long survive the Revolution, Proletkultist attitudes toward science nonetheless infiltrated the managerial philosophies of the forerunners of Glavnauka, all of which came under the jurisdiction of the Commissariat of Enlightenment. The scientific department of the Commissariat of Enlightenment of the Union of Communes of the Northern Region, for example, sought to do away with the Academy of Sciences and its *sharlatanizm* (here, a self-conscious and pretentious attitude toward science), and to create a communist science compatible with "collective ideals";[82] it failed in this effort. The scientific department gave way to the Petrograd Department of Scientific and Higher-Education Institutions, organized in 1919, which in turn became the Petrograd Administration of Scientific and Artistic Institutions of Akadtsentr, and later still the Leningrad Department of Glavnauka (which functioned until 1926), under M. P. Kristi.[83] Finally, in December 1921 Glavnauka arose under the chairmanship of F. N. Petrov, as a result of the reorganization of Akadtsentr (the major organizational and ideological organ of Narkompros, responsible for general education, higher education, and several research institutes).[84]

At the start, Glavnauka faced criticism from such moderate leaders of the Bolshevik party as Gorbunov. In the spring of 1918, Narkompros had proposed that a Russian association of science be established under its jurisdiction to supervise all scientific research in Soviet Russia,[85] a move opposed by Vesenkha's Scientific-Technical Department. Directed by radical intellectuals who were not as radical as the Proletkultists, Glavnauka lacked a following and had sympathizers neither in the leadership of the Party nor among the rank and file. While Narkompros managed to keep the universities, libraries, theaters, and scientific research institutes open with its subsidies, it was criticized for "lack of Bolshevik tough-mindedness."[86] The practical result of the criticism was that it had severe problems in following through on its programs, especially because of budgetary shortfalls.

This lack of political support, criticism of its performance, and financial weakness affected Glavnauka's early activities and the performance of scientific institutes under its jurisdiction. The criticism centered on two interrelated issues: the alleged lack of planning of its institutes' research activities; and its inability to compel institutes to do more research of an applied nature. Regarding the former, a plethora of new scientific institutions were formed soon after the seizure of power by the Bolsheviks in October; this opened bureaucracies such as Glavnauka to the charge that many of these institutes were superfluous, having little relation to government economic programs. In 1922, officials of Glavnauka granted that the number of new institutions had increased rapidly in the first years of the Revolution, and that existing ones had expanded "completely ignor[ing] the real potentialities" of these new institutions. As a result, some institutions had been closed and others were being cut back. But the officials warned that Glavnauka's institutions "stood at the crossroads where further reduction would be the ruin of many of them and dangerous to the Republic."[87] Nevertheless, the Sovnarkom seems to have ordered Glavnauka to close some institutions and dismiss members of its own staff: between January and November–December of 1922, Glavnauka's national staff shrank from 14,000 to 7,275 persons, and its Petrograd staff fell from 6,424 to 3,933 (of whom 1,749 worked within the Academy of Sciences and other scientific centers). Funds were also very short.[88]

Neither criticism of Glavnauka was entirely fair. As the case of LFTI demonstrates, a great deal of planning and discussion went into the creation of that institute. Furthermore, the institute's physicists in fact undertook applied research as well as production of scientific instruments. On the other hand, Glavnauka's financial difficulties often intruded upon the life of the scientists, forcing them to turn to NTO and other organizations for funding. In the second half of the 1920s, however, as a result of NEP, Glavnauka recovered from its early growing pains and its loss of staff; it employed 8,400 persons of whom 5,700 were "scientific" personnel. The number of scientific research institutes under its jurisdiction was estimated to have increased fourfold by 1926.[89] Judging by another source, the rate of growth was not quite so rapid (see table 2). As can be seen, during the 1920s Glavnauka's importance for institutes of fundamental research, and especially for physics organizations, increased markedly.[90]

Glavnauka had four departments: science, museum administra-

Table 2. Scientific Institutes under Glavnauka RSFSR, 1922–1927

Kind of Institute (by discipline)	1918	1919	1920	1921	1922	1922–23	1923–24	1924–25	1925–26	1926–27
Physics and mathematics						17	15	27	28	30
Natural sciences						7	9	21	21	22
Humanities						6	6	16	16	15
Kraevedy societies	200	248	317	372	410	477	606	663	87[a]	120[a]
Artistic						8	8	8	7	7
Other (libraries, etc.)						23	29	22	23	27
Total scientific research institutes	21	26	30	34	42	79	81	81		90
Total						538	673	757	182	221

Sources: *Itogi desiati let* (Moscow-Leningrad, 1927). p. 24; *Pervaia otchetnaia vystavka glavnauki narkomprosa*, pp. 8–11; and F. N. Petrov et al., *Desiat' let sovetskoi nauki* (Moscow-Leningrad, 1927), pp. 23–24.
[a]Only those on state budget are included for 1925–26 and 1926–27.

tion, art, and nature conservation. The scientific department had five subdivisions: physics and mathematics, natural history, humanistic sciences and pedagogy, libraries and publishing houses, and scientific and *kraevedy* societies, which were organizations that studied local customs and lore. The Party assigned Glavnauka four basic functions: (1) scientific research in the natural sciences and humanities; (2) creation of a network of museums for educational purposes; (3) research in the arts in support of economic and cultural "construction"; and (4) art and nature conservation.[91]

More specifically, the tasks of the scientific department were the planning of scientific activity to establish the "highest correlation" with the needs of "government construction"; development of scientific techniques; the recruitment of scientists and the bringing of scientific institutions into its jurisdiction.[92] Like the Scientific-Technical Department, Glavnauka administered and supported publication of research, the purchase of instruments from abroad, and foreign travel.[93] In 1922, for example, it supported, eighteen different journals of which five covered medicine; four, geography and biology; and five, the physical sciences, including *UFN* and *ZhRFKhO*.[94]

CONCLUSIONS

On the eve of the First Five-Year Plan, most fundamental physics, chemistry, and biology institutes in the Soviet Union were under the jurisdiction of Glavnauka. Nevertheless, despite its power and its ambitions, Glavnauka was in the end displaced by the Scientific-Technical Department and most of its scientific research institutes were transferred to the latter's jurisdiction. There had been increasing criticism of Glavnauka for failure to coordinate the activity of its affiliate research institutes and to use its funds to advance more applied research, not research which was perceived as pursuing knowledge for knowledge's sake. The battle against the bourgeois specialist and the introduction of the five-year plans and Stalinist policies of science, which required rationalized, comprehensive, centralized plans and scientific research that contributed to the growth of industrial production, all ensured the decline of Glavnauka's influence in matters of fundamental research. But in the heyday of the NEP, Glavnauka had protected fundamental scientific research in Soviet Russia from encroachment by unfriendly government officials.

A major factor in the success of Soviet scientists in founding scientific research institutes after the Bolshevik Revolution, paradoxically enough, was the absence of a national science policy. The two major government science organizations, Glavnauka and NTO, failed to coordinate their activities, and in fact competed with each other for influence. Scientists were thus able to secure support in the hard times of revolution, civil war, and War Communism, if not through Glavnauka or Vesenkha's Scientific-Technical Department then through contracts with factories or other bureaucracies, and many of the research centers created after the Revolution arose primarily as a result of the initiative of physicists, chemists, and biologists, not as a result of a master plan. Such is the case of the Leningrad Physico-Technical Institute.

3

The Russian Association of Physicists and the Founding of the Leningrad Physico-Technical Institute

The Russian Revolution created conditions under which physics would grow as a discipline; at the same time, physicists suffered from extreme conditions of famine and cold, international isolation, and shortages of research materials and equipment. The conditions of physical hardship slowed the attempts of physicists to resume normal research and instructional activities. They approached these problems through newly established research institutes and the Russian Association of Physicists (Russkaia assotsiatsiia fizikov, or RAF), which was founded in 1919.

Political turmoil and economic uncertainty belabored Russian physicists' efforts to create an environment suitable for the conduct of research. Until the fourth All-Union Congress of Physicists in 1924, the physicists endured trying financial circumstances and organizational headaches because of the failure of the government to support the association in a consistent fashion. The political challenges of the Workers' Opposition and Kronstadt Rebellion diverted government attention. The Revolution and Civil War destroyed *smychka* (the trade and economic ties between city and countryside) and led to a decline in industrial production to levels lower than in 1913, creating economic chaos. The "Red-Specialist" conflict, based on deep mistrust between the communist and the expert who had been trained in the Tsarist period, often interfered with efforts to establish cooperation between the government and scientists. In the period 1918–1921, the government responded to these challenges with War Communism, which was characterized by strict state control of the economy, mobilization of labor, and forced requisition of food supplies from the countryside, and which did little to promote normal conditions for science and technology.

In this environment, not surprisingly, the RAF proved inadequate to the task of rejuvenating the discipline. Physicists therefore turned to such central research facilities as the Leningrad Physico-Technical Institute (LFTI) with their professional and scientific demands for regular funding, insulation from the vagaries of policy changes, and support for increasingly expensive "big science," which required collective effort and sophisticated equipment.

This chapter examines physicists' self-perceptions and professional activities, and how they struggled with persistent problems facing their discipline after years of war, revolution, and international isolation. Its concluding discussion of the formation of LFTI and its research program highlights the haphazard paths by which scientists and governments agree to establish large-scale research institutes that provide scientists with organizational and financial stability while simultaneously serving the government's interests.

THE FORMATION OF THE RUSSIAN
ASSOCIATION OF PHYSICISTS

The Russian Association of Physicists, founded through the efforts of O. D. Khvol'son, A. F. Ioffe, and others in 1919 as an adjunct to, and later in place of, the physics section of the RFKhO, sought to develop conditions propitious for the normalization of physics research activity. Through the RAF, physicists lobbied the government to acquire basic necessities—above all else, money, but also reagents, equipment and machinery, and textbooks; maintain such professional activities as publication and conferences; and reestablish international contacts.

Physicists did not fare well initially with either the Provisional government or the Bolsheviks. But their few successes were due to the efforts of Orest Danilovich Khvol'son (1852–1934), the senior physicist of the Petrograd community. Khvol'son played a major role in the popularization of physics in Russian and Soviet society. In 1869 he entered Petersburg University, where he began work on magnetism and electricity. He continued this work in Leipzig, returning to Petersburg in 1876 as a privatdocent and defending his doctoral dissertation in 1880. He became a professor at Petersburg in 1891, and taught at several other institutes, and at Higher Women's Courses. After the Revolution he taught workers at the Communist University. He is best known for the encyclopedic five-volume *Kurs* of physics written over a twenty-year period, and for over thirty popular books on science, including the first book published in Russian on X rays (1896). Khvol'son's efforts to popularize science earned him the enmity of Marxist philosophers who suggested that he was hostile to materialism; after all, Lenin had attacked Khvol'son in *Materialism and Empiriocriticism*. Not known as an original thinker and considered a pedestrian lecturer, he nonetheless inspired the respect of his peers. He was president of the RFKhO, honorary president of the RAF, and an honorary Academician. He even received in 1926 the title "Hero of Labor" and the Order of the Red Banner of Labor, but it is not clear if these awards were as significant personally as those from fellow scientists.

During World War I a special commission on educational supplies, with S. I. Sozonov representing chemistry and O. D. Khvol'son representing physics, was established to help meet the pressing needs of the exact sciences. Seeking to build on the work of the commission, Khvol'son turned immediately to the Provisional government after the February Revolution, but the government addressed the need for a subsidy for physics only in September 1917, and the subsidy failed to materialize.[1] In a parallel effort, throughout 1917 the RFKhO asked the Provisional government for publication subvention. But these funds, too, were not forthcoming.[2]

Petrograd physicists at first barely took note of the Bolshevik uprising, but as the winter dragged on they had no choice but to recognize the government at hand. On December 12, 1917, fifty members of the physics section gathered to discuss business as usual and elected O. D. Khvol'son president, with an executive committee which included A. F. Ioffe, N. A. Gezekhus, and V. F. Mitkevich. By the time of their next gathering, the coup and continued street fighting had intruded upon the physicists' activities. The danger of meeting after dark led the executive committee to vote "to gather only on Sundays or other holidays, following the example of many other societies and institutions."[3]

Their ever-more-precarious financial situation forced all scientific professional groups and institutions to consider formal discussions with the new government. Khvol'son reported on two meetings he had attended at the Academy of Sciences, chaired by its president, A. P. Karpinskii, where scholars discussed the creation of a "union of scientific and higher-educational institutions and scholarly societies for the defense of science." The physics section thereupon selected A. N. Krylov as its representative in meetings with the government.[4] There is no evidence that Krylov met with government officials, however.

By the spring of 1918, with attendance at periodic meetings dropping to less than twenty-five and sources of support dwindling, physicists commenced a long, frustrating search for government backing for the RAF and its journal. A series of government bureaucracies approved funding, but none provided funds. At the end of March, Khvol'son approached commissar of enlightenment A. V. Lunacharskii and his assistant, Z. G. Grinberg, to discuss the physicists' precarious situation. It was the first time in his life he had met face-to-face with a government representative. He was surprised to receive funds "on the spot" for the Special Commission on Educational Supplies (a commission later subsumed within Narkompros) and the physics section.[5]

In May, the Small State Commission of Narkompros voted to give the *ZhRFKhO* 10,000 rubles for 1917, and 50,000 for 1918.[6] In June, in response to another petition to Narkompros, the RFKhO received 25,000 rubles, of which 10,000 went to the physics section, enabling *ZhRFKhO* for 1917 finally to be delivered to press. In October 1918 the Physico-Chemical Society registered with the Soviet of Workers' and Peasants' Deputies of the Vasilevskii Island district, and began to submit regular, formal budgetary requests to the government.

The physicists' organization demanded funds for publication subvention and the establishment of international and all-union contact among physicists. More than anything else, severe paper shortages handicapped publication efforts. Since physicists had received few journals from abroad since the onset of the war, domestic publication assumed monumental importance. The major Soviet physics journals—*ZhRFKhO, VRiR, UFN, Trudy GOI, Izvestiia IFBF, UFN*, and the *Doklady* and *Izvestiia* of the Academy of Sciences— were published regularly only after 1924. Between 1917 and 1923 journals appeared infrequently (see app. C, table 1) or, in the case of *Vestnik rentgenologii i radiologii*, for a handful of issues before disappearing forever. The receipt of fifty tons of paper from the United States in 1921 helped reestablish publication of some Academy journals, although problems persisted into 1924.[7]

The case of *ZhRFKhO*, the physicists' most important journal, reveals the extent of the problem. In 1919, the editorial board of *ZhRFKhO* apologized to its readers for tardiness in publication over the past year; only through the efforts of O. D. Khvol'son could they publish their journal at all, Khvol'son having secured yet another special award from Glavnauka's Grinberg.[8] In 1921, the journal again failed to be published. Regular publication of the journal resumed only in 1924.

Uspekhi fizicheskikh nauk (UFN), like *ZhRFKhO*, struggled through publication at irregular intervals for the first six years of its existence, with only three volumes published between 1918 and 1923. *UFN* was important as a forum for Western research, devoting up to two-thirds of its pages to translations of papers by Bohr, Eddington, Jeans, Rutherford, and other Western physicists. The journal also published several fundamental Soviet articles, including the first article published in Russian on the general theory of relativity, by V. K. Frederiks in 1921.

Problems in publication contributed to the international and domestic isolation of Russian physicists. A. F. Ioffe reached the conclusion that it was necessary to create a national organization of physicists. On December 1, 1918, he reported to his Petrograd colleagues that the participants of an October congress in Moscow on pyrometry, radiology, and roentgenology had sanctioned the creation of a more active national society. The Petrograd physicists thereupon called an organizational meeting for the RAF. The congress would coincide with a celebration of the fiftieth anniversary of the Mendeleev periodic chart in February 1919. The organizational committee—Khvol'son, Ioffe, Rozhdestvenskii, I. V. Obreimov, later of Ioffe's institute, and others[9]—secured emergency funding from Narkompros enabling over one-hundred physicists to meet in early February in a "melancholy Petrograd of iced-over buildings without food or heat or the promise of reimbursement of expenses."[10] The rector of the Third Petrograd Pedagogical Institute, A. P. Pinkevich, promised free housing in institute dormitories and meals in the dining hall, but participants were warned to be prepared for the harsh life of Petrograd: "It is necessary to bring with you some foodstuffs, e.g., sweet crusts, sugar, tea, jam, sausage, meat, and so on."[11]

At the organizational meeting of the RAF, physicists debated the proper relationship between science and government and measures needed to resume professional life. Within the government in 1919 there were three basic positions concerning the proper relationship between the government and scientists. The most radical position was espoused by the Proletkultists, who saw the Bolshevik Revolution as an opportunity to advance notions of "proletarian culture," to subject the former Tsarist scientist to class war, and to create purely proletarian institutions of science, government, and culture. Officials of Vesenkha believed that the scientists should be subject to state control through centralized administration of R and D to ensure accountability to Bolshevik economic and political programs. Narkompros represented a more moderate position, permitting the relative autonomy of the so-called bourgeois specialists in the formulation and administration of research programs.

The congress opened amid mutual declarations of the need in principle to support scientific and technological research. A. I. Kaigorodov, who represented A. V. Lunacharskii and M. P. Kristi for Narkompros at the meetings, expressed the government's interest, in principle, in supporting the development of physics, partly through the creation of a broad national association of physicists.

A. F. Ioffe, speaking next, drew a pessimistic picture of the problems facing Soviet physicists, but one in keeping with his view of the importance of the research "school" for modern physics. First, he called for the creation of an all-union association to disseminate professional information among scholars in a country whose communications had been destroyed by war and civil war. He emphasized the need for the establishment of broad-profile, government-supported research institutes along the lines of

Kamerlingh Onnes's Leiden laboratory for low-temperature physics. He outlined possible approaches to their organization and
administration, and offered the research program and structure of
LFTI as a case in point. The construction or purchase of complex
research equipment had not been feasible under the Tsarist system
because of their high cost and technological sophistication. Under
Soviet rule how would this change? In Ioffe's view, the founding of
several major research institutes to facilitate equipment-intensive
physics research on the structure of atoms, molecules, and crystals
was the key to ensuring that physics research in Russia measured
up to the world level.

Without a steady flow of well-trained researchers, however,
there would be no one to operate the new apparatuses. Ioffe spoke
of the need to set up graduate programs in newly established *vuzy*
(higher-educational institutions) and university physics laboratories. Courses would familiarize students with recent advances
in physics and new methods of research which were crucial to contemporary molecular and atomic physics; this program was realized in the establishment of the physico-mechanical department at
the Petrograd Polytechnical Institute in 1919. In sum, Ioffe argued
that the rejuvenation of Russian physics could be guaranteed only
by the creation of an all-union association with sufficient resources;
the creation of new research institutes; access to modern instrumentation; a training program for young physicists; and the resumption of publication.[12]

By and large, other leading physicists shared Ioffe's views. It is
true that such scholars as V. A. Anri, P. I. Lukirskii, and O. D.
Khvol'son expressed concern about the danger of excessive concentration of scientific forces within a handful of large-scale institutes. Khvol'son suggested that university laboratories alone might
suffice.[13] Others expressed the fear that "provincial physics"
would be left behind, a fateful comment on the true state of affairs.
But as talks by L. S. Kolovrat-Chervinskii on the Radium Institute,
Iu V. Vul'f on the Solid State Institute, D. S. Rozhdestvenskii on
the State Optical Institute, P. P. Lazarev on the Institute of Physics
and Biophysics, and A. E. Fersman on KEPS demonstrate, Ioffe's
call to resurrect Russian physics through government support of
these institutes found broad response.

The congress concluded with resolutions announcing the formation of the RAF, and setting forth the physicists' demands in no uncertain terms: a call for government support of scientific research and freedom in the selection of research topics. In exchange, the physicists promised to assist the state in the matter of "government construction." "The congress welcomes and recognizes the increase in the number of research institutes [whose] sum total now covers almost all [areas] of physics," the resolution read. These institutes were "the best means in order to ensure the rise and flowering of physics." Through the organization and concentration of scientific forces, physicists hoped "to create new possibilities for research" in well-equipped, modern facilities.[14] Turning to education, the physicists considered it their "obligation to inform the government about the dangers of the insufficiency of young people in higher schools which threatens Russian science. . . ." Without this, "the present generation of physicists will be without successors."[15]

The council of the RFKhO physics section became the executive organ of the Russian Association of Physicists, with Ioffe as chairman and Rozhdestvenskii, V. F. Mitkevich, and D. A. Rozhanskii as members. The governing body of the RAF would consist of a council of physics institutes with representatives of the Academy of Sciences and nine institutions including LFTI, the State Optical Institute, the Radium Institute, and the Institute of Physico-Chemical Analysis of KEPS. The RAF established two committees to advance its programs, one consisting of Ioffe, P. P. Lazarev, A. N. Krylov, and V. A. Anri to coordinate the receipt of foreign literature, instruments, and equipment, to reestablish and organize professional ties between the RFKhO and other societies, and to look into ways of increasing *komandirovki*, or research trips, abroad; and a second to consider the role of scientific research institutes in Soviet Russia, with Anri, N. S. Kurnakov, and L. A. Chugaev as member.[16]

The RAF acquired enough financial stability to fulfill its national responsibility only with the fourth congress in Leningrad in September 1924. In the interim, the RAF met in Moscow in 1920, in Kiev in 1921, and in Nizhni Novgorod in 1922. The first meeting of the RAF, called by members of the Moscow Physical Society (MFO), was held in Moscow in September 1920. Leading Petrograd physicists assisted in its organization. The RAF faced head-on the problems of budgetary and equipment shortfall, international isolation, and interruption of publication. The physicists again requested government funding for a physics journal as the organ of the association, and an insert or separately published abstract of foreign scientific literature; and they pressed for foreign contacts and trips abroad, as well as government permission for institutes and *vuzy* "to acquire books and instruments from individuals," an appeal to the government to allow some private trade during War Communism. Difficulties in shaking funding loose from Glavnauka to cover personnel, publication, and operating costs continued throughout the year.[17]

Lack of funds had not, however, precluded the development of a broad national research program. At the organizational meeting of the RAF in February 1919, physicists had discovered that their discipline was vital, and perhaps not as isolated from European achievements as they had thought. Scholars delivered over sixty scientific papers on such diverse topics as optics and spectrography, the strength of materials, rectification of alternating current, relativity theory, and quantum theory. A talk by V. A. Anri on the Bohr model of the atom generated discussion by Iu. A. Krutkov and A. F. Ioffe. Ioffe later delivered a paper on quantum theory.

At the 1920 Moscow congress there were sectional meetings on such topics as spectral analysis, the physics of crystals, thermodynamics and the kinetic theory of gases, stereoroentgenoscopic measurements, and physical chemistry, with over 100 papers given in all. Ioffe, Lazarev, and Bursian also discussed the activities of their respective institutes. There were 500 participants (with 400 from Moscow).[18]

During the year before the next congress, the RAF severed its official ties with the physics section of the RFKhO (although overlap of most leading personnel moderated this independence). Its membership had expanded to include sixteen scientific institutions and societies in ten cities.[19] But this institutional growth masked severe organizational problems which became apparent at the second congress of the RAF, held in Kiev in 1921. The Kiev Physico-Mathematical Society, the host for the congress, and the council of the RAF had hoped that physicists throughout the country would attend, seeing that Narkompros Ukraine had funded the gathering. Unfortunately, the Civil War disrupted plans, and of the nearly 500 individuals interested in attending, only 278 managed to make their way to Kiev; very few physicists came from Moscow and Petrograd.[20] Furthermore, of 125 papers proposed for the conference, only 59 were delivered.[21]

The position of the RAF continued to deteriorate. On October 1, 1921, the RAF ceased receiving funds from the state, although the Bolsheviks repeatedly called on the physicists to help in "government construction." Throughout 1922, Ioffe and Vul'f engaged Glavnauka in lengthy correspondence over the precarious existence of the association.[22] The RAF turned to its now seventeen institutional members for support. Only four institutions agreed to provide funds, while three declined and the rest remained silent. LFTI gave 10.5 million of the nearly 25 million rubles the association acquired. Only because of this was the association able to enter the 1922–1923 year with a surplus.[23]

The government was so strapped for funds that the RAF sought out private sources to pay for their third annual congress at Nizhnyi Novgorod in 1922. V. K. Lebedinskii, his Nizhegorod Radio Laboratory (NNRL), and the Nizhegorod Association of Physics sponsored the congress. The RAF struggled with the authorities over the issues of housing, food, and pocket money. The organizers reduced their original budget from 1.8 million rubles to 800,000, 600,000, and then 500,000 rubles, but were unwilling either to cut the budget further or to invite only those with the wherewithal to attend. Glavnauka apologized for the difficult circumstances: "Recognizing that it is absolutely desirable to call the congress, Glavnauka . . . does not have the wherewithal to subsidize it" because of a decision of Narkompros that "all congresses of a theoretical character have been turned down until after the new harvest."[24]

At the last moment, in mid-August, the Nizhegorod Fair Committee gave 200,000 rubles to the RAF, and the organizational committee sent telegrams by wireless to announce the convention date. The Commissariat of Ways of Communication gave half-price rail tickets to all attendees; the provincial organization of health, Gubzdrav, located housing for 200 participants; the provincial military committee, Gubvoenkom, and the radio laboratory gave support to the leadership of the RAF; and Lebedinskii's staff acted as chauffeur, service center, electrician, and rapporteur.[25] In all, 239 physicists representing twenty-two cities were able to attend, with 94 from Nizhnegorod, 62 from Petrograd, 45 from Moscow (and only one from each of eleven cities), including 44 professors and 58 college teachers. The organizers considered the conference successful in view of "external conditions" and evidence that scientific work, at long last, was "returning to normal conditions."[26] The physicists selected Ioffe as president of the RAF, with Vul'f as his deputy and Khvol'son as honorary chairman.

In fact, however, normal research conditions had yet to be established. The RAF protested inadequate government support, having failed to secure a long-term commitment for the physics enterprise. The situation was particularly precarious in the provinces and with respect to new university and *vuzy* training programs. The absence of a regularly printed journal prevented a large number of completed works from being published. When would the Bolsheviks make good on the promise to support the RFKhO, RAF, and their journal?[27] Until governmental bureaucracies themselves sorted out which organs were best suited to fund physics research or how much independence scientists should have, the RAF would find it difficult to achieve its goals.

THE ESTABLISHMENT OF THE
LENINGRAD PHYSICO-TECHNICAL INSTITUTE

Soviet physicists faced the organizational, physical, and psychological problems of revolutionary Russia with vision and resolve. While encountering continued uncertainties in the funding of the RAF, they convinced Bolshevik leaders to support the creation of several research institutes. These facilities—the Leningrad Physico-Technical Institute, the State Optical Institute, the Radium Institute, and the Institute of Physics and Biophysics, to name the most important—became the locus of physics research in Soviet Russia, and soon displaced universities and the Academy of Sciences as the center of scientific activity.

The reasons for LFTI's success in creating a basis for its expansion as the preeminent physics institute in the USSR highlight the fact that Russian physicists encountered problems both peculiar to the Soviet situation and common to physicists in other countries. First, scientists took advantage of growing state interest in supporting scientific R and D. Since the turn of the century, governments had recognized the importance of science and technology for economic development, international security, and prestige. Germany, Great Britain, and the United States began to establish such state-supported research institutes as the Kaiser Wilhelm Gesellschaft, and to fund universities at higher levels. Industry also supported R and D in extensive laboratories at firms such as Siemens AG, Westinghouse, and General Electric.

In Soviet Russia, too, the state recognized the crucial role science could play in meeting the national priorities of applied research, instrument building, and even basic science. It was no easy task, however, to secure government support and maintain it at levels necessary to expand research programs, hire adequate staff, and produce machinery, equipment, and instruments. LFTI would be forced to contract with a number of governmental organizations to provide supplementary funding, and to develop its own facilities for the production of scientific instruments.

Second, just as in Europe and America, farsighted scientific leadership was a prerequisite to the organization of modern Soviet research institutes. Kamerlingh Onnes in Leiden, Niels Bohr in Copenhagen, Ernest Rutherford at the Cavendish Laboratory, and Ernest Lawrence in Berkeley had proved to be first-rate scientists and able administrators. A. F. Ioffe provided this leadership in the Soviet Union. As president of the RAF, representative on several government scientific advisory boards, and editor of major physics journals, he was in a position to advance policies which promoted the expansion of the physics enterprise. Within his institute, this leadership included establishing the crucial foci for physicists' efforts.

Unlike their colleagues abroad, Soviet physicists expended a great deal of energy reestablishing activities considered "normal" to the conduct of physics. They faced civil war, the physical decay of their cities, and a government inattentive to their needs. Even with the help of the RAF, several years would pass before physicists resumed regular publication, reopened dialogue with their European counterparts, or created a graduate program for young scholars.

As a first step Ioffe joined with M. I. Nemenov to create the institutional basis for his research program. Nemenov, a famous roentgenologist and founder of the Russian Society of Roentgenologists and Radiologists who worked in the Women's Medical Institute, believed that questions concerning the application of X rays in medicine could be answered only within the framework of a new institute. He had sought government support since 1910. Ioffe, of course, believed that establishing research institutes would put Russian physics on the right path. Primarily because of financial obstacles and shortfalls of staff, Ioffe and Nemenov agreed they should marshal forces to create one institute in which the interests of both medicine and physics could be concentrated.

In April 1918, Khvol'son, V. I. Vernadskii, L. S. Kolovrat-Chervinskii, Nemenov, Ioffe, and others met at the request of V. T. Ter-Oganesov of Narkompros to consider Nemenov's proposal for the organization of the new institute. This committee initially recommended that the institute be formed within the Petrograd Homeopathic Hospital, and then suggested that it be housed within the Women's Medical Institute (ZhMI) because of shortages of space, funding, and staff.[28] Only by the end of September did the Small Regional Commission of Narkompros approve the project, first as an autonomous unit within ZhMI and then as the independent physico-technical and radiological division of the State Roentgenological and Radiological Institute (Gosudarstvennyi rentgenologicheskii i radiologicheskii institut, or GRRI), after a brief battle with B. V. Verkhovskii, the director of the Women's Medical Institute, who opposed the separation for financial reasons. In late October, Narkompros's Z. G. Grinberg gave Ioffe permission to take on "all matters touching upon the organization of the Roentgenological and Radiological Institute from . . . Petrograd ZhMI." On May 6, 1919, Narkompros officially incorporated GRRI under its jurisdiction.[29]

GRRI consisted of four departments: physico-technical (fiziko-tekhnicheskii otdel or FTO), headed by A. F. Ioffe; M. I. Nemenov's medico-biological sector; the optical division, which soon became the State Optical Institute (Gosudarstvennyi opticheskii institut, or GOI), headed by D. S. Rozhdestvenskii; and the radium division, under L. S. Kolovrat-Chervinskii. The government gave GRRI a broad mandate within which it found the leeway to undertake a substantial fundamental research program, including scientific research on radioactivity, other kinds of radiant energy, and their applications in medicine and technology; dissemination of information among specialists and the "broad masses" through lectures, courses, and the publication of journals and books; the founding of museums; and the development, testing, and production of physico-technical instruments.[30] While the statutory authority for the creation of the institute existed, however, physicists now faced the problems of securing facilities, personnel, and regular government support. Unlike Nemenov, who succeeded in finding a home for the medico-biological department at the former site of the Homeopathic Hospital, Ioffe encountered protracted difficulties, and for almost five years was forced to occupy several laboratory rooms at the Petrograd Polytechnical Institute which were available only thanks to the generosity of its director, V. V. Skobel'tsyn.

Toward the end of 1918 and throughout all of 1919, Ioffe and Nemenov were involved in the organization of GRRI, addressing the complex and time-consuming problems of structure and administration, equipment shortages, staffing, and development of a research program. GRRI's administrative difficulties were exacerbated by two factors: first, the two departments of the institute shared only an artificial connection from the start; second, Ioffe and Nemenov alternated as president of the institute, a post which itself increasingly became a formality. An annual report to Glavnauka referred to this "burdensome" and confusing problem in no uncertain terms.[31] Ioffe and Nemenov came to believe that the best solution would be to break GRRI into several independent research institutes, a view shared by N. Kamen'shchikov, a representative of Narkompros in charge of overseeing academic institutions.[32] A commission on the reorganization of GRRI, under the chairmanship of Glavnauka's M. P. Kristi, with N. Kamen'shchikov, A. F. Ioffe, V. I. Vernadskii, L. V. Mysovskii, and M. I. Nemenov, was established to consider the proposal.

The commission learned about the purely formal connection between the departments and their virtual scientific and administrative-economic independence. Ioffe expressed the concern that the creation of three autonomous institutions might interrupt ties between scholars and weaken educational functions. After a two-hour meeting on November 23, 1921, the commission approved the formation of the new institutes, creating a joint academic council to meet Ioffe's concerns, and called upon the government to help locate facilities for them.[33] The three institutes were the Roentgenological and Radiological Institute, the Radium Institute, and the State Physico-Technical Roentgenological Institute (Gosudarstvennyi fiziko-tekhnicheskii rentgenologicheskii institut, or GFTRI),[34] which was referred to variously as GFTRI, GFTI, and LFTI in documents throughout the 1920s. Ioffe became the director of GFTRI, a position he held until December 8, 1950; A. A. Chernyshev became deputy director.

The charter of the State Physico-Technical Roentgenological Institute gave it broad responsibilities and wide latitude in controlling its own affairs. Its research program included investigation of X rays, electronic and magnetic phenomena, and the structure of matter; the application of this work in technology; the production and testing of instruments; and even the dissemination of scientific information among the "broad masses" through the establishment of experimental stations, museums, and libraries. It had the right to convene commissions, hold public lectures, print scientific and educational works, call congresses, organize expeditions, and even send its members on domestic and international komandirovki.[35] But the new charter did not resolve the persistent problems of budget, equipment, paper shortages, and staffing. In an annual report to Narkompros in 1920, physicists complained that "the organization of the institute is still far from complete, for example the equipping of the building for the physico-technical department has not been finished, so that the entire department is located within the walls of the polytechnical institute which [itself] is experiencing great difficulties in heating. [This] tells upon the work of the department."[36] The institute proposed installing an office representative in Moscow to press the government over these problems, and requested over ten million rubles for the representative and his staff;[37] the government did not approve this project.

Before beginning normal operation, LFTI had to secure a suitable building. Within the center of Petrograd few buildings were appropriate, even with repair and renovation, to serve as physics laboratories. There was a housing shortage, and the urban problems of noise, road vibration, and dirt would interfere with complex research. More to the point, LFTI was closely tied to the Petrograd Polytechnical Institute, located roughly ten kilometers from the city's center. Scientists at the Ioffe Institute often taught at the polytechnical institute, whose students—numbering in some semesters thirty to forty—turned to LFTI for their laboratory practica. With the help of M. P. Kristi of Glavnauka, Ioffe found a two-story brick building which had served as a military psychiatric hospital before the Revolution, across the street from the polytechnical institute and located in the Sosnovka, a park on the outskirts of Petrograd. The building required extensive renovation and repair, but its location and size were ideal.

Narkompros had few resources. Materials, electrical equipment, pipes, laboratory equipment, and office furniture were in short supply. But the institute was able to take advantage of Glavnauka's jurisdiction over and inventory of existing museums and scientific centers in Petrograd. It obtained from the Winter Palace items "without artistic value," including a solid oak table, lamps, a clock, curtains, a piano, padded chairs for the library, and a rug; chemicals with accompanying shelves and containers, other laboratory materials, instruments, books and journals, and marble slabs were removed from the Agricultural Institute.[38]

Unfortunately, Narkompros could do little to forestall the instrusion of cold into the institute's world. Heating fuel shortages persisted throughout the winter of 1921–1922; Ioffe was obliged to petition Narkompros for almost fifteen cords of wood for heating and production of gas.[39] Removal to the new quarters was also hindered by deficiencies of funds. The institute's physicists contributed long hours to the new laboratory, combing the neighborhood and region to locate materials and equipment. While engaged in moving furniture, cleaning, and repairing, physicists managed to continue their research during these months, and often worked through the night on their householding and menial tasks in order to return to research in the morning. Often, especially on Sundays, scientists gathered to tackle the dirtiest work of fixing equipment and cleaning the new building. The amount budgeted for the move itself was also small.

By January 1923 the move to the new building had begun: the heating system was operative; electrical construction was nearly complete; the telephone system worked; furniture was in place; piping for the lighting and laboratory networks was soon completed; and finally, a new transformer from Moscow's Glavelektro was put in place, as was a thirty-kilowatt generator. Throughout

Table 3.   Structure of LFTI, 1924

Department and Laboratory	Director
Physics department	A. F. Ioffe
Laboratory of electronic phenomena	N. N. Semenov
Laboratory of molecular physics	I. V. Obreimov
Laboratory of heat engineering	M. V. Kirpichev
Laboratory of theoretical physics	V. R. Bursian
Technical department	A. A. Chernyshev
Laboratory of electrical oscillations	L. S. Termen
Electrovacuum laboratory	Ia. R. Smidt
X-Ray laboratory	N. Ia. Seliakov
Laboratory of roentgenographic analysis	M. M. Glagolev

Sources: *A LFTI*, f. 1, op. 1, ed. khr. 17, ll. 8–9; and Sominskii, p. 253.

the move, V. V. Kuibyshev, the director of Glavelektro Vesenkha, and later head of Vesenkha, who had met Semenov in Samara before the Revolution, assisted in securing materials and expediting requisitions.[40] On February 4, 1923, the new institute held a gala attended by M. P. Kristi, representatives of the Petrograd scientific community, and employees of the institute. A. F. Ioffe praised the physicists' efforts and the government's help in opening the institute in a talk entitled "Science and Technology." V. R. Bursian discussed the institute's organizational history.[41] Ioffe was moved by the relative "luxury" of his institute, which was equipped "like a European scientific institute." He confided to his wife his plans to expand research, facilities, and production, even before his physicists had settled into a new routine, with funds from new contracts totaling fifty thousand rubles.[42]

LFTI consisted of physics and technical departments (with a production division and workshops), each with four laboratories (see table 3), most of which were under the direction of individuals from Ioffe's Petrograd physics circle. Ioffe formed a laboratory of general physics about this time to conduct research in the areas of his greatest interest: the strength of materials, the development of new kinds of accumulators for high voltage, the study of the electrical properties of insulation and of the photoeffect in insulators. From its first day, LFTI maintained a formal structure which emphasized its dual nature as a basic and an applied research institute. In this way physicists sheltered theoretical endeavors from charges that institute activities lacked a close link with "practice."

The success of LFTI in creating a firm institutional basis was due to the development of a diverse institutional research program of basic and applied physics, and production activities of interest to government and physicists alike. The institute's physicists sought to develop a healthy environment for the rapid transmission of research results from basic science into production and back. They embarked upon theoretical endeavors, while concentrating their efforts in basic science on investigations of the electrical, mechanical plastic, and elastic properties of the solid state. They also sought applications in the physics of conductors and insulators, X-ray techniques, communications, and heat engineering for problems of government construction. An original program of applied research enabled the Ioffe Institute to assume prominence among Soviet physics institutes in terms of both the development of a national research agenda and political relations with the new regime, while efforts in theoretical physics assumed importance in creating a sense of community among Petrograd physicists.

Theoretical pursuits initially took place in a suprainstitutional environment consisting of two specially funded commissions for atomic and molecular physics whose members cut across several institutes. The Molecular Commission was set up by Ioffe to support research on the physics of crystals and the properties of molecules in the solid state. D. S. Rozhdestvenskii organized the Atomic Commission at the State Optical Institute in Petrograd in 1919 to promote investigations into Bohr's theory of the atom.

Dmitrii Sergeevich Rozhdestvenskii was born in 1876 in Petersburg. His father was a middle-school teacher. Upon finishing gymnasium with a silver medal in 1894, he entered the physico-mathematics department at Petersburg University. His early interests included the biological sciences, but he quickly found a home in physics, especially under the influence of the experimentalist N. G. Egorov, a professor of the Military Medical Academy who taught a course on spectral analysis at the university. He became a laboratory assistant to Egorov at the academy in 1900. Like most promising young Russian scientists, Rozhdestvenskii went abroad twice to study, first with O. Wiener in Leipzig in

1901 and then with Paul Drude in Hessen in 1903. He returned to Petersburg University in 1903 at the invitation of Khvol'son and Borgman, and began to participate in scientific sessions of the RFKhO in 1909.

Like A. F. Ioffe, Rozhdestvenskii was an experimentalist whose experiences with the backwardness of Russian industry convinced him of the need to reorganize Tsarist science and technology, in this case instrument building and optical glass, an interest that grew out of his work in the analysis of spectra. Also like Ioffe before him, Rozhdestvenskii created a scientific school. Beginning in 1907, the year of Ehrenfest's arrival in Petersburg, Rozhdestvenskii began to equip an optics laboratory at the university, and followed the example of Ioffe and Ehrenfest by forming a physics circle. By 1915 he had six assistants. Later that year he became the director of the Physics Institute of Petrograd University, and in 1916 an ordinary professor.

In 1912 Rozhdestvenskii completed his master's thesis, which provided a new method for investigating the anomolous dispersion of light. World War I interrupted further work in this area. There were only a handful of enterprises in the entire empire which manufactured optical glasses, including the Obukhovskii works. Before the war Germany monopolized almost all production. During the war and after, Rozhdestvenskii worked to develop an indigenous industry through KEPS. The facilities of the optical department of KEPS and Rozhdestvenskii's laboratory at the university were combined to form the State Optical Institute (GOI) in 1918. The optical institute initially was one of four departments of the State Roentgenological and Radiological Institute, created at the suggestion of Ioffe who recognized that Russia had in Rozhdestvenskii one of the world's outstanding specialists for the study of radiant energy and the spectrum. Before moving to its own quarters on Vasilevskii Island, GOI occupied fourteen rooms at the university. Rozhdestvenskii served as its director from 1918 to 1932. In the mid-1930s, after he and his institute came under attack for failure to combine theory and practice in the work of its physicists, Rozhdestvenskii became increasingly depressed, and in 1940 he committed suicide.[43]

The activities of Rozhdestvenskii's Atomic Commission centered on the electrical fields of the nucleus, magnetic interactions between electrons, X-ray spectra of nuclei, and atomic spectroscopy. Meeting first on January 21, 1920, the commission provided a forum where such scholars as Rozhdestvenskii, A. N. Krylov, V. R. Bursian, Iu. A. Krutkov, and A. A. Friedmann discussed their research.[44] At the first session Ioffe delivered "The Structure of the Atom as Shown by X-ray Spectra" and Krylov offered "Several Observations on the Motion of Electrons in Helium." A week later the commission heard a paper by Bursian entitled "The Problem of Quantization of the System of Electrons in the Module of the Atom." Weekly meetings continued throughout 1920,[45] supported by 1.1 million rubles from Glavnauka which acted on Ioffe's advice that the commission's efforts must continue for "the benefit of science and society."[46]

In addition to contributing to the sense of community among Petrograd physicists, the commission had an important social function. The physicist and specialist in quantum mechanics V. A. Fock, who began his career at GOI, credits Rozhdestvenskii with turning his interest toward Bohr's theory of atomic structure in June of 1919, propelling him into a career as a theoretician, and saving his life with an "atomic ration."[47]

Ioffe's Molecular Commission also served a social function. At LFTI a group of young theoreticians gathered around Ia. I. Frenkel' (and his work on electron theory and theory of conductivity), V. K. Frederiks (relativity theory), and Iu. A. Krutkov. They recognized the opportunity to build upon their work as Rozhdestvenskii had at the optical institute. Ioffe petitioned Narkompros to support the creation of a molecular commission for his research on the physics of crystals and the properties of molecules in the solid state (specifically the mechanism of plastic deformation and texture during cold tempering of steels, using monochromatic X-ray machines and applying the methods of Ehrenfest, Debye, and Scherer). He claimed, "We have [in X-ray methods] the objective method for the determination of the position of atoms in the molecule and electrons in the atom." This simple solution to molecular structure "would be the best proof of the vitality of the central research institutes created by Soviet power," and considering Russia's isolation and food crisis would carry "world, scientific," and "important political significance."[48] While the government approved funding for the commission in March, and endorsed additional individual support for Ioffe, Chernyshev, Kapitsa, Semenov, and others to hire personal staff in May, it appears that only in October, after additional delay, did Narkompros give final approval to the molecular commission with funding.[49]

Scholars at LFTI supplemented the activities of the Molecular Commission through basic investigations of the structure of matter with X-ray methods, and undertook research of a more applied nature in such areas as the ionization of gases, the physics of electricity, heat engineering, and physical properties of metals. According to the academic secretary of the institute and head of the theoretical department, V. R. Bursian, LFTI had in essence four basic foci, which centered on quantum theory and investigations of interatomic processes using X-ray methods.[50] The first area involved work in theoretical physics: Iu. A. Krutkov's study of quantum theory; the efforts of A. F. Ioffe, G. A. Grinberg, and V. R. Bursian on the motion of the electron in the Bohr model of the atom; and activities associated with the Molecular Commission. This work led the authors to believe that the Bohr model was in need of some modification.

The physics of amorphous bodies was the second focus of the LFTI program. Ia. R. Smidt, N. Ia. Seliakov, M. M. Glagolev, and M. V. Kirpichev, who produced a thin liquid film for research on the structure of amorphic bodies, particularly liquids and gases, undertook research in this area, along with E. N. Goreva and A. A. Gorev, who solved the problem of building a rectifying apparatus. As a third focus, LFTI physicists examined the phenomenon of ionization in gases, which was closely tied to the construction of evacuated X-ray tubes in the institute's production department. The theoretical and experimental work of N. N. Semenov and V. M. Kudriavtseva on the construction of a probe for the measurement of potential in a vacuum, and the efforts of M. M. Bogoslovskii, A. A. Chernyshev, and L. S. Termen to produce Coolidge tubes, vacuum ovens, kenotrons, generators, and high-voltage oscillographs were also aspects of this research activity.

The fourth area, the major focus of LFTI physicists, was based on the investigations of Ioffe, Kirpichev, and Termen into the mechanism of residual and elastic deformation of crystals, growing out of the work of Laue and the Braggs. While the Laue-Bragg method was applicable only to entire monocrystals of sufficient size, the LFTI physicists used the work of Debye to show that with monochromatic X rays it was possible to produce and then interpret the interference pictures of polycrystalline aggregates and even amorphic bodies. Ioffe and Kirpicheva undertook the study of deformation of crystals in order to clarify the mechanism of plastic deformations which occasionally appeared even during small loads applied to monocrystals. This led to the discovery of asterism. The theory of crystal lattices could not explain the phenomenon of asterism—sudden division of pivots or abutments in what were, according to theory, perfectly elastic crystals. In a classic work, conducted in 1919 but published in 1922, Ioffe and Kirpicheva set forth the first complete interpretation and description of asterism, and a detailed picture of plastic (residual and elastic) deformation.[51]

Research on plastic deformation led to the study of the problem of strength of solid bodies, and of the methods of their yielding or failure by means of plastic flow and brittle rupture. Ioffe, Kirpichev, and M. A. Levitskaia published a continuation of this research in *Nature* in 1924 after Ioffe had made special arrangements with Rutherford to publish the work. In this work, the Ioffe group established the relative character of the difference between plasticity and strength of a given material, showing with rock salt that one and the same material is plastic at high temperatures and brittle at low temperatures.[52]

The study of the strength and plasticity of solid bodies raised questions regarding the theoretical versus real strength of crystals. Theoretical research on the interaction between ions in the crystal lattice, developed in the 1920s by Max Born, first permitted the calculation of the strength of ionic crystals under tension. However, in calculations by this method of the limit of strength, for example of sodium chloride, the strength in theory turns out to be 200 kg/mm^2, whereas in experiment for brittle rupture it never exceeds 0.5 kg/mm^2. It was known by the end of the nineteenth century that the observable strength of ionic crystals depends to a significant degree upon the surface conditions of the material.[53] This led the British physicist A. A. Griffiths to the conclusion that the decrease in strength observed in real crystals is the result of the presence of microscopic surface fissures or cracks, and using glass he demonstrated that these fissures really can significantly lower strength.[54] But Griffiths's experiments did not prove that a solid body will possess strength in conformity with theoretical calculations in the absence of fissure. Ioffe and Levitskaia directed their research toward rock salt, using artificially created crystals produced by submersion in hot water, where the limit of strength reached 160 kg/mm², which approached the theoretical limit.

A number of physicists rejected the Ioffe-Levitskaia conclusions, asserting that the method used to produce artificial crystals not only "washes away" fissures but also makes the crystals more plastic, so that the observed increase in the limit of strength may be conditioned by plastic deformation. However, Ioffe demonstrated that producing artificial crystals by submersion and creating conditions under which the surface defects cannot be formed has no real effect on plasticity.[55]

The creation of production facilities proved to be a double-edged sword. "Production" included the repair of radio and electrical equipment belonging to various governmental organizations and factories under contract, the outright sale of various instruments of which LFTI was the sole manufacturer in the USSR, and the construction of apparatuses for the institute itself. These activities took place in the institute's well-equipped technical department and the State X-Ray Tube Factory (formerly the Fedoritskii factory), which came under LFTI's technical supervision in December 1919 and included the production of monochromatic X-ray tubes and vacuum spectrographs for soft X rays, Langmuir pumps, kenotrons, rectifiers, Coolidge tubes, powerful cathode relays and high-intensity lamps, and other kinds of vacuum tubes.[56] During 1922, production capacity for all types of equipment expanded, although capacity far exceeded production.[57] In 1922–1923, through contracts with the Scientific-Technical Department of Vesenkha, the military, the Commissariat of Post and Telegraph (Narkompochtel), the Committee on Inventions, and a whole series of Leningrad factories including Krasnyi putilovets and Bolshevik, the technical department increased production and ultimately received a number of patents for its work.[58]

Because of increased demands on the technical department, its close supervision of the X-ray tube factory, and a desire to formalize relations with funding agencies, the academic council of LFTI decided in April 1923 to expand the technical department and create a production facility "for the development of instruments and apparatuses which are invented and developed during work of the institute connected with its basic scientific-technological research." The income received through the production department was "its own special means" to be used to expand scientific research, produce new instruments, and to award prizes for work completed within LFTI. A. A. Chernyshev, head of the technical department, had the authority to conclude contracts, sign checks, and undertake promotional and administrative activities in the production department.[59]

While providing additional income, however, the creation of the new facility led to the criticism that the institute was misusing scarce resources and ignoring basic research. Contract orders strained production capacity, occasionally causing the institute to overlook "the development of new kinds of technical instruments and apparatuses without a direct link with life." In addition, the costs of R and D frequently exceeded the institute's capability to undertake it. While Glavnauka permitted the income from the production department to revert to LFTI, this was not enough in a time of budgetary shortfalls. Paradoxically, some physicists complained that "the highest organs of the government" were "insufficiently attentive" to the institute; the government failed to let an adequate number of interesting contracts with LFTI, "preferring to conclude contracts abroad." In spite of all this, because of the "energetic activity" of A. A. Chernyshev and his staff, the institute managed to organize four fully equipped production projects for electrovacuum instruments; radio and telephone; X-ray tubes; and radio signalization. Chernyshev, in particular, was responsible for these successes; he had used personal monies to support production activities.[60]

The benefits of the production department far outweighed its costs. Its activities demonstrated that the institute shared government interest in resurrecting Soviet industry through participation in applied research. Through production contracts, LFTI secured badly needed infusions of capital to expand its basic research program. Finally, the Leningrad Physico-Technical Laboratory, which grew out of the technical and production departments, ensured the institute's continued vitality throughout the 1920s.

By 1926 the institute had thirty-four scientific personnel, sixty-two technical personnel including fifty-three *sverkhshtatnye*, and twenty-three administrative personnel including ten *sverkhshtatnye*.[61] The staff had grown to over thirty physicists including many, of the future leaders of Soviet physics: Frenkel', Kapitsa, A. F. Val'-ter, D. V. Skobel'tsyn, B. M. Gokhberg, and V. A. Fock (see app. B, table 1).

Thus, in the six-year period between 1917 and 1923, physicists in Petrograd, working in close association with A. F. Ioffe, joined forces to establish the Leningrad Physico-Technical Institute, one of the major research institutes created immediately after the Revolution. They embarked on an ambitious research program which focused on atomic and molecular structure, in particular the physics of crystals. Ioffe took advantage of competition among several governmental bodies for control of science policy, and of the needs of the Party for science and technology in matters of "government construction" to get support from Glavnauka, Vesenkha, and a number of other bureaucracies and enterprises. Beyond finance, LFTI kept contacts with government organizations to a minimum; it assured Glavnauka that it was in constant contact with the local Party committee and workers' and peasants' inspectorate on important issues. With conditions for physics research finally returning to normal, it remained to resurrect graduate training.

THE TRAINING OF SOVIET PHYSICISTS

The question of staffing of Soviet physics institutes assumed critical importance in 1918. The decay of Petrograd told heavily upon the professoriat and student body. Cold and shortages of heating oil and food shut down most higher-educational institutions, and students fled the city, cutting enrollments to ten or twenty percent of the prerevolutionary level. Even more damaging to university life was the destruction of the professoriat. A large number of instructors emigrated or moved southward, and many died, including the historians A. S. Lappo-Danilevskii and A. A. Shakhmatov, the economist M. I. Tugan-Baranovskii, the Pushkin specialist P. O. Morozov, and A. S. Famitsyn, the agronomist. To combat these problems and to speed up the training of specialists, Glavprofobr (the Main Administration of Professional Education) was organized;[62] Glavprofobr experienced limited success and the shortage of faculty persisted throughout the 1920s.[63]

Furthermore, just as publication of journals was delayed severely, so textbook publication was interrupted. Only old, out-of-date textbooks were available and then only on the black market. Tens of thousands of books were needed for some courses but were nearly nonexistent. For example, O. D. Khvol'son's *Kurs fiziki*, the major physics text, was not to be found in the provinces and had to be bought collectively by students in cities. Sometimes twenty or more shared the text. Some 300,000 copies were needed but unavailable. In December 1920, Gorky reported to the eighth All-Russian Congress of Soviets about the serious book famine. There were almost no new books, while older editions were rapidly wearing out, were obtainable in numbers of three or five thousand, or had disappeared.[64]

In the minds of many, the reorganization of university education raised the specter of the continuing domination of institutions of science and technology by Tsarist specialists. Some recognized that individuals with specialized knowledge would play a vital role in "government construction." Lenin himself, while suggesting on one occasion that every cook could master the complexities of government, recognized the central role of Tsarist specialists in the construction of communism. These individuals, he argued, were materialists at heart by virtue of their proximity to modern science and technology. Over time, they would come over to the cause of socialism.[65]

But the more radical, primary among them the Proletkultists and Workers' Opposition, called for class war against "bourgeois remnants." They alleged that these specialists would strive to keep the peasant and worker from improving his lot by using outdated and unegalitarian admissions standards. The Tsarist professor would inculcate the students with counterrevolutionary political and social ideology. And finally, the specialist would serve the interests of the capitalist for profit at any expense, leading to exploitation of the worker. Was it not time to allow the victorious working class to control all social institutions to advance the cause of the Revolution? When would the worker wrest authority from the bourgeois specialist in science, technology, and education?

These notions were at the root of the "Red-Specialist" conflict and became crucial again during the cultural revolution of the late 1920s.

Among its first steps, the Soviet government attempted to introduce universal education from the elementary school level upward, and to open higher-educational institutions to the disadvantaged—those who by virtue of social origin or inability to pay could not receive an education under the Tsarist system. As a result, like those of other universities and *vuzy*, the student body of the polytechnical institute changed drastically in the first years of revolution. In 1920 demobilized soldiers joined the more traditional student body; in 1922 younger students who had finished the first Soviet schools entered; and from 1923 to 1924 *rabfaktsy* (students who had finished special "workers' faculties," a kind of remedial school for the less fortunate) and the Communist-Thousands (*kommunisty-tysiachniki*) descended upon the institute.[66]

LFTI physicists participated in pedagogy for workers, for example in the opening of Communist University in 1921. Many of its students—demobilized Red Guards, Siberian partisans, and the Communist-Thousands—were simply illiterate; the program of Communist University was geared to teach the basics of contemporary exact science. Its courses included lectures by Ia. I. Frenkel' on the electrical nature of matter and relativity theory, and various lectures by O. D. Khvol'son.[67] But it must be noted that only with the return to economic normalcy associated with the NEP; the organization of *rabfaky*, special schools to handle demobilized soldiers and workers; and the reinstatement of entrance exams based more on academic qualifications than on social origin did life in higher-educational institutions return to normal.

In the meantime, how did LFTI face the problem of training physicists, which required educational reform? How could Narkompros be prevailed upon to support the new endeavor? And since at first many school administrations were hostile to the new Soviet power and refused to recognize the Sovnarkom or have contact with Narkompros, when would normal university life resume? Physicists ignored the Red-Specialists debate. They saw in the revolution in education an opportunity to establish new programs of matriculation and to push through curriculum reform. This would end the Tsarist system's emphasis on mathematical competency for all but theoretical physics. A. F. Ioffe was prominent in this effort, creating the physico-mechanical department (Fiziko-mekhanicheskii fakul'tet, or FMF) in 1919 to meet the twofold needs of curriculum reform and training of physicists.[68] At Petrograd University, Rozhdestvenskii would lead a similar charge against the old-guard physicists to lessen the traditional emphasis on mathematics.

After he had been rejected by the physics department at Petrograd University for a professorship in 1916, Ioffe and a childhood friend from Romny, S. P. Timoshenko, set out to organize a physico-mechanical department at the polytechnical institute with Ioffe responsible for physics and Timoshenko for mechanics.[69] Ioffe, Timoshenko, and several other professors met throughout the fall of 1918 to discuss the creation of an engineering-physics department, roughly along the lines of the Massachusetts Institute of Technology. They believed that modern industry required factory laboratories and testing bureaus to stay current with recent advances. At the end of the nineteenth century, institutes in Petrograd, Kharkov, and Tomsk, and such technical higher schools in Moscow as the Institute of Ferrous Metallurgy had trained a "universal engineering scientist" to work in the broad area of technology. The shift at the turn of the century to polytechnical education with experimental and laboratory practica represented an attempt to train specialists with a narrower focus and more practical skills. The physico-mechanical department grew out of this latter trend away from broad specialization.[70] Narrower training with practical skills would also soothe government concerns that new educational programs be commensurate with the needs of "government construction."

Bureaucratic politics on both a governmental and an institutional level belabored the organization of the physico-mechanical department (the so-called *fizmekhfak*). At first, the physicists were confused about where to turn for support. When founded in 1903 at the instigation of Count Sergei Witte, the Petrograd Polytechnical Institute had naturally fallen under the Ministry of Finance. Should Ioffe approach the ministry's successor, the Commissariat of Trade and Industry, or would Narkompros be more forthcoming? Ultimately, it was decided that Narkompros would endorse

the effort with the support of the rector of the polytechnical institute. In July 1919, A. F. Ioffe, A. N. Krylov, M. V. Kirpichev, and
student representative P. L. Kapitsa met to approve a list of twenty
potential faculty.[71] Several more-conservative members of the
polytechnical institute, who believed that the *fizmekhfak* would turn
out neither physicists nor engineers, opposed its creation. The entire faculty approved the new department, but because of delays
associated with the approval process only six students entered in
the fall of 1919. By 1921, another sixty students had entered the
program.

The fact that those wishing to enter the polytechnical institute
took entrance examinations for well-established departments before taking those for the physico-mechanical department indicates
its early lack of reputation. Those not admitted to other departments turned to the physico-mechanical department and hoped to
transfer later on. By 1922, only twenty or so of the sixty students
remained, but many were replaced by students from other
departments.[72] By 1925 the physico-mechanical department had
grown to include fourteen full-time professors, thirty-eight instructors, and two hundred students and engineers.[73] Three-quarters of
the faculty worked simultaneously at the Ioffe Institute, including
Ioffe, Semenov, Bursian, Frenkel', Obreimov, and Levitskaia.[74]

The program of the physico-mechanical department was
oriented more toward training than toward fundamental research,
with its four-year program closely tied with instruction in the
electro- and radiotechnology, mechanics and hydromechanics,
metallurgy, and shipbuilding departments.[75] First-year students
took courses in mathematics, theoretical mechanics, physics,
and chemistry. A. F. Gavrilov lectured on mathematics, V. V.
Skobel'syn on "electrical current" and oscillating motions, and
Ioffe on molecular physics, using his *Molecular Physics* in place of
Khvol'son's *Kurs*. In the second year, practica supplemented such
coursework as Semenov's lectures on electronic phenomena,
Obreimov's on optics, and Frenkel''s on thermodynamics and statistical physics. A. A. Friedmann, known today for his work in
relativistic cosmology, also gave several second- and third-year
courses on theoretical mechanics (the dynamics of points and
monosystems).[76] In the third year, students turned to independent
research in preparation for the final year of specialization in one of

the following eight groups: theoretical mechanics; applied and building mechanics; hydro- and aeromechanics; "pure" physics; optical technology; heat engineering; electrotechnology; and X-ray physics and technology. Later, molecular and magnetic phenomena were added to the list of specializations.

The cold and the general devastation that affected all scientific institutions of Petrograd also disrupted activities at the polytechnical institute. Shortages of fuel and electricity were endemic; heating and lighting were provided only sporadically. The cold was especially debilitating during the winter of 1921–1922, when the main building of the polytechnical institute was unheated. Classes were postponed to the coming summer months, transferred to professors' apartments, or canceled outright, and many students simply went home.[77] Selected practica were held in the chancellery of the faculty, a relatively small room in the middle of which was a brick fireplace with an exhaust pipe through a window. One student would arrive early to stoke the fire; depending upon the moisture of the wood, the students often conducted experiments in dense smoke.[78]

These difficulties notwithstanding, the physico-mechanical department quickly became a source of physicists for the Ioffe Institute, other research centers, and industrial laboratories. The Sunday physics circle, so important to prerevolutionary Petrograd physics, began again to meet regularly at the polytechnical institute, with Ioffe frequently sharing his knowledge of developments on the European front. Ioffe's talks were of particular interest to faculty and students alike since he had returned from *komandirovka* abroad in the fall of 1921. The circle provided Ioffe with the opportunity to know his students well, and he selected and transferred the better among them to LFTI, including A. I. Leipunskii, later the head of the Ukrainian Physico-Technical Institute in Kharkov and director of the Soviet program for the development of the fast breeder reactor in Obninsk. One student, V. N. Kondrat'ev, recalled:

> Studying in the polytechnical institute and simultaneously working in FTI, we were introduced to science, giving all our free time to it. Such a combination of study and scientific work was made easier by the fact that [LFTI] did not have its own quarters in the first years of its existence, and took shelter in several rooms of the main building of the polytechnical institute. In this very building at that time

almost all of the basic work of the FMF was conducted. The fact that a large number of teachers of this faculty simultaneously were scientists at [LFTI] also played a not unimportant role. This facilitated scientific communication of junior with senior comrades, which occurred both in the classrooms of the FMF and in the scientific laboratories and seminars of [LFTI].[79]

Petrograd physicists had successfully turned to the training of physicists and curriculum reform, creating a department that was "closely tied with [LFTI] where a significant number of the students in senior courses, under the leadership of the physicists of the institute, participated in scientific-research work."[80] Between 1922 and 1930, sixty-one scientists finished the physico-mechanical department in specialties from physics, mechanics, and radiotechnology to heat engineering and materials testing. Five-sixths of these physicists graduated to scientific work, and over one-half worked in LFTI (see app. B, table 2). Ioffe's position as dean of the physico-mechanical department and director of LFTI facilitated the symbiotic relationship between the two institutions—one devoted to research, the other to education.

CONCLUSION

In the first years after its organization, the Leningrad Physico-Technical Institute conquered a number of challenges peculiar to Russia, and others shared by representatives of their discipline throughout the world. Difficulties encountered in reestablishing activities considered "normal" to physics—publication, international exchanges, and training—plagued Soviet scientists until 1924. The increasing specialization and complexity of physics research demanded expensive equipment and funding from the state to build it. Political and economic uncertainties made it hard to generate government support. But owing to the efforts of a recently established professional association and its leaders, physicists were successful in creating the conditions for the flowering of Soviet physics in the mid-1920s.

Within the Soviet government, Glavnauka and the Scientific-Technical Department of Vesenkha competed with each other for control of the scientific enterprise, the former allowing more autonomy for scientists and the latter pushing for centralized administration of R and D, as well as more research of an applied nature. Physicists took advantage of the lack of agreement about national science policy, and seized upon the Bolsheviks' need for their advice and respect for science and technology to establish before 1920 a series of scientific institutes, notably the Leningrad Physico-Technical Institute. The organizational, pedagogical, and financial handicaps of the first years of Soviet rule notwithstanding, Petrograd physicists quickly resumed research in such areas as atomic and molecular physics.

The next period, beginning with the transfer of LFTI into its own building and lasting until the First Five-Year Plan, is perhaps the most glorious for the institute and the discipline in general. The number of physicists grew; their institutional bases and research programs expanded; and international and domestic cooperation were reestablished as a spirit of corporatism spread among Russia's physicists.

4

The Flowering of Soviet Physics: National Achievements and International Aspirations During the New Economic Policy

The mid-1920s mark a turning point for Soviet physics. The New Economic Policy (NEP) had taken hold and was in full swing. In 1921, in the face of overwhelming obstacles—famine and a poor grain harvest, industrial decline brought about by fuel and other material shortages, and a breakdown of the transport system, as well as the challenges of the Workers' Opposition and the Kronstadt Rebellion—the Bolshevik, at Lenin's urging, had abandoned War Communism for a return to small-scale retail and industrial trade and the institutions of a money economy. Scientific research institutes benefited, not only because of a significant increase in the level of government support and an end to food, fuel, and other economic shortages that had plagued them until 1923, but also because of farsighted leadership in the physics community.

This leadership, provided by A. F. Ioffe, members of his school, and the Russian Association of Physicists, promoted policies which in many cases overlapped with those of the government. Scientists wished to embark on research in a number of new directions, expand institutional bases, and go about the business of science with a minimum of political interference. Their research efforts included programs of interest to "government construction": electrification, communications, metallurgy, and so on.

The government hoped to promote economic stability by funding R and D. Officials in Vesenkha and Glavnauka believed that through its funding policies, the government needed to indicate its preferences for the general directions of research but that scientists best understood how to carry out that research at the institute level. The Scientific-Technical Department of Vesenkha provided capital for the expansion of the national scientific enterprise in concert with Glavnauka. It relied heavily upon scientists who served on its advisory boards to determine where funding should go. The number of institutes and scientists, and the level of funding grew as a result of these policies.

With this largesse, physicists founded a series of new institutes, most notable among them the Leningrad Physico-Technical Laboratory (LFTL), a quasi-independent laboratory closely associated with the Ioffe Institute. In fact, while Bolshevik administrators treated them as independent entities, the laboratory and the institute shared personnel, equipment, and space. Physicists ably turned this to their advantage when requesting funding for the expansion of research programs.

The successes of the Leningrad Physico-Technical Institute went hand in hand with those of the Soviet physics community, both in the domestic arena and abroad, as it became the preeminent Soviet physics institute. LFTI physicists commenced new programs of basic and applied research, modernized and enlarged laboratory facilities, and assumed leadership in publication in Soviet physics journals, in terms of both quantity and quality. And its physicists were among the first to reestablish contacts with European and American scholars, at long last ending years of isolation.

It was no easy task to reestablish ties with Western scientists, because of problems with visas, bureaucratic foul-ups, and shortages of hard currency. But until pressures for "autarky" in science became pronounced under Stalin in the 1930s, Soviet physicists participated actively in the international arena. The Russian Association of Physicists would celebrate nearly a decade of uninterrupted achievements at their sixth congress in 1928, a congress that was truly national and international in flavor. The result was the development of a corporate spirit among physicists within both the USSR and the world community of science.

THE REESTABLISHMENT OF SCIENTIFIC RELATIONS WITH THE WEST

In 1920, with the end of the Civil War, scientists began to press for approval to travel to the West with *valuta*—hard currency—to purchase instruments, equipment, books, and journals. While the

RAF had formed a commission to look into the reestablishment of ties with the West, individuals turned out to be far more important in securing such ties. The physicists D. S. Rozhdestvenskii, P. L. Kapitsa, and A. F. Ioffe, and the mathematicians A. N. Krylov and V. A. Steklov all organized lengthy and fruitful stays abroad. However, they encountered significant diplomatic, bureaucratic, and financial obstacles in normalizing relations with the West.

Since nearly all European nations, with the exception of Weimar Germany, hesitated to grant Soviet scientists official status until after the signing of the Treaty of Locarno in 1925, and Germany and Russia were the pariahs of Europe, it was difficult for any Soviet scholar, ballet dancer, or musician to secure a visa.[1] The situation was made no easier by the actions of the Communist International, or Comintern. An unofficial arm of the Soviet government, it sought to foment revolution throughout Europe with the complicity of the Soviet diplomatic corps. This generated mistrust of and distaste for the USSR among all Western diplomats.

The diplomatic relations with Germany proved to be the saving grace. The Soviets set up a number of joint scientific, technological, and military endeavors with the Weimar Republic which were in the interests of both during a period of international isolation. For scientists the most important endeavors involved research travel to Germany (and from there, with some luck, to England, France, and Holland) and the establishment of Russian-language publications houses there.

Trips abroad, or *komandirovki*, were administered by government bureaucracies, although physicists determined the specifics of each case. This often led to delay in approval of visas, and tied up letters of credit needed for purchases abroad. Vesenkha and Glavnauka instituted formal procedures for scholars wishing to go abroad. The collegium of Glavnauka considered applications for funding and approval of *komandirovki* and expeditions. Voting was not a formality. In June of 1920, M. N. Pokrovskii, V. T. Ter-Oganesov, and M. P. Kristi decided to postpone A. F. Ioffe's first proposed trip to Estonia and from there to Europe. Only in 1921 did the collegium approve Ioffe's *komandirovki*.[2]

In addition to an application procedure, the presidium of the collegium of the Scientific-Technical Department of Vesenkha established basic standards to be met, including demonstrated knowledge of literature in the field, the impossibility of accomplishing the research project without going abroad, a need to become familiar with new technology or patents, and so on. After some years the procedures became formalized, which signified to physicists tacit approval to go abroad and generated more applications. Starting from a handful of scholars in 1920, by the 1927–28 academic year there were almost 400 foreign trips under the auspices of the Scientific-Technical Department, although by the next year, on a budget of 120,000 gold rubles, the number had fallen to 140 scientific *komandirovki*, of which 11 were to the United States.[3]

In addition to its support of foreign travel, the Scientific-Technical Department set up the short-lived but successful Bureau of Foreign Science and Technology (Biuro inostrannoi nauki i tekhniki, or BINT) to combat a disastrous situation in the area of scientific publication. Few technical books or magazines made their way to Russia. Paper shortages interrupted domestic journal publication. BINT was to support foreign research trips, gather scientific and technological information abroad, and publish Russian-language materials for shipment back to the USSR. Although an arm of the Scientific-Technical Department of Vesenkha, BINT assisted Glavnauka in resurrecting scientific publication as well. By the mid-1920s institutes of Glavnauka received between twenty and thirty thousand books each year from abroad, while sending ten to twenty thousand back in exchange. In 1927 these institutes maintained 874 German and 385 American scientific journal subscriptions.[4]

In 1921 Lenin criticized the performance of BINT, saying it must have "fallen asleep" in its efforts to acquire for Moscow "one copy of all of the most important machines from among the newest to study and learn." BINT most likely was an arm of the Soviet secret police for purposes of industrial espionage; branches were organized in all large manufacturing centers of Europe and tied to each Soviet mission. But BINT served scientists well, supporting publication through its Berlin office, and facilitating the travel of such scholars as V. N. Ipatieff, A. N. Krylov, P. P. Lazarev, A. F. Ioffe, and countless others abroad.[5]

Physicists from LFTI in particular seem to have benefited from the normalized contacts. Among the almost forty Soviet physicists who visited Europe and the United States, Iu. A. Krutkov, L. D.

Landau, V. A. Fock, Iu. B. Rumer, P. L. Kapitsa, George Gamov, K. D. Sinel'nikov, Ia. I. Frenkel', and A. F. Ioffe represented the Physico-Technical Institute. Although officials within the Academy of Sciences had approached Narkompros as early as the summer of 1918 about the need to open contacts with the West, the successful normalization of scientific contacts among physicists was mainly attributable to Ioffe and his friendship with Paul Ehrenfest.[6]

Ehrenfest tried to maintain contact with Soviet colleagues during the war and Civil War. He had worked closely with Iu. A. Krutkov in Leiden just before the war, and after the war broke out gave considerable support to the physicists V. K. Frederiks and V. M. Chulanovskii, who were with Hilbert in Göttingen and who faced the threat of internment. But Ehrenfest received little information from his colleagues in Petrograd until the summer of 1920, when correspondence with Krutkov and Ioffe started again. He was thus prepared to offer advice and assistance in securing visas when Rozhdestvenskii, Ioffe, Krylov, Kapitsa, Chulanovskii, Krutkov, and others made their way abroad. By 1921 scientific relations had returned somewhat to normal, and Ehrenfest established many contacts between European theoreticians and their young Soviet colleagues.[7]

In 1920 physicists at LFTI sent a petition to Glavnauka on "the necessity of foreign travel for scholars for scientific purposes." They emphasized the international character of science, and the fact that it was a "product of the collective experience of all mankind which demanded uninterrupted interaction" of scholars of all countries.[8] The academic council of the institute voted to ask Glavnauka to send Nemenov and Ioffe abroad to purchase instruments, reagents, apparatuses, and journals, and to renew the exchange of ideas.[9] Ioffe simply would have to go abroad.

In summer of 1920, Ioffe wrote Paul Ehrenfest in Leiden, asking him to help LFTI, the State Optical Institute, the Institute of Physics and Biophysics, and Moscow University to catch up with European research by sending the main recent books and journals to an address in Finland which would forward them to him. Ioffe would see to it that money was sent no more than two weeks later. In his letter Ioffe also described his work on atomic and molecular structure, and that of Rozhdestvenskii, V. R. Bursian, Krylov, Krutkov, and A. A. Friedmann. While Ehrenfest gladly provided what articles and monographs he had at hand, he could not single-handedly see to it that Soviet physicists were reacquainted with Western physical achievements.

The international isolation of Soviet Russia and the government's typically inept handling of funding authorization dogged the physicists' every step. First, it was no easy task to put visas in order for the scientists who wished to go abroad. The Soviet bureaucracy processed the visa applications slowly, perhaps training for the legendary standards of inefficiency achieved in later years. More important, refusal of all countries except Germany to recognize Soviet power waylaid many a scholar at the Estonian border, including Petr Kapitsa, who spent months waiting. Second, the government failed to provide hard currency in a timely fashion. These difficulties interrupted Ioffe's journey on several occasions.

Ioffe dealt with Narkompros's Z. G. Grinberg, commissar of education of the northern region, and M. P. Kristi, director of Leningrad Glavnauka, in continuing discussions about a research trip. In spite of rampant inflation and limited hard currency reserves, Narkompros saw fit to approach the Sovnarkom (Council of Peoples' Commissars) with a request to approve several hundred thousand gold rubles for Ioffe's *komandirovki* through the Commissariat of Foreign Trade. Sominskii says that Lenin's direct intervention secured approval.[10]

In mid-January 1921 Ioffe traveled to Moscow to get his papers in order; on February 12 he left for Estonia, where he languished for two weeks waiting for a visa and wondering where the letters of credit were.[11] Ioffe was abroad from February until August 1921, and then again on a second *komandirovki* from April until September 1922, and corresponded with his wife in Leningrad during these periods. The letters are extant, and reveal Ioffe's repeated frustrations in arranging even simple agreements during his travels in Europe.[12]

For his first trip, Ioffe had hoped for visas to Holland, Sweden, and Germany, but received only the latter and arrived in Berlin at the end of March. He immediately ordered 391 journal subscriptions for three- to five-year periods and 350 books, but had to await Kapitsa's arrival with the letters of credit to secure the purchases. Ioffe passed the time by sending his wife small purchases of cocoa, chocolate, cheeses, and medicines, pestering N. N. Semenov for reports from the institute, and demanding that someone be sent to Moscow to light a fire under the bureaucrats at the Commissariat of Foreign Trade.[13] It seems that a large number of the books and journals ordered never arrived; the number of journal subscriptions received at the institute's library, in fact, dropped from fifty-four in 1922 to forty-two in 1925 and thirty-six in 1926;[14] perhaps the credits were not as large as initially approved.

Several personal successes somewhat buoyed Ioffe's spirits despite the continued funding difficulties, which Ia. I. Frenkel' was hard at work trying to solve. Ioffe gave lectures at several Berlin physics colloquia to which Planck, Nernst, and others responded respectfully.[15] To Ioffe's delight, Ehrenfest arrived in Berlin in April to renew their friendship, discuss issues in physics, and plan how to normalize Soviet scientific exchanges with the West. Ioffe finally received the letters of credit in May, completed his transactions, and sent instruments and machine tools worth DM 2.4 million, chemical products worth DM 60,000, and a spectrograph to LFTI.[16]

In June, Ioffe continued to England and meetings with Ernest Rutherford and the Braggs, and finally met up with Kapitsa. Once again, however, uncertainties over letters of credit dogged his every step. Ioffe expressed particular displeasure at how L. B. Klassen, a career diplomat and at that time trade representative in London, had delayed receipt of funds. When would the credits clear?[17] Ultimately, Ioffe secured another eight boxes of books and instruments, including five Coolidge tubes and kenetrons costing £7,738. His meetings with Rutherford and the Braggs were particularly successful, and he made arrangements for Kapitsa to stay in Cambridge if he was willing, and to publish an article in *Nature*. He also dined at H. G. Wells's house with several British scholars, and no doubt talked with him about the activities of TsEKUBU.[18]

In July, Ioffe returned to Petrograd by way of Berlin, Hamburg, and Leiden. Ehrenfest had succeeded in securing a visa for him, and in organizing colloquia and meetings with Lorentz and Kamerlingh Onnes. In spite of several scientific achievements, he remained gloomy over his failure to receive as many gold rubles as promised.[19] Yet, as the institute's 1921 annual report noted, "The scientific activity of [LFTI] was facilitated as a result of the *komandi-*

rovki of the chairman of the department, A. F. Ioffe." In addition to the "numerous instruments and materials for experimental work which have already been received," the report continued, the quality of library holdings had improved manyfold.[20]

During his second trip abroad in 1922 Ioffe succeeded in procuring another one hundred boxes of equipment, including two powerful 200 kV transformers, string galvanometers, and a Siemens oscillograph.[21] While in Berlin, Ioffe met with the specialist in solid-state physics Karl Wagner, the theoretician Arnold Sommerfeld, Wilhelm Wien, Walter Gerlach, and others. A meeting with physicists in Göttingen proved even more fruitful to Ioffe's efforts to reestablish normal scientific exchanges with his European colleagues: the German physicists agreed to send several copies of reprints of all German articles that had appeared in the main physics journals dating back to 1914, where possible, to the Petrograd Dom uchenykh; Ioffe requested from twenty or so Petrograd physicists lists of two to five books each which they would want sent to them in the USSR from abroad free of charge; and he asked for bimonthly reports on LFTI. Another German scientist agreed to organize a center for the selection of individual reprints of scientific work to send to Russia. Finally, a number of scholars expressed an interest in publishing the short-lived *Vestnik rentgenologii i radiologii* in German translation.[22] These efforts may have been connected with BINT.

By 1924 foreign contacts were an integral part of Soviet research activities and an indispensable source of scientific equipment. In that year, the Optical Institute's D. S. Rozhdestvenskii traveled with $80,000 to Europe, where he purchased instruments which "opened the possibility of beginning systematic production of optical glass in [a Russian] factory." A special customs point was opened for the hundreds of boxes of scientific materials received for the institute.[23] The fact that the government included a monthly allowance of 1,000 rubles for foreign purchases and authorized a one-time allocation of 7,500 rubles for LFTI early in 1924, after lobbying by N. N. Semenov in Moscow, indicates that the government recognized the importance of regularized scientific foreign contacts.[24]

In 1924 alone, twelve Academicians received *komandirovki* to Europe,[25] and during the 1923–24 academic year, Ioffe, A. A.

Chernyshev, L. S. Termen, Ia. R. Shmidt, and M. V. Kirpichev all represented the Ioffe Institute abroad. By 1926, one-fifth of the Ioffe Institute physicists had visited Western physics facilities (although it must be noted that the younger scholars, who might have benefited the most from travel abroad, were excluded from it.[26]

Kapitsa, the future Nobel Prize winner and specialist in low-temperature physics and magnetism, established himself securely in Rutherford's Cavendish Laboratory in Britain in 1921, and spent the next thirteen years there.[27] Beginning in 1928, I. V. Obreimov, G. A. Gamov, and K. D. Sinel'nikov, all of the Ioffe school, and the latter two with funding from the Rockefeller Foundation, spent time in Cambridge as part of the famous "Kapitsa Club."[28]

Soviet theoreticians were able to participate in the European and American physics communities in part with the support of the International Education Board (IEB).[29] The IEB was funded by the Rockefeller Foundation under the umbrella of the General Education Board. It operated between 1923 and 1938; the main period of its activity was 1923–28. The IEB gave out funds for agriculture and science, and in each of those categories provided support for (1) traveling fellowships; (2) traveling professorships; (3) grants to such institutions and organizations as Niels Bohr's Institute of Theoretical Physics and Kamerlingh Onnes's low-temperature laboratory in Leiden; and (4) "miscellaneous."

No Soviet institution received funds from the IEB; only "scientists" did. From its earliest days, the directors of the IEB were hesitant to have extentive contacts with the Soviet government, apparently because of mistrust of the "Bolsheviks." C. B. Hutchison, director of Agricultural Programs in Europe in the mid-1920s, did not wish "to close the door on Russia at this time. . . . But until conditions improve in Russia and we have an opportunity of visiting . . . , we ought to proceed very slowly." Beyond fellowships, the burden of proof lay with Soviet Russia.[30]

In addition, IEB officials feared that Soviet scholars who received fellowships might not return to Russia, using the opportunity to emigrate. This would defeat the purpose of the award, which was to support the development of a particular science by training its younger members for service to their national communities. There was also concern that those who stayed in the West might infect Western scientists with seditious bolshevik ideology.

Finally, there were the usual visa problems for IEB scholars. Those who wished to study in the United States had to apply for visas at the American consulate in Riga in person. The U. S. Department of State assured IEB officials that all consular officers had been informed in August of the previous year that the IEB had been placed "on the list of institutions approved by the Department of Labor for alien students,"[31] but this did not guarantee a smooth trip.

Most IEB officials recognized the role that the IEB could play in helping the Soviet scientists recover from the strains of war and isolation. Dr. Wickliffe Rose, president of the IEB in the mid-1920s, believed the IEB should undertake to supply scientific literature from a distribution point in Paris; $3,500 would apparently buy new journal subscriptions, with another $20,000 needed for back issues.[32]

More important for Soviet Russia were IEB fellowships. Twenty-three Soviet scholars and eight "Russians" (including the emigres Otto Struve and George Kistiakowsky, later a Harvard professor and science advisor to Eisenhower) received IEB funding. The "Soviets" included seven mathematicians, three zoologists, three agricultural specialists, and seven physicists. The first Soviet fellows, in September 1924, were Paul Aleksandrov, a mathematician, and Eugen Gabrishchevskii, a zoologist, both from Moscow University. Over the next five years such important figures as the geneticist Theodosus Dobzhanskii and the physicists V. A. Fock, Ia. I. Frenkel', George Gamov, later a professor in the United States and father of the "big bang" theory, Iu. A. Krutkov, K. D. Sinel'nikov, and D. V. Skobel'tsyn received IEB fellowships. According to Soviet records, the Nobel Prize–winning theoretician L. D. Landau was also a fellow, but there is no record of this in the archives. In addition, there was some discussion of the participation of another eventual Nobel Prize winner, I. E. Tamm, in the IEB program, but he apparently never applied.[33]

Through these travels, the physicists strengthened their ties with foreign scholars and brought back scientific instruments and ideas for their laboratories. The sabbaticals of Frenkel' and Ioffe demonstrate the richness of the foreign research experience for theoretician and experimentalist alike.

In March 1925, Dr. Augustus Trowbridge, director for Europe of the IEB, met with Paul Ehrenfest in Leiden concerning Ehrenfest's plans to get "a few of the more gifted young Russians into Western Europe."[34] Specifically, Ehrenfest recommended Iu. A. Krutkov and Ia. I. Frenkel', maintaining that "one will be doing the most that one can for this branch of science in Russia at the present time since these men will go back with new ideas gained by researching for a year in the West." Trowbridge supported the candidacy of Frenkel', citing the latter's impressive publication list of fourteen articles and five books since 1912. Max Born also supported Frenkel', saying that he showed "in his work a very complete mastery of the mathematical methods of physics, a deep insight into physical relations."[35]

Iakov Il'ich Frenkel' (1894–1952) was a leading theoretician, important in the development of relativity theory and quantum mechanics in Soviet Russia. He was born in Rostov-on-Don, but his family moved to Lugansk, Kazan, and Minsk before settling in Petersburg in 1909 where he finished gymnasium and entered the university in the mathematics division of the physico-mathematics department. He graduated in 1916, about the same time he began to attend Ioffe's seminar in physics at the polytechnical institute. In 1917 he left for the Crimea where he taught in the university in Simferopol, then served the Bolshevik government as head of the department of higher schools and technical education of Narkompros Crimea. In 1921 he returned to Petrograd to work with Ioffe in the Physico-Technical Institute, where he remained for the rest of his life, from 1932 as director of its theoretical department.

Frenkel' had a biting sarcasm and wit, and activities which kept him in the public eye caused significant professional distress. His love of music and facility with the violin also created the image of someone more in tune with bourgeois sensibilities than with the demands of Soviet socialism. Frenkel' was active in the RAF, in the Academy of Sciences as a corresponding member from 1929, as a propagandist of modern physics in popular science journals, and as a member of the international physics community. He came under attack for idealism in the 1930s in spite of his loyalty to many Bolshevik causes, because of his outspoken rejection of dialectical materialism and his ridicule of strictly mechanistic explanations of electromagnetic phenomena in any workers' or students' forum. For this reason, and because of anti-Semitism, he never became an academician.

In his application to the IEB, submitted apparently on Ehren-fest's advice, Frenkel' expressed the desire to work in Göttingen with Born, beginning in the autumn of 1925, to work on "the electronic theory of solid and liquid bodies" and "general electro-dynamics in connection with the quantum theory," before travel-ing to Cambridge for a few months. He hoped to obtain results "regarding the nature and magnitude of attractive and especially repulsive forces between atoms, and, further to extend the theory [to] liquid bodies, the liquid state of matter . . . being much nearer to the solid than the gaseous one." Frenkel' proposed to "show that the electron must be considered as spatially unextended force centers." He argued that "this conception requires a reform of the current views on electromagnetic energy and radiation. It also leads to new, hitherto unnoticed solutions of the field equations which seem to correspond to the nonradiative (stationary) notions of the quantum theory. . . . " Rose approved the physicist's can-didacy, a monthly allowance of $182 for Frenkel' and his wife, and additional funds for travel.[36]

When Frenkel' arrived in Berlin on November 11, however, he found that Born had already departed to deliver a course of lec-tures at MIT.[37] After consulting with Ehrenfest and Ioffe, Frenkel' left instead for Hamburg to study with Pauli and Stern, with plans to return to Berlin to work with Einstein, and finally in the spring or summer to Göttingen and Born. This proved to be time well spent. At a seminar early on in his stay, attended by Nernst, Planck, Laue, Einstein, and others, Frenkel' delivered a talk on his theory of electroconductivity of metal which was enthusiastically received. Later he had the opportunity to consult in Paris with Leon Brillouin and with Paul Langevin, who was head of the France-USSR Society and a member of the French Communist party. And finally, he consulted with Born from April until June, when he finished research on a book on electrodynamics and awaited the arrival of Lukirskii, Bursian, and Semenov and his wife. After a month in Oxford with Kapitsa, Semenov, and Seliakov, he returned by way of Göttingen in late October to Leningrad.[38]

Frenkel''s *komandirovki* had a twofold importance for Soviet phys-ics. First, it served to cement ties with European theoreticians, to convince them of the high quality of research going on in the Soviet

Union, and to pave the way for such scholars as L. D. Landau and George Gamov to follow him to their laboratories. Ehrenfest reported to Rockefeller's Dr. W. E. Tisdale that Frenkel' "was a very good appointment, that he uses his time well . . . that it is extraordinary for a man to be able to receive the new impressions which Frenkel' has received and at the same time get so many good results in a short time."[39] Second, when Frenkel' returned to LFTI, he was better prepared to provide first-rate leadership to a new generation of theoreticians in Leningrad: V. A. Fock, G. A. Grinberg, and D. D. Ivanenko, and in the 1930s, M. P. Bronshtein and L. V. Rozenkevich. Frenkel' had the opportunity to go abroad again in 1930–31, primarily to the University of Minnesota, and upon his return contributed to the development of quantum mechanics in the USSR.

A. F. Ioffe's travel in Europe and the United States had a profound influence on his views of the appropriate relationship between industrial laboratories and physics research, its organization, scale, and desirable level of funding. Trips to General Electric, Westinghouse, and American universities convinced him that education, research, and production might find a symbiotic relationship in LFTI and the physico-mechanical department of the polytechnical institute. Ioffe was nominated for an IEB fellowship but turned it down for other sources of funding. European scholars received Ioffe as an honored guest; they were relieved to see international scientific relations return to normal. His stories of the achievements of Soviet physics in isolation in less than a decade impressed them. Ioffe, for his part, gained confidence that his institute was moving along the right path, and welcomed the recognition for his work.

Ioffe was abroad for two major trips in the mid-1920s. From November 1925 until February 1926 he visited Berlin, Paris, and the United States, where he stopped at Columbia University, the Rockefeller Institute, the cities of Chicago, Madison, Boston, Kansas City, and Pasadena, and en route toured General Electric and Westinghouse facilities. Kamerlingh Onnes and Trowbridge of the IEB intervened with the American embassy in Paris to get Ioffe's U.S. visa.[40]

In January 1927, Ioffe again went abroad, for his most successful *komandirovka*. He arrived in Boston on January 12. He gave a series of lectures at Harvard and MIT, before traveling to Schenectady, Washington, D.C., and Pittsburgh, where he participated in colloquia. He was especially impressed by 400 kV Coolidge tubes with forty-centimeter spheres at the General Electric facilities in Schenectady, where he also discussed shears and displacements with Langmuir. Ioffe traveled to Berkeley in February for a semester of teaching. He developed a close working relationship with Gilbert Lewis and Leonard Loeb and wrote *Physics of Crystals* during this time. In the introduction to the book Ioffe noted that

> the University of California, in inviting me to give a course of lectures for a semester, has given me the opportunity to organize into a consistent system the results of a number of investigations on the physics of crystals, which I and my collaborators have been carrying on for the last twenty-five years.[41]

The University of California awarded Ioffe an honorary Ph.D.,[42] and paid him the compliment of offering him a professorship, a position Ioffe chose to decline. While university officials were concerned about Ioffe's communist sympathies, the curriculum vitae which he provided them clearly stated, "The present high position which he holds in Russia is despite the fact that he is there known to be unsympathetic to the Communist cause. Thus the confidence and reverence felt by the communists for Ioffe is shown by his election to the leading positions despite his disbelief in their political views."[43]

His semester in the states led Ioffe to pursue funding which would generate an independent source of hard currency for his institute. The institute had signed the first of several contracts with foreign firms for research and development.[44] On his return to Leningrad in 1928, Ioffe stopped in New York and Boston where he negotiated another agreement, this one with Vannevar Bush and L. Marshall to develop and manufacture thin-sheet insulation with potential backing of up to $50 million from T. J. Coolidge and J. P. Morgan. This was one of many new ventures for the newly established Raytheon Corporation, through which Ioffe would get a salary of $12,000, certain patent rights, and royalties in exchange for development of the insulation. Ioffe also negotiated a contract with a consortium of German electrotechnical companies including Siemens AG to create a laboratory for the production of insulation, which guaranteed Ioffe DM 10,000,000 and a seven percent royalty.[45] While satisfied with the terms of the Raytheon agreement for his institute, Ioffe was nonetheless concerned: "But personally it means a burden for me. Financially, it increases my expenses. . . . During about two years I shall be obliged to think and to work on the improvement of insulation which does not interest me and to neglect my real scientific work. I know this perfectly well, but I think I have to go through the same of [sic] science and physics in Russia."[46] This comment indicates that Ioffe found all of his efforts to secure funds for institute and laboratory—through Vesenkha and Glavnauka, or from abroad—to be burdensome, distracting him from his work, but necessary to ensure the financial security of his institute. Ioffe used the funds received through the summer of 1928 "for my collaborators and pupils being sent to Europe. All together there are twenty-five physicists from my institute abroad."[47]

Ioffe failed to develop thin-layer insulation, a material of greater dielectrical strength than existing materials, because of miscalculations of several physical parameters that he, I. V. Kurchatov, and others carried out. This failure would result in political fallout in the 1930s, but at the time the commercial ventures attracted the attention of some Western scientists and firms and gained Ioffe respect at home. What happened to the Raytheon contract is unclear. It seems that Bush traveled to Leningrad to see Ioffe's laboratories and efforts, was disappointed, and, as the writer of popular-scientific books Otto Scott argues, "came to realize that Ioffe's theory was fallacious."[48]

It is true that Ioffe and his institute failed to deliver on a contract, something that at the very least must have caused him some discomfort. But what matters here is not that thin-layer insulation never materialized but that Ioffe, Frenkel', and other Soviet scientists succeeded by the end of the 1920s in reestablishing scientific ties with their European and American counterparts. They were surprised by the wonderment with which they were received, and warmed by the hospitality and cordiality of their hosts. Through the dogged and often frustrating pursuit of international contacts and persistent lobbying of the government, Soviet physicists had achieved one of their major goals of the postrevolutionary period. They participated in international conferences, for example the Solvay meetings; they traveled abroad to study, deliver lectures, consult, and to teach; they published extensively in foreign journals; and gained insights into the organization of R and D in industrial research laboratories. This latter experience had a significant impact on the organization of research in the USSR, as the case of the Leningrad Physico-Technical Laboratory demonstrates.

APPLIED PHYSICS RESEARCH IN THE
SOVIET UNION IN THE 1920s

In the 1920s Soviet scientists and government officials went beyond the framework of research institutes to establish organizations whose goals lay closer to the production end of the "fundamental research-technology" continuum. In some cases, this involved the creation of laboratories within factory enterprises. In others, it involved the establishment of research facilities which served directly one branch of industry, or *glavk*, of Vesenkha. In the case here, it involved the formation of a seemingly autonomous physics laboratory whose research served a whole range of bureaucracies, enterprises, and fields of technology.

The Leningrad Physico-Technical Laboratory (Leningradskaia fiziko-tekhnicheskaia laboratoriia, or LFTL) was a hybrid fundamental physics research institute and applied laboratory. Like Rozhdestvenskii's State Optical Institute and its glass instrument production facilities, the laboratory combined the production of vacuum tubes and X-ray equipment with applied and fundamental research. In a brilliant sleight of hand, it used the same facilities, equipment, and researchers as the Leningrad Physico-Technical Institute, but gave them different names, thereby generating additional income. Unlike present-day American industrial laboratories, it supported a broad range of industries, not just a single corporation. It also preserved education and training functions normally associated with a university or institute of technology. But more importantly, it enabled physicists to stay on the good side of the government as the pressure to conduct research of an applied nature in service of state economic and political goals became pronounced.[49]

Since the founding of LFTI, A. F. Ioffe, N. N. Semenov, A. A. Chernyshev, and others had pursued R and D contracts with other government agencies and industrial enterprises in areas important to rebuilding the Soviet economy, such as radio and telecommunications, electrification, heat engineering, and metallurgy. The institute took advantage of its special X-ray equipment and glass and machine shops to contract with several district government offices to undertake needed repair work in radio and telecommunications, and electronics. It repaired a transformer for the Leningrad District Health Office, built radio-signalization systems for the Hermitage, the Commissariat of Finance, and the State Bank, and produced electronic vacuum tubes and lamps for a number of different contruction firms including Volkhovstroi and Shagurstroi, and for the Donbas region. The institute also manufactured electronic equipment for the military academies and bureaucracies. The contracts for this work ranged from 4,000 to 80,000 rubles, and in the 1924–1925 fiscal year contributed almost 240,000 rubles to the institute's budget.[50]

Contracts with a number of Leningrad factories (and some as far away as the Ukraine and Siberia) contributed to the economic well-being of the institute and enabled it to demonstrate that its research efforts were closely linked to the "productive process." At first, the agreements were limited to the production of special electrical equipment and tools, and the repair of vacuum tubes, generator lamps, and the like for the Krasnyi putilovets (Red Pathfinder), Krasnyi vyborzhets (Red Vyborger), Kabel'nyi zavod (Cable Factory), and Bolshevik factories.[51] Later, they expanded to include exchanges of personnel between industry and the institute for one- to three-month periods, extensive consultative work, and even theoretical physics research, especially during the years of the first five-year plans.

Since its contracts and consultancies did not provide a regular and reliable source of funds, physicists sought out permanent increased funding for an applied physics laboratory. On one level both Glavnauka and physicists at the institute were satisfied with their relationship. The physicists often expressed thanks to Glavnauka for its support, for "attentive and gracious" relations. While Glavnauka funding was crucial to the day-to-day operations of the institute, however, it was insufficient to support the new activities on which the institute was anxious to embark.[52]

Glavnauka was hardly flush with funds; it supported the LFTI's basic budget but did not have the wherewithal to contribute to the expansion of the institute's fundamental and applied research. In addition, there may have been a restriction on the kinds of research which Narkompros was willing to fund. Ioffe therefore turned to the Scientific-Technical Department of Vesenkha for supplementary funding and to factories in and around Leningrad for research contracts. The establishment of the Leningrad Physico-Technical Laboratory in 1924, based in part on the European and American example, was the crowning achievement of this effort.[53]

In Ioffe's view, the laboratory was the first link between physics research, technology, and production, and part of the threefold hierarchy between physics research and industry—from the physics institute as the closest link to science, to the factory laboratory, and finally to the trust or enterprise itself. In establishing this hierarchy Ioffe hoped to protect scientists' autonomy in developing research programs, meet government demands for applied R and D, and placate those scientists who thought he placed too much emphasis on applied pursuits at the expense of fundamental.

Ioffe believed that two main problems interfered with the creation of a healthy relationship between science and production. Both were a matter more of worldview than of organization or finance. The first was that many "practical engineers" considered it improper for research institutes to interfere with their efforts at the laboratory or in the productive process in general. Similarly, they and their enterprises considered expenses for scientific research an unnecessary obligation. This worldview was reinforced by the rise of the Stalinist system of centralized planning, which emphasized such short-term goals as plan fulfillment rather than long-term, initially expensive but ultimately more efficient technological improvements.

For their part, physicists, in Ioffe's view, all too frequently saw production and technology as dirty work or something beneath them. Their task was to discover new processes or materials; the engineer should do the rest. In the presence of an overly centralized planning system without market demand, physicists' discoveries merely withered on the vine. There was no common ground for engineers and scientists to discuss the problems of applied R and D.

Ioffe proposed two solutions. One was the strengthening of an intermediate instance between institutes and factories: factory laboratories. Factory laboratories were concerned with the "essence of the matter." They were close to the productive process and hence knew "all defects and demands of production." By linking the laboratories more closely to research institutes, the latter would then "know better the needs of industry and could best satisfy them." Since the initiative for the organization of factory laboratories often came from the factories themselves, it was necessary to avoid the danger of linking research too closely to factory production targets. As Ioffe wrote:

> The leading elements of trusts and factories sometimes attempt to a certain degree to transform these laboratories into research institutes which are fully subordinate to a given trust. Such a path, which is necessary in conditions of competition, that is, under the capitalist order, loses all advantages in the socialist economy.[54]

Another danger was the fact that too many factory laboratories were involved in product testing, whereas their main goals, Ioffe argued, should be to ensure that scientific results found their way into production and to indicate to research institutes where more fundamental work ought to be carried out.[55]

The second path toward an improved connection between science and production was through education and retraining to improve qualifications, through both formal coursework and periodic assignment of scientists to factories, and engineers and workers to scientific research institutes. The Leningrad Physico-Technical Institute adopted this latter approach. It regularly sent its scientists to such factories as Krasnyi putilovets, Bolshevik, the Izhorsk Works, and the Kharkov Electromechanical Works, to consult on questions of metallurgy, motors and transformers, and production problems. In exchange, the institute accepted factory specialists to work in its laboratories for two- to three-month periods. This method met with some success. P. P. Kobeko and I. V. Kurchatov reported that the workers of the Krasnyi treugolnik (Red Triangle) factory did not understand the insulating properties of ebony although they were engaged in its production. Workers were brought to study ebony at LFTI with the help of *praktiki* from the polytechnical institute, and production miraculously improved,[56] although it may have been the attention given to the workers that made the difference.

Unfortunately, Ioffe's approach suffered from a number of shortcomings. For example, his view of the correct relationship between research institutes and industry was mechanical and almost simplistic. He forgot how hard it had been for him to establish a leading research center. Economic decline, international isolation, and backward technology handicapped the efforts of industrial laboratories to build modern research facilities. Ioffe had succeeded through the confluence of hard work, experience, and fortuitous good luck. In addition, he did not fully comprehend how to ensure that physics activity found rapid application. The tension between planning and individual initiative, between Red and specialist on the shop floor, between engineer and scientist, between the policies of Glavnauka and Vesenkha, and above them all, the presence of an increasingly bureaucratized state and economy provided significant barriers to the application of scientific advances to the production process. In the 1930s Ioffe would come under attack from three groups: physicists who thought his interests were synonymous with those who favored increased expenditures on applied R and D, planners who believed he supported fundamental research at the expense of applied research, and Communists who mistrusted his "bourgeois" roots.

But Ioffe's position as laboratory and institute director, consultant for various factories, and policymaker on various Glavnauka and Vesenkha boards gave him the authority to make decisions with implications for the entire physics enterprise. Already in 1924 he had chaired a special commission for the reorganization of 113 Leningrad factory laboratories. His plan was adopted for the Leningrad region by the Scientific-Technical Department and later applied to other areas of the USSR.[57] Leningrad served as the focus for this plan because of its importance to the Soviet economy as the "means of production" for the rest of the country, with its machine tool, metal working, textile, electrical, and other industries; in the mid-1920s industrial growth in Leningrad was nearly fifty percent faster than in the rest of the USSR.[58] In 1926 the Scientific-Technical Department again consulted Ioffe about the organization of a network of factory laboratories to speed up the introduction of scientific advances into the economy. Ioffe subsequently criticized Vesenkha's inadequate support (1.3 million rubles) for the system of factory laboratories, as well as its insistence on creating the network as quickly as possible. He believed that up to three years were needed to introduce the plan.[59]

Ioffe turned to the creation of the Leningrad Physico-Technical

Laboratory with the goals of a closer tie between physics research and industrial production and regular funding from sources other than Narkompros in mind. The Scientific-Technical Department of Vesenkha first considered the organization of this laboratory in October 1924, and voted to give no less than 3,000 rubles per month to this endeavor. Iu. N. Flakserman of Vesenkha looked into the prospects for additional funding for future years. The department also authorized the hiring of six new employees to staff the laboratory.[60] A month later the department gave final approval to the new physics laboratory and authorized funding.

The laboratory was responsible for "physico-technical research" and its applications to the needs of industry, primarily for the northwest region of the RSFSR; completion of special projects put forth "at the request of various institutions of the Republic"; development of closer ties between problems of physics and technology; familiarization of industry and corresponding technical circles with the scientific achievements of the laboratory; and "assistance in the training of qualified specialists of different regions of technology, primarily workers of factory laboratories."[61] The department required that research results be published; fifteen volumes of *Trudy Leningradskoi fiziko-tekhnicheskoi laboratorii VSNKh NTO* (hereafter *TLFTL*) appeared over the next five years.[62] A. F. Ioffe assumed directorship of the new laboratory; A. A. Chernyshev and N. N. Semenov were his deputies.

The newly formulated relationship with Vesenkha was advantageous to Ioffe's institute for fiscal and programmatic reasons. Even though funding came from "industrial bureaucracies," it was intended to support work that not only "has the character of applied physics" but which also "takes shape during the process of laboratory work," that is, of an experimental or even theoretical nature. In fact, approximately one-half of all articles published in *TLFTL* concerned theoretical research. Vesenkha's support was lavish enough to permit the institute to expand its research program, and included annual appropriations, unrestricted funds for research of physicists' choosing, and research contracts for which up to 25 percent of the contract's sum was put at the disposal of the director's office.[63]

The Leningrad Physico-Technical Laboratory was located in part in a three-story building that had formerly been a kindergarten on

Table 4. Organization of Leningrad Physico-Technical Laboratory, 1928–1929

Group of Physical Departments	*Group of Materials Testing Laboratories*
Laboratory of A. F. Ioffe	
Magnetic laboratory (Dorfman)	Laboratory of mechanical prop-
Physics department (Ioffe)	erties
Physico-chemical department (Semenov)	Physico-mechanical department (Davidenkov)
Theoretical department (Fock)	Roentgenotechnological depart-
Group of Communications Departments	ment (Seliakov)
	Metallurgical department (Davidenkov)
Technical acoustics (Andreev)	Crystallography department (Obreimov)
Electrical oscillations (Rozhanskii)	
Scientific radiotechnology (Papaleksi)	*Heat Engineering Group*[a]
Radiotechnical applications (Chernyshev)	Physical chemistry (Semenov)
High-frequency machinery (Vologdin)	Department of study of high-pressure steam
Laboratory of special applications (Termen)	Laboratory of factory research
	Gas laboratory (M. V. Kirpichev)
High-voltage laboratory (Chernyshev)	Combustion engineering
Electromechanics (Liust)	Thermal laboratory
Electroinsulation department (A. F. Val'ter)	(A. A. Gukhman)

Source: *Piatiletnii plan GFTL*, pp. 12–32.
[a] Personnel not provided in *Piatiletnii plan GFTL* for heat engineering group, but likely include those listed.

Priiutskaia ulitsa; it shared other facilities and staff members with the institute. The joint facilities grew rapidly: by the end of 1925 there were fifteen departments, and by 1929 there were four groups comprising twenty-four departments (see table 4), as well as 151 personnel of whom 107 were scientists and graduate students, 14 were involved in administration, and 30 in service and support.[64] The overlap in facilities and staff resulted in a synergistic relationship between the laboratory and the institute. As noted in the annual report of the Ioffe Institute for 1924–1925, "the results of scientific research of the institute which to one degree or another may be utilized for technical goals will be handed over for furthest possible development to the laboratory, which has as its

task to serve industry by searching for new productive methods
. . . and also new methods of research and testing. In this way, a
tie between science and production is achieved."[65]

The laboratory maintained close contacts with the physico-
mechanical department of the polytechnical institute. This ensured
a steady stream of qualified young scientists into the laboratory, its
assumption of pedagogical functions, and a closer tie to industry.
A large number of polytechnical students worked under laboratory
faculty on industrial R and D. Examples of these projects include
A. K. Val'ter's study of electrical fields in cable under the direction
of Semenov with funds from Glavelektro; Ia. G. Dorfman's study
of the heterogeneity of materials, and the testing of sheet metal for
defects by magnetic methods, a project for the Izhorsk factory; and
A. F. Val'ter's research for the Scientific-Technical Department and
Glavelektro on electrostatic fields.[66] Ultimately, "the vast majority" of senior theses in the physico-mechanical department were
conducted "under the direction of responsible scientists of [the
laboratory] and in part even within its [walls]."[67]

The expansion of LFTL and the creation of new laboratories
within it was based on three patterns. One was to identify an area
of interest to a funding agency and request funds to acquire the
necessary material, equipment, and publications for research of an
applied or theoretical nature. A second path rested upon the initiative of scholars to move into a new area of research and later to
justify continued receipt of funds in that area. A third path was to
reorganize existing laboratories or subdepartments into new ones.
The heat engineering department (*teplotekhnicheskii otdel*, or TTO)
was formed in this way, although the simple reorganization took
several years to accomplish.

The TTO, one of seven departments founded in 1925, consisted
of three laboratories and the subdepartment of industrial research.
Funding for its creation came primarily from the *glavk* of the metal
industry. Professor V. E. Grumm-Grzhimailo, who originally conceived of the TTO, felt to a great measure indebted to the chairman
of Vesenkha, F. E. Dzerzhinskii, who "was very sympathetic to the
idea of the organization of a special laboratory for the study of
the motion of gases in industrial furnaces" and provided its
wherewithal.[68] Soviet physicists modeled the TTO on European
and American laboratories.[69]

Research in the department centered on the physical processes of industrial furnaces and boilers, control and testing instruments, and insulation, including laws of heat transfer, and the motion of gases in pipes and tubes. In addition to an optical experimental method for the study of heat transfer and exchange of liquids and gases, physicists employed rather detailed theoretical modeling methods to develop exact methods of measurement and control.[70] The basic focus of the TTO was not "testing of machines but the study of separate physical processes occurring in different heat machines and apparatuses," that is, the study of the physical process in laboratory conditions "in its pure form, not obscured or disturbed by other working processes of the machine."[71] By organizing theoretical research in conjunction with applied, physicists were able to defend the former from charges that it was either a fruitless waste of time and funding or a "needless remnant of bourgeois science." Indeed, theoretical research completed within TTO was tested in Leningrad factories, which worked closely with the Ioffe Institute.

By 1929, the LFTL had grown into an internationally renowned physical research center. It had expanded year by year, adding new facilities and equipment. Its researchers published widely in Soviet and Western journals. It served both the physics community and the Soviet economy. The vital connection between the laboratory, such facilities as its heat engineering department, and graduate education at the polytechnical institute convinced most physicists and policymakers alike that the relationship between science and industry envisaged by both groups had largely been realized by the end of the 1920s.

THE RESEARCH PROGRAM OF THE IOFFE
INSTITUTE: PURE AND APPLIED SCIENCE
IN SERVICE OF PHYSICS AND GOVERNMENT

In the 1920s the research program of LFTI set off in a number of new directions, matching the growth of its facilities. From a central focus on the physics of crystals, research now covered such regions as the physics of metals, heat engineering, and theoretical physics. With funding from Glavnauka, Vesenkha, and a number of other government agencies, the research program matched the interests of the state and physicists: the physics of dielectrics and dielectrical breakdown; the physics of metals; abrupt and plastic deformation; electrical and mechanical strength of crystals; and communications.[72]

Research on the electrical and mechanical properties of crystals continued in the institute throughout the late 1920s in several major areas: the physics of dielectrics; plastic and puncture deformation; torsion, twisting, and other aspects of mechanical strength; dielectrical breakdown and loss; piezoelectricity; and photoeffect in crystals. Ioffe focused his attention primarily on the study of so-called "high-voltage polarization" in dielectrical crystals.[73] He gained the interest of several other physicists to work in this area, including K. D. Sinel'nikov, an IEB fellow who had worked at the Cavendish Laboratory and later became director of the Ukrainian Physico-Technical Institute, and I. V. Kurchatov, father of the Soviet atomic bomb project.[74] A large group of institute physicists assisted in conducting careful measurements on the distribution of charge in the layer of the cathode and its dependence on crystallographic direction, temperature, and the size of the charge which is applied to the difference of potentials and force of current.

The existence of large fields in dielectrics led to attempts to apply the phenomenon practically. However, to do this it was necessary to have a sufficiently full picture of the phenomenon of electrical breakdown in a dielectric. At that time, according to Wagner, breakdown sets in as a consequence of the heating of the dielectric when current passes through it. As the work of N. N. Semenov and others showed, the thermal mechanism of breakdown definitely plays a role in certain circumstances (e. g., in rock salt and glass at high temperatures). In normal conditions such as room temperature, however, when electrical conductivity of dielectrics is very small, the mechanism of breakdown operates in another way. The materials-testing department of LFTL actively investigated this phenomenon in search of more complete understanding and potential applications, for the most part related to the electrification of the country—the development of cable insulation, transformer oil, and so on.

Semenov, Lydia Inge, and A. F. Val'ter examined the theory of breakdown at low, room, and high temperatures and at high frequency. In 1924 they published a paper on the thermal theory of breakdown, starting from Wagner's theory, and proved its inapplicability at other than high temperatures, especially for such materials as rock salt and glass; at room temperatures, Wagner's theory demanded much higher breakdown voltage. Working with glass, they concluded that breakdown had a thermal character at intervals between 120° and 140°C, whereas at lower temperatures the mechanism was ionic. The decrease in resistance of insulation during breakdown also pointed to an ionic mechanism.[75]

Later, focusing on low-temperature breakdown, institute physicists introduced a new theory of purely electrical breakdown where greater importance was attached to local heterogeneities of electrical fields at different points of the dielectric which materially effect dielectrical strength. What was more, heterogeneities rendered breakdown dependent upon a series of such other factors as frequency. It was well known that at radio frequencies solid dielectrics break down at significantly lower stresses than during normal technical frequencies. In light of the rapid and ongoing development of the radiocommunications and telegraph industries of Soviet Russia, this problem acquired great technical significance for the question of the production of good insulation. Several highly heterogeneous materials (e.g., resin and other organic compounds) turn out in general not to be good insulators,[76] while other, more homogeneous ones, like glass and porcelain, require great care in use.[77] Val'ter and Inge attempted to establish whether at high frequencies the phenomenon could be accounted for by the thermal and ionization processes which led to it at low frequencies.[78]

In the period 1927–1933 LFTL continued applied research on dielectrical breakdown under contract with several industrial enterprises and government organizations. The high-voltage department of LFTL contracted with Sevkabel' (Northern Cable) in 1925 to apply a model of thermal breakdown to electrical fields in cable using a capacitor probe. V. A. Fock ran calculations on thermal resistance in cable, "an example of fruitful application of higher physico-mathematical analysis to questions of technical practice."[79] Later, for Sevkabel', Val'ter and Dmitriev conducted research on the breakdown of impregnated cable paper. The researchers tested various kinds of cellulose paper to determine the effect produced by degree of moisture, thickness of paper, temperature, and character of current (alternating, direct, or impulse current) on dielectrical strength. They determined that the breakdown tension values varied directly with the compactness of the paper:

the breakdown gradient somewhat decreased as thickness increased, while temperature had little effect on dry, impregnated paper; direct and impulse current had greater breakdown tension values than alternating current; and finally moistness produced the greatest effect.[80]

The research of institute physicists on the physics of metals covered a wide range of properties, applications, and methods of testing. The period witnessed the development of new carbon steels; the testing of conductivity;[81] hardening and tempering, and ferromagnetism; and the application of acoustical, optical, and X-ray analysis and stress tests to codify and explain the properties which technologists had long mastered.[82] N. N. Davidenkov, head of two departments in the materials-testing group of LFTL, conducted a series of experiments in metallurgy at the request of the Scientific-Technical Council of the northwest region. He developed an acoustical technique for measuring stress by the string method,[83] and a sclerometer for repeated impact testing (under contract with the Commissariat of Ways of Communication and the Ingostal' factory).[84] Davidenkov sought to determine the modulus of elasticity of steel during processes of thermal working—cold hardening, annealing, and alloying. Using acoustical and mechanical methods, Davidenkov concluded that his work confirmed experimentally and theoretically the modulus of Sears.[85]

Efforts to understand the properties of crystals, including elastic, mechanical, and abrupt deformation; flow and glide; and strength and cohesion, remained at the center of the institute's research program. In the area of plastic and abrupt deformation, M. V. Klassen-Nekliudova achieved several experimental and theoretical advances in a 1927 paper.[86] Klassen-Nekliudova presented results of her work on abrupt deformation in another paper published in 1928 on displacement in monocrystalline (zinc, bismuth, and brass [70 percent copper]) and polycrystalline (selenium, aluminum, zinc, and brass) metals.[87]

THE RAF AND THE FLOWERING OF
SOVIET PHYSICS

During the NEP the RAF enjoyed growth in membership, vitality, national influence, and international reputation. The association assumed all-union lobbying and representational functions, helped to return publication and *komandirovki* to normal, and contributed to the development of a strong sense of corporate spirit among physicists. Such physics institutes as the Leningrad Physico-Technical Institute, the State Optical Institute, and the Institute of Physics and Biophysics remained the locus of research activity and the source of financial stability. The association was thus an autonomous and powerful ancillary agent that maintained independence from Party control until 1931 and complemented research institutes as a power base for physicists.

The RAF held three national congresses in the mid-1920s, each more international in character and better attended—up to 800 delegates. The fourth, in 1924 in Leningrad, focused the paradox of the wave-particle nature of radiant energy; the fifth, in 1926 in Moscow, considered the quantum mechanics of Heisenberg and Schrödinger; and the sixth, in 1928, examined the experimental results in support of quantum mechanics. The 1928 congress began in Moscow and then traveled by steamboat from Nizhni Novgorod down the Volga to Saratov. The steamboat trip symbolized the strong feelings of professionalism which physicists had developed in the 1920s, and the growing vitality of the discipline.

The fourth congress of the RAF opened in mid-September of 1924 in Leningrad, with A. F. Ioffe, president of the association and chairman of the congress, calling the 426 delegates to order; over 200 participants were from Leningrad, 80 from Moscow, and 124 from Kiev, Tashkent, Tbilisi, and as far as Vladivostok. Participants were housed in the European Hotel and a TsEKUBU dormitory. By the end of the congress nearly 625 were in attendance. Ioffe had contacted the Scientific-Technical Department about the necessity of inviting Ehrenfest, Langevin, and Einstein to the congress, and received 500 gold rubles toward that end, but only Ehrenfest could attend. Before getting down to the business of physics, Ioffe thanked Glavnauka, the Leningrad regional government, and NTO for their support, and welcomed the RAF's single foreign guest. Ehrenfest acknowledged his welcome, noted the zeal with which Soviet physicists worked, and called for continued support of theoretical physics in Russia.[88]

The main issue for physicists during the Leningrad meetings was the controversy over wave and corpuscular theories of light which was coming to a head and solution in the development of quantum mechanics. O. D. Khvol'son highlighted this controversy in his opening address, "On the Contemporary Struggle of Two Theories of Light." Physicists subjected quantum theory to a detailed discussion later in the week after Ehrenfest read a paper entitled "The Theory of Quanta," and Ioffe followed with a presentation entitled "Atoms of Light." There were, in addition to the general meetings and discussions on quantum theory, twenty-nine other sectional meetings on questions of theoretical, experimental, and applied physics.

Leningrad physicists dominated the conference proceedings. This should come as no surprise since two of the major physics institutes of Russia and a plurality of its physicists resided there. Of 162 papers presented for which author and city can be determined, 61 (37.8 percent) were given by Leningrad physicists (20 from the State Optical Institute and 17 from LFTI), and a bit more than one-fifth by Moscow representatives.[89] What was surprising was the fact that the number of papers proposed was two times higher than anticipated, with nearly two-thirds coming from newly established institutes. This lead to the establishment of three additional subsections in optics, physical chemistry, and geophysics.[90]

The RAF also held an organizational meeting to discuss four issues of primary importance to its members: the livelihood of the RAF; the connection between physics research and technology; physics "schools" in Russia; and the establishment of the priority of the discoveries of Russian physicists. While there is no record of these discussions, it is clear from the resolutions passed by the association that physicists were concerned about continuing organizational and financial difficulties. They voted to look into measures to extend their activities throughout the empire and develop provincial physics, to work more closely with Glavnauka, and to consider how to improve ties between industry and physics, with the General Electric and Westinghouse laboratories possibly serving as models, although stressing the importance of theoretical physics and the need to improve training of theoreticians. The importance of industrial research was clear from the fact that the RAF organized a series of excursions to several institutes, Leningrad-area factories, Volkhovstroi, the GOI glass factory, and the Pavlovsk observatory.[91]

The fifth congress of the RAF, held in December 1926, in

Moscow, drew 800 individuals. As he had for the fourth congress, Ioffe once again appealed to the Scientific-Technical Department for hard currency to defray costs to enable Western physicists to attend, but only at the sixth congress were dreams of a truly international gathering realized.[92] Roughly 200 papers were delivered, with Moscow and Leningrad physicists giving 78 and 40 papers respectively.[93]

Discussions of the new physics—Heisenberg's and Schrödinger's contributions to quantum mechanics—and its implications for atomic and molecular properties of matter were central to the fifth congress, with D. D. Ivanenko and L. D. Landau presenting the results of their research on the relationship between classical and wave mechanics, in essence substantiating the Schrödinger method of operators. V. R. Bursian and George Gamov also gave papers on various aspects of quantum mechanics.

In spite of the apparent successes of the fifth congress—the breadth of its program, the high level of attendance, and the clear signals that Soviet physics was healthy and growing—the proceedings provoked discontent. As S. I. Vavilov wrote at the time, no longer was it possible to hold general congresses of doctors or scientists, attended by hundreds upon hundreds, with strict limitations on time to talk in mechanical, formally run sessions. The fifth congress of the RAF had "preserved several of the insufficiencies" of traditional congresses: there were general sessions, truncated talks, bunched papers, and short and impatient discussions.[94] What was the goal of a congress? How was it possible to ensure the vital, unconstrained exchange of ideas? At Ioffe's suggestion, an attempt would be made to hold the next conference during a steamship trip down the Volga, without the press of the crowds but with freely held discussions on controversial questions of contemporary physics. Ioffe's idea was thought at first to be "fantastic" and "impossible to realize," but after the fact, Russian physicists considered the "floating congress" to be practical, productive, and a confirmation of Ioffe's leadership.

The Sixth Congress of the RAF: An International Celebration

In spite of their general sympathy with Ioffe's plans for the sixth congress, many physicists saw them as so extravagant that they insisted on some sort of a compromise with more traditional congresses. The congress therefore opened in Moscow on August 4, 1928, in the Dom uchenykh. After four days "as usual," its members boarded a train to Nizhni Novgorod and then proceeded down the Volga on the steamship *Aleksei Rykov* to Stalingrad, stopping in Nizhni, Kazan, and Saratov (all university towns) for popular lectures and discussions with local audiences consisting of teachers and students. In Saratov, after the official close of the congress, a number of Soviet physicists, accompanied by Debye, Dirac, and others, continued on steamboat to Stalingrad, by railroad to Tikhoretskaia-Vladikavkaz, down the Georgian Military Highway to Tbilisi, and finally to Batum. The congress thereby satisfied "all tastes and desires"; it was a congress both in the "classical" sense and in the new Soviet sense, and played the role of popularizer of physical ideas in the provinces. T. P. Kravets, caught up in the excitement of the sixth congress, commented,

> It is difficult to think of better circumstances for direct exchange of opinions, for more in-depth discussion of a scientific theme, for more sincere rapprochement between workers of close regions [of physics] than the uncrowded, calm life on a steamboat, sufficiently comfortably equipped and by the force of things removed from the noise of a large city, from the broad mass of individuals who, although interested in the works of the congress, do not introduce [anything] into its labors of active creativity. Meetings on the shore along the way fully served these masses. How many new acquaintances were tied together on the Volga! How many interesting general gatherings were conducted in the salons of the ship! How many private conversations which were rich in content individual scholars held, strolling along the decks and slowly feasting their eyes on the broad, melancholy vistas flowing by![95]

The *Aleksei Rykov* thus provided the serene environment in which physicists could get away from "broad masses" and celebrate their achievements during ten years of Soviet power.

The general makeup and program of the sixth congress reflected a youthful and vital discipline, and the growing international respect that Soviet physicists commanded. While precise information is unavailable, observers agreed that the RAF membership was, on the whole, youthful, a fact not surprising in view of the then recent establishment of graduate physics education on a broad scale in Soviet Russia. Over 400 physicists attended the Volga gathering, with 143 from Moscow, 83 from Leningrad, 154 from the "provinces," and 21 foreign scholars; Glavnauka paid for the congress and the traveling expenses of its foreign participants, including Max Born, Peter Debye, Richard von Mises, Charles Darwin, and P. A. M. Dirac from England, France's Leon Brillouin, Phillip Frank from Czechoslovakia, and Gilbert Lewis from the United States.[96] Ioffe also invited Augustus Trowbridge, the Rockefeller official whose efforts had facilitated the foreign travel of Frenkel', Sinel'nikov, and the others, to the conference, but Trowbridge could not attend.[97]

The congress was thus truly international in spirit and composition and, coming three years after the international celebration of the two-hundredth anniversary of the Academy of Sciences in 1925, served to cement ties between Russian and Western physicists. Vavilov described the international flavor: "Within the walls of a former seminary in Nizhni an English speech with American pronunciation on the thermodynamics of chemical processes was heard (Lewis); in the packed conference hall of Kazan University the physics of crystals was discussed in a number of languages . . . ; in Saratov, Born drew matrix tables on a blackboard," and throughout it all Ioffe could be heard translating from Russian to English or German and back again.[98]

While the fourth congress, in Leningrad, had focused on the wave theory of matter and the corpuscular theory of light, and the fifth had considered the abstract forms of the matrix equations of Heisenberg and Schrödinger's wave mechanics, the major focus of the sixth congress was the discussion of experimental data which had accumulated in support of wave mechanics, and the physical consequences of quantum mechanics. Participants expressed surprise at the fact that so little time had transpired since de Broglie had set forth his principle that for each material phenomenon observed by us, there is a wave "picture" lying at its foundation— that is, there is a relationship between observed material motion and the oscillating parameters of propagation of waves which are equivalent to it.[99]

During the proceedings a number of scholars tried to clarify the physical consequences of Schrödinger's theory, and debated the basic significance of his equations, some focusing on the real basis of their statistical character, others still doubting their rigor. Still other physicists searched for a correlation between Schrödinger's work and Heisenberg's matrix calculations. Frenkel', for example, elucidated the relationship between Schrödinger's theory and classical quantum mechanics and statistics in a paper discussed by Landau and Darwin. He also delivered papers on the connection between the wave mechanics of rotating electrons and the classical theory of the electromagnetic field; on the wave theory of matter; and several other, more popular talks on quantum mechanics.

As they had at the previous congress, Ivanenko and Landau also spoke on quantum mechanics, specifically on the foundations of quantum statistics and indeterminacy, with a discussion of the Pauli principle and its mechanical necessity and statistical significance, as well as on the principle of causality in contemporary physics in general.[100] Ivanenko's paper entitled "On Causality and Wave Mechanics," based on the ideas of Bohr and Heisenberg on the impossibility in principle of observing intra-atomic processes, produced the unexpected conclusion that "theoretical physics should replace philosophy."[101] Since Ivanenko soon assumed the reactionary position of watchdog, guarding Soviet theoretical physics from the dangers of idealism, this conclusion is all the more surprising. On the whole, physicists agreed that results of various approaches and methods for the most part coincided, and therefore that quantum mechanics should be considered proven.

Among other papers on theoretical atomic physics, Landau added a discussion of the magnetic moment of the electron. This question of the spin of the electron, so recently introduced into theoretical physics by Uhlenbeck and Goudsmidt, was also discussed by Darwin, Brillouin, and Frenkel'. During one session on the *Aleksei Rykov*, Dirac lectured about the theory of the electron and wave mechanics which, when combined with relativity theory, led inevitably to the "spinning electron," the moment of spin not an auxiliary postulate but a consequence of wave mechanics and relativity theory. In the Physical Auditorium at Saratov, Max Born gave a talk on the different interpretations of quantum mechanics and on the advantages of the Göttingen-Heisenberg point of view, in opposition to Frenkel', Landau, and Ivanenko.[102]

Ioffe attempted to summarize these discussions and the role of experiment in verifying the recent theoretical advances in a paper entitled "Experimental Work on the Wave Theory of Matter." In his opinion the notion that theory develops independently of the participation of the experimentalist was incorrect. On the contrary, it was necessary to conduct experiments which exposed to verification the basic positions of theory, such as the experiments of Davisson and Germer on the reflection of an electron beam from a monocrystal, which produced a marked convergence between the scattering of the electron stream and the diffraction picture of the action of X rays on a crystal, and which immediately corroborated the basic idea of de Broglie on the equivalence of the electron and the wave.

Ioffe also discussed a number of moments in his work and that of his co-workers on the general issue of providing experimental verification for theoretical advances in solid-state, atomic, and molecular physics.[103] In all, there were two hundred papers read at six general and twenty-one sectional meetings on molecular physics, optics, electromagnetism, accoustics, X rays, geo- and biophysics, technical physics, physical chemistry, and theoretical physics,[104] with LFTI physicists reading twenty-three of the seventy-one papers (32.4 percent) for which authorship can be determined.[105]

The congress was not entirely uplifting for Russia's physicists, however. During the steamboat trip along the Volga, they became aware of the fact that the conditions for physics research in the provinces stood in stark constrast to the high level of theoretical discussions held on the *Alexei Rykov*. These highlighted the growing disparity between physics in Moscow and Leningrad and physics in the provinces, something which the scholars hoped to eliminate in the coming years with the establishment of regional physico-technical institutes. Physicists learned of the "wretched" material and moral conditions of provincial physical laboratories. The once well-equipped Saratov University physics laboratory could carry on research only with instruments which could be procured through the journal *Everything for the Radio*. Universities in Nizhni Novgorod and Kazan were also in critical condition, hence the slogan of the congress, which Ioffe had suggested, "Decentralization of Physics!!" The RAF from this point on began to press the government for equipment and money to establish new physics institutes in Kharkov, Tomsk, and elsewhere.[106] Toward these ends, they enthusiastically called for annual gatherings, in 1929 perhaps in the Ukraine, in 1930 in the Caucasus.

For ten years the association's meetings had grown in popular-

Table 5.   Attendance of RAF Congresses, 1919–1930

Year and Date	Location	Attendance	Number of Papers
February 1919	Petrograd	100	60
September 1920	Moscow (1)	500	100+
September 1921	Kiev (2)	278	59
September 1922	Nizhni Novgorod (3)	239	106
September 1924	Leningrad (4)	426	162
December 1926	Moscow (5)	800	200
September 1928	Moscow/Volga (6)	401	200
August 1930	Odessa (All-Union)	750	170

ity and attendance (see table 5). During the Volga conference Russian physicists gathered with European and American scholars for a two-week discussion of the latest developments in physics, and engaged in a frank and open debate over quantum mechanics. No longer were Russian physicists limited to contact with German physicists. They now took advantage of regular sabbatical stays at American and British universities, and published more widely abroad. For their part, European and American physicists in greater numbers attended Soviet congresses and recognized their Russian colleagues abroad. The RAF had achieved the goal of reestablishing Russia's scientific connections with the West. Soviet physics had indeed flowered.

CONCLUSIONS

On the eve of collectivization and rapid industrialization, Soviet physicists believed that they were part of the international community of scholars. The Russian Association of Physicists and its leaders developed a series of policies geared to creating conditions propitious for the growth of the discipline. This involved heroic efforts to reestablish scientific publication, secure the appropriate setting for research, and overcome international isolation.

International contacts were important to Soviet physicists for a number of reasons. First, foreign contacts provided the opportunity for physicists to participate in the world community, share and exchange their ideas with their Western counterparts, ensure Soviet priority in discovery, and return to the USSR well prepared to train a new generation of physicists. Second, contacts with

Western firms resulted in research contracts which provided additional hard currency for physicists. Finally, Soviet scientists often based their ideas for the organization of new research institutes on the example of university and industrial physics laboratories in the West.[107]

More than any other research center, the Leningrad Physico-Technical Institute and Laboratory stood at the center of this process and symbolized the potential of Soviet science. It had experienced ten years of uninterrupted growth in terms of employees, budget, and research program. It now consisted of four departments with over thirty different laboratories and sectors,[108] staffed by more than one hundred full-time physicists, many of whom studied and traveled in the West. Work was funded by a number of bureaucracies; contracts from industry supplemented this funding. And with all governmental and economic organizations the institute was on very good terms. This is because physicists and the state shared many goals, primary among them the organization of industrial R and D for economic development.

Moreover, fundamental research and theoretical investigations supported by various branches of Soviet industry through Vesenkha, Glavnauka, and enterprises such as Sevkabel' and Elektrosila in fact often had immediate applications in electrification (insulation, cable, and transformers), in radio and telecommunications (vacuum tubes of all sorts), the production of new steels, and the construction of new industrial furnaces.

Yet there were storm clouds on the horizon. When the Russian Association of Physicists disbanded at the end of the sixth congress in this environment of expansion and vitality, they did not realize they would come together only one more time as a national organization. The boat ride down the Volga which had symbolized the independence of Soviet physicists represented to an increasingly class-conscious Party these physicists' elitism and aloofness. The Shakhty and Industrial party affairs signaled an end to the autonomy of these scholars. Specialists—scientists, technologists, and engineers—came increasingly under attack during the introduction of the Stalinist politics of science, cultural revolution, and "class war" directed by the Party against specialists during the period of forced collectivization, industrialization, and the First Five-Year Plan.

5

Physics During the First Five-Year Plan: Industrialization and Stalinist Science Policy

Stalin's revolution from above, the so-called Great Break (*velikii perelom*), marks a turning point in the history of Soviet physics. During this time of rapid industrialization and forced collectivization, the state introduced a series of new policies aimed at making scientific research and development better serve the needs of "socialist reconstruction." Like "government construction" before it, "socialist reconstruction" was a catch-all phrase which in this case signified all political, economic, and cultural measures intended to complete the socialist revolution begun by Lenin in 1917. The goal was to create "socialism in one country" since world revolution had failed, indeed was alleged by Stalin and his supporters to be a bankrupt Trotskiite concept. In Stalin's view, the USSR was to be transformed into a socialist fortress which, in a struggle with the hostile capitalist encirclement, would be victorious.

Stalin turned the energy of an increasingly working-class Party and vigilant secret police toward enemies believed to be lurking within Soviet society: the *kulak*, the better-off peasant in the countryside; the NEPman, the small-scale, petit bourgeois trader; the Trotskiite and other traitors within the Party; and the Tsarist engineer and scientist "wrecker" in the R and D apparatus. The Stalinist system used administrative measures and terror against society to achieve its goal of unquestioned loyalty to the means and ends of "socialist reconstruction." Terror would be used increasingly throughout the 1930s in both the city and the countryside, and millions would die of starvation or execution.

The following section of this book examines how physicists reacted to the assault on their autonomy which accompanied the three major events of Soviet history in the late 1920s and 1930s:

rapid industrialization under the five-year plans, collectivization, and cultural revolution. This chapter focuses on the development of Stalinist science policy and the growth of the national physics enterprise during the First Five-Year Plan. It identifies the sources of physicists' continued autonomy in the face of economic and political pressures to conduct research in keeping with the goals of "socialist reconstruction."

Chapter 6 turns to the effect of cultural revolution on the conduct of fundamental research. Cultural revolution involved the penetration of Party cadres into institute management; advancement (*vydvizhenie*) of workers into positions of administrative and academic responsibility by virtue not of their expertise in physics but of their social origin and political views; and class war against "bourgeois specialists."

It was, however, encroachment on the philosophical front which physicists most resented and opposed. Chapters 7 and 8 evaluate the disruptive impact in theoretical physics of philosophical disputes between ideologues who saw idealism lurking behind every theoretical pronouncement, and physicists who foresaw the potential damage to their discipline caused by misguided ideological regulation of the epistemological questions posed by the "new physics"—quantum mechanics and relativity theory.

STALINIST SCIENCE POLICY[1]

From the point of view of Stalinist economic planners and Party officials, science and technology were a central instrument of "socialist reconstruction." They would bring about rapid industrialization and collectivization of agriculture by making all areas of the productive process more rational and efficient. Unlike the capitalist West, the socialist Soviet Union would tie science and technology "organically" to production. Bringing science and technology to the service of the state required class war against the bourgeois specialist and his subjugation to Party organs. It required new institutions for the administration of R and D. The introduction of Stalinist science policy was intended to meet all of these goals.

Stalinist science policy involved the centralization of the administration of R and D through the creation of new governmental

administrative and financial organs, and the emasculation of Glavnauka and its replacement by the Commissariat of Heavy Industry
as the primary organ of science policy. It required that researchers
stress application more than fundamental research, that applications have the greatest importance in industrial spheres through
vnedrenie (best translated as "assimilation"), and that researchers'
activities be encompassed by long-range, narrowly focused
"thematic" plans. And it led directly to the destruction of all
professional organizations, including the Russian Association of
Physicists, which were seen as representing the specialists' interests rather than those of "socialist reconstruction."

Physicists held that Stalinist science policy encroached on their
autonomy. Through perceptive leadership they were able to blunt
some of its impact. Rapidly increasing budgetary allocations also
permitted leeway in the selection and determination of research
programs. Since they best understood the research agenda and
served on the newly constituted science policy boards, they were
ensured that their concerns would at least be heard. It may also be
that the pressure for accountability to the state required those
directing research to make rational choices at a time when a huge
influx of funds might have encouraged waste of resources.

To what extent did physicists' new accountability to the state
counter the greater autonomy granted by increased budgets, the
expansion of research programs, and a growing network of physics
institutes? The victory of Stalin and his followers in the apparatus
ensured that the Party, and not scientists, would have final say in
determining national priorities. The Soviet government intended
to control the physics enterprise at all levels, from the individual
researcher to the institute and commissariat. It enforced new policies through the successors of the Scientific-Technical Department:
first the Scientific-Technical Administration of Vesenkha, then the
Scientific Sector of the Technical and Economic Planning Administration of Vesenkha, and, later still, the Scientific Sector of the
Commissariat of Heavy Industry (Narodnyi komissariat tiazheloi
promyshlennosti, or Narkomtiazhprom). The Physics Association
of Narkomtiazhprom was introduced at the same time to supplant
the physicists' professional organization, the RAF. The RAF soon
disappeared, as did all other professional organizations, to be

resurrected as the Physics Society of the USSR only under Gorbachev's policies of *glasnost'* in the late 1980s.

Stalinist science policy involved the victory of Vesenkha over Glavnauka. On the eve of the First Five-Year Plan, it was not clear that Vesenkha would usurp Glavnauka's role in the administration of fundamental science. Glavnauka had ambitious plans for the development of its research network. Throughout the late 1920s scientists pressured it for higher salaries and larger institutional budgets. In an effort to meet these demands and attract first-rate personnel, Glavnauka planned to raise salaries of scientific workers within its institutes perhaps threefold between 1928 and 1933, with the total amount of salaries for scientific workers increasing from 7.2 million rubles to 23.2 million rubles (of a total budget of 167.6 million rubles). Glavnauka forecast an increase in the number of scientific workers in its institutes from 3,200 to 5,000 in the same period.[2]

Reflecting the tenor of the time, the deputy director of Narkompros, V. N. Iakovleva, called on Glavnauka to improve its efforts in the matter of "socialist reconstruction" at a meeting of the Party cell of Narkompros in 1929.[3] Officials instructed all scientific institutes under Glavnauka's jurisdiction to channel their research in support of rapid industrialization and collectivization. Institutes within its physico-mathematical division subsequently included in their plans increased investment in agricultural research (realized in the founding of A. F. Ioffe's Institute of Physical Agronomy in 1934), electrification (the Leningrad Electro-Physical Institute, 1934), and the formation of a network of physics institutes throughout Siberia and the Ural provinces (UkFTI, SFTI, and others).

When Glavnauka lost out to Vesenkha, however, its plans for expansion came to naught, and a majority of its research institutes were transferred to the jurisdiction of Vesenkha. Narkompros had long been seen as soft on the issue of class war against the "bourgeois specialist," and unwilling to force the specialist to cede the right to formulate the national research agenda to Party and government organs. The Party then orchestrated the Shakhty and Promparty (Industrial party) affairs, with show trials replete with the repentant engineer confessing his "wrecking" of Soviet industry. In the atmosphere of the Great Break, and class war against the engineer and scientist as signified by the show trials, it followed that Glavnauka, too, would come under attack.

N. I. Bukharin, chairman of the Scientific Sector of Vesenkha and a moderate in terms of science policy, could do little to temper the policies of Vesenkha's V. V. Kuibyshev, Narkomtiazhprom's G. K. Ordzhonikidze, and I. V. Stalin, whose centerpiece was centralization of control of the scientific enterprise from above by the Party. Kuibyshev, earlier head of Glavelektro, then chairman of Vesenkha from 1926 to 1930, and finally head of Gosplan, the state planning commission, from November 1930 until his death in January 1935, embraced superindustrialization with fervor. Ordzhonikidze, the chairman of Vesenkha after Kuibyshev, also believed that science and technology must be put to use to push industrial production further and further. He was commissar of Narkomtiazhprom when the light and timbers industries were separated off from heavy industry in 1932. He wholeheartedly had supported the Promparty affair but in February 1937 committed suicide, apparently rather than face purges within Narkomtiazhprom over "wrecking" after members of his staff and factory directors he had appointed were arrested and his brother executed.[4]

Glavnauka quickly lost its domain of scientific research institutes. While in 1929 there had been over 230 central research institutes, 119 scientific societies, and over 700 museums under its jurisdiction, by 1932 Glavnauka administered only 76 institutes, of which 24 were museums, 18 were biological research centers, 11 were libraries, and only 5 were physics institutes: the State Radium Institute, the Siberian Physico-Technical Institute, the Nizhegorod Physico-Technical Institute, the Kazan Chemical-Scientific Research Institute, and the Physics Institute of Moscow State University, an organization central to the philosophical controversies brewing on the horizon.[5] By the middle of the next year only 32 of these institutes remained under Glavnauka's jurisdiction.[6]

Research institutes which were associated with the problem of "socialist reconstruction" became wards of Vesenkha. The Scientific Sector of Vesenkha, later of Narkomtiazhprom, took as its mandate the coordination of basic and applied scientific research. It was created in 1930 to administer scientific research institutes in twenty associations covering physics, chemistry, the oil and gas

Table 6. Scientific Research Institutes and Personnel of Narkomtiazhprom, 1931–1935

	Jan. 1931	Oct. Nov. 1931	Jan. 1932	Jan. 1933	Jan. 1934	1935
Research institutes	65–70	128		154	158	138
–physics		6		6		6
–Leningrad physics		4		4		4
–chemistry		24		31		32
–energetics		6		8		8
–fuel		10		9		7
–metallurgy		12		11		11
–machine building		16		17		17
–construction		6		24		15
–associations		6		10		12
–other		37		43		24
Personnel	19,600	29,942	31,915	45,213	42,023	33,830
–leading	1,688	2,903	9,177	13,316	13,730	
–scientific workers	6,312	8,757			11,189	
–technical	6,931	12,099	9,013	13,307	12,222	
–other (including aspirants)	4,669	7,252				
–physicists			448	657	672	560
Total budget (in million rubles)	138.2		230.8	273.7		268.6

Sources: *Sorena*, no. 2/3 (1931):282–288; no. 2 (1934):170–176; no. 4 (1935):137; *FNiT*, no. 7/8 (1932):105–109.

industries, and so on. The associations had been formed upon the dissolution of the Scientific-Technical Administration.[7] A. F. Ioffe was chairman of the Physics Association, whose organization and functions are described below. Seventeen research institutes including LFTI remained outside of the jurisdiction of the associations until late in 1931 when they were brought firmly within the control of the Scientific Sector.

For three years the Scientific Sector experienced uninterrupted growth; it received massive infusions of funds and by 1934 included over 150 research institutes and associations. The number of scientific workers also increased rapidly. Nearly 30 percent of all scientific personnel in the USSR now worked in institutions of Narkomtiazhprom, primarily in chemical, metallurgy, and machine-building institutes (see table 6), while 28.6 percent worked in Narkomzdrav (the Commissariat of Health), not quite 12 percent under Narkomzem and Narkomsovkhozov (Agriculture and State Farms), and only a little over 4 percent in Narkompros.[8] For physicists, the most pressing issue which accompanied the transfer of their institutes from Glavnauka to the Scientific Sector was the introduction of long-range planning of scientific research.

THE PLANNING OF PHYSICS RESEARCH

During the first five-year plans, the policies of Vesenkha and Narkomtiazhprom dictated the scope of scientific research and the subordination of "bourgeois experts" to the planning organs. The government at first required rather detailed plans, ordering institutes to justify research and specify anticipated results. By the mid-1930s, scientists no longer had to provide such comprehensive documents, having convinced officials that general plan outlines sufficed.

Physicists initially provided quite detailed lists ("thematic plans"), itemizing between 150 and 200 individual projects for each year of the five-year plan. The plans were upward of three hundred pages, and included a general outline of the program; finances; a laboratory-by-laboratory breakdown of research projects proposed and under way and their dates of completion; publications or other products of research, and a chart of the thematic plan. This chart comprised nearly half of the document and specified laboratory, number of topics, name of the topic, who initiated a research project, its significance for industry and a given region of the country ("for the USSR," "for Leningrad," etc.), the number of man-hours, cost, duration, and anticipated cooperation with other departments and institutes.[9]

Physicists seem to have discovered that such detail constrained their activities, however, locking them into narrowly construed topics of research and making it difficult for them to embark upon unforeseen, and therefore unplanned, topics of research. They began to complain about onerous reporting requirements, and thought it was dangerous to tie physics research too closely to narrow projects. They soon drew up more general alternative plans which provided greater leeway in project selection. In 1935, for example, LFTI listed only fifty-five individual projects. Of these, the physicists duly reported, forty-six were completed on schedule, three new projects were added during the year, ten were cancelled for various reasons, and two remained unfulfilled. Reports always stressed the close "tie with industry"—as demonstrated by contracts with a number of factories, institutes, and governmental bureaucracies including the Scientific-Technical Defense Bureau, and the Bolshevik, Kazitskii, Svetlana, Kirov, and Sevkabel' factories.[10]

By the late 1950s, planning documents had to be submitted on standardized forms provided by the State Statistical Administration. Each institute was required to submit a series of these forms to the responsible higher authority (Academy of Sciences, ministry, etc.); documents for institutes employing as many as five thousand workers amounted to less than one hundred pages.[11]

Physicists' success in tempering onerous reporting requirements could not forestall a decline in their control over the physics enterprise. Pressure for planning of R and D and accountability to state organs had been growing since the mid-1920s. At the second all-union congress of scientific workers in February 1927, Party members pushed for "more rational and planned distribution and utilization" of the resources available to science.[12] The third all-union conference, held in February of 1929 in Moscow, echoed this call.[13] It became clear that the scientist would have to confront the pressure directly and attempt to play an active role in the formulation of plans in order to maintain a degree of control over the research enterprise. In discussions of the implications of planning for physics research, A. F. Ioffe was a major spokesman.

Once Ioffe realized that physics activity would become subordinate to such central planning and economic organs as Gosplan and Narkomtiazhprom, he sought out ways to protect physicists' autonomy while addressing the Party's concerns for accountability to state economic programs. Ioffe delivered his first major public pronouncement on the subject sometime before the Sixteenth Party Congress in 1929 or 1930. This was just after the Shakhty and Industrial party affairs, which had alerted specialists that their aloofness from politics and from the economic problems facing the country was about to end.

In his address Ioffe set forth several themes which were to be repeated over the next four years: he urged flexibility in the planning of scientific research; he proposed that physics be organized in a hierarchy, from factory laboratories and branch institutes of industry to research institutes, covering the gamut of research from applied to basic and theoretical, but avoiding the subordination of the latter to industrial production; he expressed concern over the possible excesses of *vydvizhenie;* and he argued that the primary problem of contemporary physics and technology was energetics. Ioffe frequently drew on the example of LFTI to illustrate that physicists' interests were commensurate with those of the Party.[14]

At the first session of the Central Scientific-Research Council of Industry in 1929, speaking with such leading scientists as the electrotechnologist V. F. Mitkevich, the chemist V. N. Ipatieff, and S. A. Chaplygin, a specialist on aerodynamics, Ioffe addressed each of these issues. He recognized that planning scientific activity on an all-union scale was important, especially in light of the expansion of the physics research network and the creation of new physico-technical institutes. But he urged that the initiative for planning and selection of topics of research remain with physicists. The formation of an all-union association of physicists with these powers, to replace the RAF, would serve such a purpose. He argued against *vydvizhenie* in a veiled manner, calling for advancement of workers according to the nature of the work, not of the worker. And, sensing the approaching pressure for autarky in science, he called for the strengthening of international scientific relations.[15]

Many within the Communist Academy of Sciences—an organization of marxist scholars established in the early 1920s to rival the "Tsarist" Academy, and until 1924 known as the Socialist Academy—criticized Ioffe's view of physics during the reconstruction period. V. P. Egorshin maintained that organizing R and D as Ioffe had suggested would give physicists too much power. "The government," Egorshin claimed, "would play the role of financier, patron of the arts, who would award a set percentage of its budget to scientific work" while physicists decided who got what and how. Egorshin believed that "competent organs" should fix specific research tasks before consulting scientists, and, of course, finance the research and equipment acquisition.[16]

The major steps toward the introduction of formal and regular planning of scientific activity were taken at two all-union conferences. The first, held in Moscow on April 6–11, 1931, brought together all of the leading scientists and technologists in the USSR—over one thousand delegates in all. Such speakers as N. I. Bukharin, now head of the Scientific Sector, and the geologist A. E. Fersman underscored the importance of planned science in addressing the problems of "socialist reconstruction." The permanent secretary of the Academy of Sciences, V. P. Volgin, the physical chemist A. M. Frumkin, Ernst Kol'man, a mathematician active in the Communist Academy of Sciences, N. I. Vavilov, the geneticist and brother of the future president of the Academy, and others discussed issues of planning in science according to their disciplinary specialties.

In his presentation, Ioffe urged planners to recognize the danger of project-specific, long-range planning. Fixed targets might provide benefits in the short run; in many fields, however, five- to ten-year forecasts were necessary to avoid "spoiling" possible long-term results. He warned that "it is necessary to keep in mind that the tempos of development may be entirely different from those we anticipate."[17] Ioffe introduced three kinds of research tasks that planners and physicists alike might consider: (1) a group of thirty or so imminently realizable tasks; (2) long-term problems; and (3) a category best called "futuristic" physics research. Speaking about the latter Ioffe said, "We are now in a period of extremely rapid and dynamic development in science, and there is nothing impossible, so that in several months or a year it may be possible to speak completely seriously about a problem. . . . Not to work on [it] would be criminal."[18]

Ioffe's concrete examples of long-range and futuristic research, which centered primarily on energy technologies, must have bordered on the fantastic for his audience. Ioffe suggested the construction of an entire domed city; the utilization of the energy of the Arctic Ocean by tapping differences in water and air temperature; desalination of water from the Caspian Sea (accomplished today by the BN-350 breeder reactor in Shevchenko); the development of photochemistry and photoelectricity; and the design of energy-efficient buildings. The achievements of his institute in solid-state physics were an indication that the fantasies could become reality.[19]

Vesenkha responded to Ioffe's speech as he had hoped. It gave the institute a rather broad mandate, calling for it to focus on research in the areas of electrification and communications and avoid tasks of "a secondary nature," and approving a series of projects already begun: work on thin-layered insulation; technical problems associated with high-voltage transmission; and the development of an apparatus capable of going from alternating to direct current and back. Vesenkha, for its part, would contribute "shock work [*udarnoe stroitel'stvo*]" to the institute by raising operating funds substantially and awarding an additional 200,000 rubles for LFTI's experimental factory, plus 130,000 rubles and 10,000 rubles in hard currency for work on new kinds of insulation.[20]

The second all-union conference on planning of scientific research in heavy industry was called jointly by Narkomtiazhprom, the Academy of Sciences, and Gosplan in Moscow in late December of 1932. It was held simultaneously with the second conference of VARNITSO (the All-Union Association of Workers of Science and Technology Building Socialism), a leftist organization which hoped to attract the specialist to the program of "socialist reconstruction" by agitating to raise the political consciousness of scholars and mobilizing them for "shock work."[21]

The conferees were concerned with the Second Five-Year Plan. In physics this meant that research institutes should focus on solutions to pressing technological problems in the physics of metals, fuels, electrification, and communications. Physicists had convinced planners of the need to support fundamental research in ten major areas from nuclear physics to biophysics.[22] The conferees thus endorsed a Second Five-Year Plan which included all major foci of LFTI: quantum mechanics and atomic and nuclear physics, where harnassing the energy "would provide greater practical possibilities for defense, medicine, agriculture, and chemistry"; research on the solid and liquid states, including the physics of metals, glass, plastics, and colloidal systems, viscosity, and the theory of the solid-state and liquid crystals; optics (the nature of light and its interaction with matter); electrical phenomena; heat engineering; radio- and electrotechnology and acoustics; and geo-, bio-, and agrophysics.[23]

The conference concluded with resolutions on Bukharin's speech, "Technological Reconstruction and Current Problems of Scientific Research," which recognized the achievements of the scientific research institutes in the USSR but drew attention to shortcomings in their organization and performance. Surprisingly, he criticized the recent practice of creating new institutions without real guarantees of manpower or budgetary allocations and "the simultaneous, premature splitting of new [institutes] from huge ones, as a result of which we have several tens of institutions, which do not justify their appointments."[24]

This criticism could have been directed at the leadership of LFTI and its burgeoning physics empire, but wherever directed, it was paradoxical, since the establishment of new research institutes had occurred largely at the direction of the state, or with its approval, and was part and parcel of the entire industrialization effort. Naturally, the growth of the R and D apparatus was accompanied by overlap and duplication of effort, shortages of personnel and equipment, and bottlenecks. Ironically, all of this in turn generated pressure for greater coordination of R and D, and the centralization of administration.

It may indeed be that the state and scientists shared certain goals in physics research during the first and second five-year plans. For example, both wished to expand the research enterprise, to introduce scientific achievements more rapidly into production, and to take advantage of the rationalization and economy promised by planning. However, planners in Gosplan, bureaucrats in Narkomtiazhprom, and Party officials emphasized heavy industry at the expense of other sectors of the economy, put too much store in "the unity of theory and practice," and saw centralization and administrative fiat as a solution to any problem which seemed to slow down the penetration of scientific advances into the economy. Party officials sought concrete, yearly specification of research goals and the development of technology. They were satisfied with the performance of physicists in such areas as electrification, communications, and heat engineering, but did not abandon the incessant calls for even greater application and more rapid assimilation by industry of scientific achievements.

Leading physicists were concerned that the new system of planning and administration would needlessly constrain research and actually lead to a decline in their productivity. They feared premature identification of targets, arguing that it was often impossible to predict future fruitful areas of research. Furthermore, they opposed shortsighted planning which bound physicists and their resources too closely to industrial production targets.

Physicists differed with the government in one area in particular: the role of specialists and their professional organizations as policymakers and advisers in the national scientific enterprise. Now that the Russian Association of Physicists was subsumed by Narkomtiazhprom, physicists had lost a crucial battle to the Party.

THE CREATION OF THE PHYSICS
ASSOCIATION OF NARKOMTIAZHPROM

During the introduction of Stalinist policies toward science, the Communist Party infiltrated and subjugated virtually all organizations, from professional societies of physicists and architects[25] to such centers of research as the Leningrad Physico-Technical Institute and the Academy of Sciences.[26] The Party intended to eliminate the last vestiges of independence which scientists, engineers, and other specialists had achieved in national associations. This was part of an attack on any movement perceived to be technocratic by the Stalinist leadership.[27]

Soon after the RAF met in their final congress in 1930, the group disappeared without a trace, and in its place several advisory boards and planning organizations arose within Narkomtiazhprom and the Academy of Sciences. Physicists attempted to head off the dissolution of their association by demonstrating to the Scientific Sector of Vesenkha and the government's Central Executive Committee how they shared national goals of "socialist reconstruction."

Narkompros abandoned its professional physics organizations at this time. The activities of the Moscow Physics Society and the Russian Physico-Chemical Society were suspended by the physics leadership. The two societies had long met infrequently and had become social clubs more than physics societies. Narkompros repeatedly refused to reconvene them until twenty professors from the physics department of Moscow State University voted unanimously in Narkompros's recently reorganized academic committee of physics to support their reestablishment. The Moscow physicists also wished to take over the editorship of the major Soviet physics journals. They saw the current editors, physicists active in the RAF, as "idealists." But Narkompros's academic committee of physics was unceremoniously disbanded in 1938, no doubt at the behest of the physics leadership, which saw Stalinist philosophical views as dangerous to the discipline.[28]

At the first All-Union Congress of Physicists, held in August 1930 in Odessa, physicists addressed the issues of cultural revolution and planning in physics head-on.[29] This was the largest gathering of physicists in Russia to date, with over 950 members in attendance. The 15 foreign participants who made their way to Odessa included Arnold Sommerfeld and Fritz Houtermans. (Houtermans worked at the Ukrainian Physico-Technical Institute in 1935–1937 and fell prey to the purges.) The congress had four plenary and thirty sectional meetings during which physicists delivered 170 papers.[30]

Judging by the content of the first two plenary sessions, the Great Break had a direct impact on physicists' activities. At the first session, physicists examined "the question of the connection between physics and dialectical materialism." At the second they discussed "the problem of cadres and the tie between physics and production." The two remaining plenary sessions focused on wave mechanics and the question of the orientation of dipolar molecules. In addressing questions of organization and planning, physicists passed a resolution designed to ensure their continued input into science policy. They called for the creation of an all-union plan of scientific research, and for the establishment of an all-union association of physicists (*Vsesoiuznaia assotsiatsiia fizikov*, or VAF), which would assume the responsibility for national policy and maintain scientific initiative among physicists.[31]

Following the congress, Ioffe wrote the academic committee of the Central Executive Committee (TsIK) about the proposed professional association and tried to deflect the kinds of criticism of the "bourgeois specialist" to be expected in the 1930s. Ioffe assured the TsIK that Soviet physicists were aware of a "major flaw both in the organization and in the direction of the very work" of the Odessa meetings. They had devoted inadequate attention to the rapid growth of cadres, the establishment of physics institutes in the provinces, and the development of both pure and applied physics occurring in the USSR. There was, to be sure, "no connection between the plans of institutions and laboratories throughout the entire Soviet Union and the general plan of the national economy.[32]

This did not mean that the initiative for planning should come only from industry and leading government organizations. Through the VAF, Ioffe explained, physicists would provide the general direction of and exercise control over scientific research. It would take on the following responsibilities: (1) the formulation of a general plan of research for the entire Soviet Union, including "thematic plans" of a technological character which were "advanced by the economy," meaning industry itself, and discussed among scientists and representatives of industry; (2) the training of cadres, both for research institutes and factory laboratories; and (3) the strengthening of ties with Western science. Ioffe closed his letter to the Central Executive Committee with a request for financial support for the VAF. The VAF council, chaired by Ioffe, included all leading Soviet physicists: Semenov, Ia. I. Frenkel', D. S. Rozhdestvenskii, S. I. Vavilov, L. I. Mandel'shtam, I. E. Tamm, and P. P. Lazarev.[33]

In keeping with its past view of the role of the specialist in Soviet society, Narkompros supported the creation of the VAF and vied with the Central Executive Committee for its jurisdiction. The VAF's educational, training, and ideological purpose—to attract "a broad mass of workers into scientific research" and to build the "research activity of the society upon a basis which guarantees the dialectical materialist elaboration of physical problems"—meant logically that it fell within the Narkompros bailiwick.[34] However, the general decline of Narkompros in the area of science policy meant that the VAF as envisaged by physicists and Narkompros would never be constituted.

Gosplan and Narkomtiazhprom opposed Narkompros on the matter of the physicist's association. Gosplan believed that the creation of an all-union association of physicists was timely but opposed its subordination to Narkompros, which it saw as being dominated by bourgeois interests. Officials at Narkomtiazhprom believed that the creation of a physics association under the umbrella of an organization, such as their own, with close ties to industry better served the goals of the five-year plans.

The head of scientific affairs for the Central Executive Committee, V. T. Ter-Oganesov, informed Ioffe that Narkomtiazhprom would be the home for a new physics association, and asked his help in creating it.[35] The new association then joined several other new bodies responsible for science policy in physics, including the Association of Physico-Mathematical Sciences of the Academy of Sciences, which was created in 1931 with groups in mathematics, physics, technology, and astronomy;[36] Ioffe served on the predecessor of this body, established in 1929 to formulate five-year plans.

The physics association was one of nine associations of Narkomtiazhprom—three others were being organized—each of which represented a major branch of science and technology for industry. These included physics, heat engineering, metallurgy, machine building, and chemistry. The institutional members of each association included *vuzy*, factory laboratories, *glavki*, trusts, enterprises, and such basic research institutes as LFTI; each association had subcommissions which represented the various fields of its discipline.[37] Physicists from LFTI continued to play a prominent role in the administration of their new association (see table 7).

The functions and makeup of the associations reflected the exigencies of Stalinist politics of science, but ensured the input of Soviet scientists in planning. Narkomtiazhprom charged the associations with keeping an eye on the needs of industry, including efforts to assimilate new foreign and domestic technology; the

Table 7.   Organization of the Physics Association of Narkomtiazhprom

---

President of Bureau of the Presidium: A. F. Ioffe
Deputies: P. P. Lazarev, S. I. Vavilov, B. M. Gessen
Academic Secretary: D. Z. Budnitskii[b] (Deputy Director of LFTI)

1. Mechanics, acoustics, and electrical oscillation: L. I. Mandel'shtam N. N. Andreev[b]	5. Magnetism: Ia. G. Dorfman[a,b] N. S. Akulov
2. Optics: S. I. Vavilov, G. S. Landsberg	6. Molecular physics: A. F. Ioffe, P. P. Rebinder I. V. Obreimov[b,a]
3. X ray, light, and electronic radiation: S. T. Konobeevskii P. I. Lukirskii[b,a] P. S. Tartakovskii[b]	7. Nuclear physics: G. A. Gamov[b] A. I. Leipunskii[b,a] I. V. Kurchatov[b,a] D. V. Skobel'tsyn[b]
4. Geophysics: P. P. Lazarev	8. Theoretical physics: Ia. I. Frenkel',[b,a] I. E. Tamm

---

Source: *Sorena*, no. 4 (1933):173–175.
[a] Members of the Ioffe "school."
[b] Physicists at LFTI or other FTI.

identification and development of individual research tasks; the creation of yearly thematic plans; the supervision of scientific research conducted among the associations' institutional members; the organization of brigades for the exchange of scientific information; the development of a data bank of scientific workers within each association; and the holding of congresses and conferences.[38]

The rapid development of subdisciplines within a field of science requires regularly scheduled conferences organized around specialized topics. But Soviet physicists wanted both academic conferences to discuss developments in subdisciplines and professional congresses to consider broader disciplinary issues. Since they were denied the opportunity to meet as an autonomous all-union professional association, physicists understood the last charge from Narkomtiazhprom as a mandate to convene frequently in conferences devoted to subdisciplines. Between 1931 and 1939, physicists gathered over twenty-five times in congresses with as many as two hundred participants. (See app. A, table 4 for a list of these major all-union conferences.)

The introduction of Stalinist science policies led to the centralization of control of the physics enterprise under the jurisdiction of Narkomtiazhprom. In word and deed this signified government intentions to put physics to work for the industrialization effort. Another result of the policies was the subjugation of the physicists' professional organization to economic organs of the Party. Yet scientists found room to maneuver within the constraints imposed from above. Their service on government science advisory boards assured them access to the policymaking process. Massive infusions of capital provided the wherewithal to maintain fundamental research programs while turning attention to applied research. And scientists and the state shared the goal of the expansion of the R and D apparatus, as is clear from the case of LFTI.

In an attempt to make science accountable to the needs of "socialist reconstruction," the government created an administratively top-heavy, inflexible system of research. Many of the policies adopted during this time nevertheless contributed to scientists' continued autonomy in the formulation of research programs and direction of research institutes. Massive increases in the budgets of those organizations throughout Soviet society which were seen as having the greatest potential for promoting economic growth contributed to the growth of the national physics enterprise. These increases also gave physicists more leeway in determining the direction of their research. In spite of repeated calls for research with direct application to the productive process, physicists found flexibility, if not in the system of administration, then in the profusion of funding, to embark on new programs in semiconductor and nuclear physics.

Under the watchful eyes of Vesenkha and Narkomtiazhprom and the leadership of physicists at LFTI, the physics R and D apparatus developed in two distinct patterns. The first involved the "spinning off" of new institutes in the Leningrad area from existing institute personnel and facilities. The second involved the establishment of physico-technical institutes in Tomsk, Sverdlovsk, Kharkov, and Dnepropetrovsk. LFTI itself continued to expand, adding new laboratories and sectors, and going through a number of reorganizations during the First Five-Year Plan. The goals of reorganization were to find greater organizational stability, achieve more rational utilization of resources to avoid duplication of effort, and identify research goals which were commensurate with Stalinist policies. The first step toward this goal was the merger of the institute with the Leningrad Physico-Technical Laboratory.

Physicists began to formalize the overlap between the institute and the laboratory sometime in 1929, and sought the approval of Vesenkha and Glavnauka. Ioffe wrote Glavnauka during the summer about these plans. The laboratory, he noted, had evolved out of "the necessity of a significant expansion of the work of the institute" and had used institute equipment from the start. The two had, in fact, similar goals, and were located in the same place—to the extent that "several analogous departments are even under the same roof." They shared the library. More to the point, Ioffe admitted that the administrative and financial departments of the two organizations had been "united almost fully from the moment of the laboratory's creation." Since "merger already exists," Ioffe concluded, it could "not be destroyed without damage to both institutions and a superfluous increase of government funding."[39] In fact, this was one institution with two different names and budgets.

The joining of the laboratory and the institute did not occur straightaway. In February 1930, the Scientific Sector of Vesenkha approved the transformation of the Physico-Technical Laboratory into the State Physico-Technical Institute of Vesenkha with an independent academic council. In a fashion, for one year, two physico-technical institutes existed, both under the direction of Ioffe, with the same personnel, located on the same grounds with scientists conducting research interchangeably between institutions, one under the administration of Narkompros, the other under Vesenkha.[40] Early in the fall of 1930, Ioffe traveled to Moscow to engage Party officials over the need to put an end to this unrealistic situation; he received V. V. Kuibyshev's support to join the two institutes.[41] In February 1931, Vesenkha and Narkompros agreed to the formation of the State Physico-Technical Institute under the former's jurisdiction and with three sectors. The physical-mechanical, under Ioffe, comprised ten departments and seven laboratories divided into three groups: physics (with Ioffe as director), mechanics (N. N. Davidenkov), and insulation (A. F. Val'-ter). The chemico-physical, under N. N. Semenov, had eight departments and one laboratory. The electrophysical sector, under A. A. Chernyshev, had twenty-eight laboratories.[42] From the start, each sector had complete autonomy in the formulation of research programs and the hiring and firing of workers. Later these sectors became three independent physics institutes.

Ioffe Institute physicists now turned to the second stage of reorganization required by Stalinist policies: internal reorganization for applied research—the division of effort according to "groups" and "brigades." This was part of a short-lived national campaign to recast educators, scientists, and common workers into closely knit collectives. Researchers in Ioffe's physical-mechanical sector were organized into eight "groups" and twenty-one "brigades" (see table 8). Scientists soon reverted to the traditional nomenclature of "departments" (or divisions) and "laboratories" when it became clear that the new terms added little to the efficiency of research.

This reorganization was accompanied by the introduction of planning of scientific research. The creation of groups and brigades involved spelling out and fulfilling short-term (one- to two-year) and long-term projects, or "thematic plans," the latter usually covering such general institutional goals as basic research. Each group had its own thematic plan, and was engaged in basic and applied research, both on state budget and under various contracts. For example, the group studying the mechanical properties of materials under N. N. Davidenkov—later simply the mechanical department—had five major research foci supported by contracts with Narkompochtel (the Commissariat of Post and Telegraph), Otkomkhoz, and other government organizations and factories interested in the areas of elastic properties of materials, plastic deformation and strength, and fatigue of metals.[43] The mathematical physics group, created in January 1931 to supplement the work of the theoretical department, also undertook research on a whole range of topics from the theory of plasticity to the theory of breakdown, and from hydrodynamics to heliotechnics to acoustics.[44]

The electroinsulation department, created out of the brigade for the study of surface phenomena, was involved in research of greater currency to the industrialization effort of the country, specifically electrification and the development of long-distance high-voltage power lines. Research focused on dielectrical loss and breakdown, the behavior of dielectrics at high temperatures and

Table 8. Organization of the Leningrad Physico-Technical Institute, 1931

Group	Leader	Brigade	Leader
Energetics	Ioffe	Photoelectricity	D.N. Nasledov
		Thermoelectricity	N. I. Dobranravov
		Radiant energy	V. P. Zhuze
		Boundary regions of energy	M. A. Levitskaia
		Surface phenomena	A. P. Aleksandrov
		Solid state	A. N. Arsen'eva
		Diffusion in solid state	B. M. Gokhberg
Heliotechnics	V. P. Veinberg	(with three brigades)	
Scattering of X rays and electrons	P. I. Lukirskii	Nuclear structure	D. V. Skobel'tsyn
		Emission	V. M. Dukel'skii
		Scattering	V. E. Lashkarev
Structure of matter	I. V. Kurchatov	Amorphic bodies	P. P. Kobeko
		Ferroelectricity	I. V. Kurchatov
		Liquid crystals	V. K. Frederiks
		Physics of discharge	I. V. Kurchatov
Mechanical properties of materials	N. N. Davidenkov	Plastic deformation	M. V. Klassen-Nekliudova
		Dynamic properties	N. N. Davidenkov
		Roentgenography	N. Ia. Seliakov
		New methods	N. N. Davidenkov
Biophysics	G. M. Frank		
Theoretical physics	Ia. I. Frenkel'		
Methodology of physics	A. V. Vasil'ev and Iu. P. Shein		

Source: Sominskii, *Ioffe*, pp. 272–273.

frequencies, and the study of mechanical breakdown and dielectrical loss in such liquids as transformer oil. Showing sensitivity to the increasing pressure for applied research, A. F. Val'ter defended the fundamental nature of research in the electroinsulation department, noting that its program embraced "a broad circle of electrophysical questions which arise in the insulation industry. Questions of technology are touched upon only lightly . . . , but this is done consciously, since this circle of questions should be elaborated in institutions of the chemical industry."[45]

Stalinist policies contributed to the tremendous growth of LFTI, especially with regard to capital construction and personnel. Drawing on the rapid increase in budgets for institutes of science, technology, and industry, LFTI undertook a massive construction program from 1928 to 1932 with funds coming from both Vesenkha and Narkompros. The program included completion of a new wing of the main building and a building for the communications department, each about four stories and one hundred feet square, new apartments and living quarters for workers, a new kitchen and dining hall, renovation of Ioffe's apartment, and repair of the existing plant. While I could not find information on the actual amounts spent in 1931 and 1932, it is possible to judge the size of the capital construction and rebuilding effort on the basis of planning documents (see app. B, table 10).

Capital construction exceeded all other categories of expenditure during the First Five-Year Plan. There was so much funding for construction and salaries that the percentage of budget going to research never exceeded twenty percent of the total, unless new equipment is included; nonetheless, in absolute numbers, the amount allotted to experimental research increased tenfold from 1928 to 1932, from slightly over one hundred thousand to more than one-and-a-quarter-million rubles.

The First Five-Year Plan had similarly ambitious targets for personnel. While targets were never reached, these projections reflected two points of emphasis. One was interest in a sheer increase in numbers of scientific workers, with a commensurate increase in salaries. The second was interest in expanding the ranks to include more personnel of the "appropriate" class origins, through *vydvizhenie* (see chap. 6). According to the plan, the staff of LFTI would grow from 151 to almost 600, including nonscientific support staff, librarians, secretaries, janitors, and so forth, between 1928 and 1932 (see app. B, table 11).

While construction, staff, and salary figures are approximate, it is clear physicists anticipated that the institute would grow rapidly during the First Five-Year Plan, reflecting the general dynamism of industrialization and the belief that physics institutes would play a major role in making the USSR a modern industrial state.

THE CREATION OF THE LFTI "EMPIRE"

The Leningrad Physico-Technical Institute assumed a prominent role in the industrialization of the Soviet Union in two other ways. One was the creation of new physico-technical institutes throughout the European USSR, from the Ukraine to the Urals. The other involved the creation of independent institutes, spin-offs from the staff and equipment of LFTI. Both processes were a matter of state-institute relations and involved lengthy negotiations, although physicists seem to have taken the lead in these efforts.

Within a few years after the merger of the institute and the laboratory, LFTI went through another metamorphosis when N. N. Semenov's and A. A. Chernyshev's laboratories became independent institutes: the Leningrad Institute of Chemical Physics (Leningradskii institut khimicheskoi fiziki, LIKhF) and the Leningrad Electrophysical Institute (Leningradskii elektrofizicheskii institut, or LEFI).[46] From the time of their formation in winter 1931, the chemicophysical and electrophysical sectors had been virtually autonomous with regard to research programs and administration. The sector directors, N. N. Semenov, A. A. Chernyshev, and A. F. Ioffe, did not strive long to keep them under one roof; they found it rather cumbersome to administer physicists, chemists, and technologists and realize a rational research program between them.

The plan to create new institutes from LFTI had to be approved by several governmental bureaucracies. Ioffe brought the matter before the Central Control Commission of the Party, the collegium of the Workers' and Peasants' Inspectorate, and Vesenkha. After deliberating throughout the summer of 1931, the Party approved the formation of three institutes. Fearing that the division of LFTI would weaken Ioffe's influence and perhaps even dilute the research effort, Vesenkha created the Association of Physico-Technical Institutes (Kombinat fiziko-tekhnicheskikh institutov), with responsibility for planning and administration of the physico-technical, electrophysical, and chemicophysical institutes in Ioffe's able hands.[47] Thus, while approving in principle the division of LFTI, the central government was as yet unwilling to free Ioffe from administrative responsibilities.

The second path by which the LFTI empire expanded was the formation of new physico-technical institutes. Both physicists and the government saw the need to extend the network of physics institutes from a handful of major urban research centers—LFTI, Moscow State University's Physics Institute, P. P. Lazarev's Institute of Physics and Biophysics, the State Optical Institute, and a few others—to a nationwide system with scientific research institutes in the provinces. From the point of view of the Party and the government, the new system would include institutes located in burgeoning industrial cities, with physics research centered on the needs of the major local industry: electrification and construction (strength of materials) in Dnepropetrovsk, solid-state physics and metallurgy in Tomsk, and so on. Physicists also endorsed the system, having advanced the slogan "Physics to the provinces!" since the time of the sixth congress of the RAF.

The creation of four provincial physico-technical institutes in Kharkov, Tomsk, Sverdlovsk, and Dnepropetrovsk between 1928 and 1933 had both a positive and a negative impact on LFTI. On the one hand, the creation of an empire of institutes confirmed LFTI's reputation as the most innovative Soviet physics institute. It ensured Ioffe's influence over the development of the discipline, an influence that went far beyond his "school." In addition, at least on a formal level, it demonstrated the unity of theory and practice. On the other hand, the formation of a national network created several problems for the leadership of LFTI. First, it was only natural that the power and influence of the institute would decline as new institutes were formed and other physics institutes "spun off"; personnel, research programs, and funding were dispersed to new centers. Second, when the Ioffe Institute came under criticism from the Party in the mid-1930s, one of the charges against it would be of "empire building" or "aristocratism." That is, many ideologues, Party officials, and even physicists saw the new institutes paradoxically as an attempt by Ioffe to ensure his control over the physics enterprise for selfish, and not national, reasons. However, while LFTI maintained influence over planning in physics through its connections with the new institutes, the latter were important in their own right; each had its own foci of interest based on the strength of indigenous physics cadres and local industry, and therefore served local as well as national research interests.

Since its founding in 1929, the Ukrainian Physico-Technical Institute (UkFTI) has been a leading research center in low-temperature, theoretical, and nuclear physics. It brought together scholars from Kharkov University and the local polytechnical institute and physicists from LFTI (and several from Moscow) sent to Kharkov on long-term *komandirovki*. By 1932 the institute already had 47 physicists and 180 other personnel, as well as five major laboratories: atomic physics under A. I. Leipunskii, K. D. Sinel'nikov, and A. F. Val'ter; low-temperature physics under L. V. Shubnikov and M. Ruheman; the physics of crystals under I. V. Obreimov; ultra-short waves under A. Shitskin; and theoretical physics under L. D. Landau and L. Rozenkevich.[48] All but two of the laboratory directors, Shitskin and Rueman, came from LFTI to Kharkov.

Following the example of LFTI, the Ukrainian Physico-Technical Institute served as an important center for physics conferences, international contacts, and publication. The institute called a series of national conferences on theoretical physics in 1929 and 1934, low-temperature physics in 1937, and nuclear physics in 1940. It attracted foreign scholars to consult and lecture: Bohr, Dirac, Ehrenfest, Podolsky, Houtermans, and Weissberg. And, it was home to the German-language Soviet physics journal *Physikalische Zeitschrift der Sowjet Union*.

Organization of UkFTI began in 1928. In February, Ioffe wrote V. M. Sverdlov, then chairman of the collegium of the Scientific-Technical Administration of Vesenkha, suggesting the organization of a branch of LFTI in Kharkov. The reasons for this were the cultural and industrial significance of Kharkov and the presence of a strong group of local physicists "who participate actively in the economic life of the [Ukraine]."[49] The tie to LFTI would ensure the continued leadership of its personnel. Ioffe acknowledged he had been influenced by examples of scientific research institutes in Germany and the United States, most of which he had visited, in the formulation of these plans.[50]

The new institute would serve two major purposes: it would strengthen the Kharkov group by bringing in both established and young scholars from different disciplines at LFTI and institutes in Moscow, lured there with money from Glavnauka Ukraine; and it would ensure that the development of local industry went hand in hand with advances in physics. Party organs had instructed physicists to establish close relations with Ukrainian industry. The Party had issued the slogan "All hands to Production!" which for UkFTI meant that all possible assistance ought to be given to the development of low-temperature physics. Very low temperatures in the chemical industry are used to separate gases. With funding from Glavnauka Ukraine and Vesenkha, and under contract with various industrial trusts and planning organizations, UkFTI would establish the physical constants of gases and gas mixtures, and train specialists and technologists for this burgeoning branch of industry.[51]

In May of 1928, the government approved in principle the organization of a physico-technical institute in Kharkov, authorized funding, and gave permission to secure a site for the new institute. The Scientific-Technical Administration appointed Ioffe as chairman of the academic council of the new institute.[52] Scientific workers were to increase from ten in 1928–1929 to nearly eighty by 1933; the budgets for research, development, and capital construction were to grow sixfold for the same period.[53] None of these targets were met. While organization of the new institute began in the summer of 1928, and Vesenkha and LFTI were lavish with financial support and personnel, the problems of housing workers and attracting outside physicists continued to plague the scientific activity of UkFTI through its first years.

When K. D. Sinel'nikov returned from England in June 1930, he moved immediately to UkFTI. Sinel'nikov had spent two years in Cambridge, supported by the International Education Board. He had worked on the "magnetic properties of crystals to measure conductivity in crystals and the magnetic influence on them." He had received the second-year extension at Rutherford's urging, since the "[Kharkov] Research Institute [was] not ready, and this [would] interrupt his research" if he returned to Russia without an extension.[54]

Upon his arrival in Kharkov he noted that only the apartments had been finished. The main building, only partially complete, had a few equipped laboratories already working within it.[55] The enthusiastic young physicists who gathered in Kharkov—L. D. Landau, A. K. Val'ter, A. I. Leipunskii, I. V. Obreimov, and L. V. Shubnikov—happy to be away from the dominance of Ioffe, quickly overcame the delays in construction and built first-rate programs in such fields as low-temperature and nuclear physics.

The Siberian Physico-Technical Institute (SFTI) was smaller, had less-ambitious plans, and never assumed the importance in national physics that the Kharkov center did. Following the same pattern in formation, it was created on the base of local physicists from Tomsk (under the leadership of V. D. Kuznetsov and N. V. Gutovskii) and served the local steel industry. As early as 1922, the Tomsk physicists organized a technical institute to work in metallurgy and material sciences. This became the Institute of Applied Physics of Glavnauka. According to an official history, it was "more a scholarly association than an institute," having neither the funding nor the staff to be considered the latter.[56] Tomsk physicists were nonetheless quite productive, carrying out intensive work on the mechanical properties of the solid state, including the technical properties of metals. Tomsk physicists presented fully one-tenth of all papers read at the fourth congress of the RAF in Leningrad in 1924. By the late 1920s, the institute had developed a close working relationship with the physics department at Tomsk State University, where Kuznetsov was chairman.

Ioffe realized that the industry and the physics community of Tomsk made it an ideal setting for his blueprint for a new institute, one which could be created without great capital expenditures or the transfer of a significant number of physicists. The institute would not only conduct research of an applied character but also serve "as a center of the scientific and cultural life of the region." Accordingly, SFTI would be subordinate to Narkompros, not to Vesenkha. Ioffe stressed, however, that this did not mean that the new institute would be "hostile to the inquiries of industry." Its proposed structure ensured focus on the development of Siberian industry; there would be several laboratories of applied physics, including metallurgy, medical physics, and roentgenography.[57] Vesenkha approved the creation of SFTI on December 1, 1928, with funding during the first years of its existence channeled through both the local Party committee and LFTI.[58] Initially, SFTI had five laboratories and a theoretical department, but by 1932 it had grown to ten departments and laboratories, with commensurate increases in budget and staff.[59] In the first five years of activity, SFTI researchers focused on the structure and electrical and mechanical properties of the solid state, including semiconductors; electrochemistry; electromagnetic radiation; and absorption and surface phenomena. This work resulted in two monographs, four popular books, and eighty-one published articles.[60]

These successes notwithstanding, organizational problems and criticism by Party officials plagued the institute from its inception. In the environment of the first five-year plans it was a foregone conclusion that SFTI ultimately would be transferred from the "soft" Narkompros to the "hard" Narkomtiazhprom. In the early 1930s the institute participated in the requisite self-criticism: its physicists acknowledged that research programs lacked coordination between laboratories and were "divorced from practice." This charge held force in particular, but not suprisingly, for the theoretical department. Its efforts were "unplanned [*stikhiinyi*]" in the area of chemical physics. Institute physicists preferred affiliation with Narkompros; the organizations which served as the kernel of the institute had been under Narkompros's jurisdiction. It was decided, however, to transfer the institute to Narkomtiazhprom, over the objections of several representatives of local industry who believed that relations with Narkomtiazhprom were often of a "purely bureaucratic character."[61]

Other physico-technical institutes, including centers in Dnepropetrovsk (the Dnepropetrovsk Physico-Technical Institute, or DFTI)[62] and Sverdlovsk (the Ural Physico-Technical Institute, or UrFTI), were founded under the leadership of A. F. Ioffe, so that by 1935 LFTI and its network spanned the European USSR. A pattern for the creation of new research institutes had been established: in all, fifteen institutes have been formed as spin-offs from the Ioffe Institute during its illustrious history (see app. B, table 4). This pattern for the establishment of new research institutes has been repeated in such noteworthy examples as Akademgorodok in Novosibirsk. For the case at hand, the creation of physico-technical institutes was the realization of the slogan "Physics to the Provinces!" and fulfilled the need for applied physics research in support of local industry as an aspect of Stalinist science policy. The extension of research programs and publication accompanied the growth of the physics enterprise.

SUCCESSES IN PUBLICATION AND RESEARCH

The Soviet R and D apparatus is the largest in the world in terms of numbers of scientists, engineers, and institutions. Yet, by many standards, Soviet science has failed to perform effectively. The system proved capable of pioneering efforts in space (Sputnik), nuclear fusion (tokamaks) and fission, and elementary particle and theoretical physics, but proved incapable of maintaining a lead or catching up in areas where it was behind. Many of the reasons for these failings are found in the Stalinist system of administration of R and D. It is top-heavy, stifles individual initiative, and provides inadequate funding for fundamental research. Its centralized system of supply leads to bottlenecks in the purchase of the simplest equipment. And yet, during the period of the introduction of the first five-year plans, Soviet physicists were engaged in research which was often on the cutting edge of world physics. Of the eight Nobel prizes given to Soviet physicists, six were awarded for work completed before 1941 (see app. A, table 5).

The vitality of the Soviet scientific community can be seen in an analysis of physics publications and a discussion of two new fields of research, solid-state and nuclear physics. In both cases the prominence of Leningrad physicists is notable, the expansion of the physics enterprise throughout the European USSR notwithstanding. One measure of the dynamism of a scientific research center is the quantity of articles published in leading journals. As a rule, in scientific, scientific-technical, and popular-scientific journals in the USSR in the late 1920s and 1930s, publication in the exact and physical science fields expanded, reflecting the importance of science and technology during the five-year plans. In the popular-scientific *Sorena* (*Socialist Reconstruction and Science*) alone, in the period 1931–1932, scholars published 34 articles on physics, 21 on chemistry, 35 on metallurgy and machine building, 27 on heat engineering, and 18 on various aspects of construction.[63] By July 1935 there were over 2,000 journals being published in the USSR, of which 1,638 were in Russian, 500 on technical subjects, 250 on agriculture, and 170 each on mathematics, the sciences, medicine, and economics.[64]

The major Soviet physics journals were *Zhurnal prikladnoi fiziki* (*ZPF;* from 1931 onward, *Zhurnal tekhnicheskoi fiziki*, ZTF), *ZhRFKhO* (from 1931 onward, ZETF), and *Physikalische Zeitschrift der Sowjet Union*. Among foreign journals, Soviet physicists published most widely in *Zeitschrift für Physik*.

Physikalische Zeitschrift der Sowjet Union (*Phys. Zeit. SU*) was published from 1932 until 1937, and would counter some of the isolation physicists still felt. At the Ukrainian Physico-Technical Institute, where it was housed, it was known as "Sovfiz." A. I. Leipunskii was the editor; D. D. Ivanenko, L. Rozenkevich, and Alexander Weissberg were his assistants. Weissberg, a German physicist who, after spending the early 1930s in Kharkov, was arrested, imprisoned, tortured, and then released, claims to have originated the idea for the journal. He apparently talked directly with N. I. Bukharin, then head of the Scientific Sector, to get approval for its publication.[65] The Leningrad physicists Ioffe, Frenkel', Rozhdestvenskii, Semenov, Vavilov, and Chernyshev, and the Moscow physicists L. I. Mandel'shtam and B. M. Gessen served on the editorial board. The majority of articles published in *Phys. Zeit. SU* were written by Ioffe "school" physicists, or physicists of other physico-technical institutes, and also frequently appeared in Russian-language Soviet journals.

The importance of the journal lies in its role as a vehicle for foreign-language publication. Throughout Russian and Soviet history, Russian-speaking scientists have published journals in French, English, or German to communicate their ideas to the European community and to ensure their priority in discoveries. For example, the Academy of Sciences published its proceedings initially in Latin; by the time of Catherine the Great, in Latin and French; and by the middle of the last century, only in French, with

Russian and French publications through 1929. The onset of "socialism in one country" and autarky in science sharply curtailed the number of foreign-language scientific publications.

Ioffe explained the purpose of *Phys. Zeit. SU* in its inaugural issue. It would become "the central publication organ of Soviet physics," he wrote. "We hope that it will become, beyond this, an organ of international cooperation of physicists of all countries."[66] Tremendous strides had been made in Soviet physics in the fourteen years since the Revolution, from small university laboratories to large-scale, modern scientific research institutes. However, the work of Soviet physicists remained largely unknown or inaccessible to foreign scholars. *Phys. Zeit. SU* would overcome both of these problems, publishing in German to ensure Soviet priority and recognition, and carrying summaries of physics conferences and articles on the planning of science and even on the advantages of dialectical materialism.[67] In practice, only a handful of articles in its six years and twelve volumes of publication focused on the planning of science or dialectical materialism.

Of all foreign journals, Soviet physicists published most widely in the German *Zeitschrift für Physik*. Soviet physicists did not rely upon *Physical Review, Proceedings of the Royal Society, Philosophical Magazine, Nature, Die Naturwissenschaften,* or any other Western scientific journal to any large extent. Until 1923 when Soviet physicists finally ended their international isolation, almost no Soviet physicist published abroad. But after the reestablishment of regular scientific contacts with Europe and the United States, the number and percentage of articles by Soviet physicists in *Zeitschrift für Physik* increased rapidly, reaching a peak of 16 percent in 1926, and averaging almost 12 percent for the period 1924–1931. Between 1920 and 1936, of nearly seven thousand articles published in *Zeitschrift für Physik,* 8.5 percent were written by Soviet scholars. The Ioffe Institute produced a remarkable three-eighths of the articles written by Soviet Physicists between 1923 and 1936; that is, one institute, founded only in 1918, was responsible for the publication of one of every thirty-nine articles (see app. C, table 4).

Three factors led to a decline in the number of articles submitted to *Zeitschrift für Physik*. First, the pressure for autarky in science discouraged Soviet physicists from publishing abroad. Second, many physicists, no doubt encouraged by the government, protested the rise of National Socialism in Germany by suspending efforts to publish in the journal. Third, the establishment of *Phys. Zeit. SU* in Kharkov provided an outlet for foreign-language publication.

The formation of new physico-technical institutes and such spin-offs as LEFI and LIKhF meant that LFTI would no longer dominate publication in Soviet journals as it had in the 1920s. Nevertheless, the institute's performance shows the acumen of its physicists in generating state support while avoiding excessive government interference in research programs. Physicists at LFTI wrote one of every six articles published in ZETF (*Zhurnal eksperimental'noi i teoreticheskoi fiziki*, the continuation of ZhRFKhO, which ceased when the RAF was disbanded), the major Soviet theoretical and experimental journal between 1931 and 1939. Two trends become apparent in the mid- to late 1930s. First, physicists in the provinces began to publish a greater share of articles, reaching one-fifth of all articles published in 1939. Still, for the nine-year period under consideration, nearly two-thirds of all articles were written by physicists at LFTI and other institutes in Moscow and Leningrad, which, according to at least one quantitative measure, reflects the failure of physics to thrive in the provinces. Second, while physicists at other physico-technical institutes contributed a greater absolute number of articles in the period 1935–1939 than in 1931–1934, the percentage of articles declined from 46.1 percent to 31.1 percent, due to the expansion of ZETF to a journal averaging 1,100 to 1,300 pages per volume and an increase in the number of articles overall and from other institutions. It is clear, however, that LFTI retained its leading position, and that a handful of physics institutes dominated publication in the 1930s (see app. C, table 5).

LFTI physicists' record in ZTF, the major journal of technical and applied physics, is not nearly as impressive, although the same trends are visible as in ZETF. That is, from 1931 to 1939, one in eight articles was authored by Ioffe Institute physicists, one in five by physicists at other physico-technical institutes, and one in four if LEFI is included. Moscow physicists were responsible for an increasing share throughout the 1930s, and wrote 30 percent of all articles published for the decade (see app. C, table 6). Physicists at institutes in Moscow and Leningrad contributed over three-quarters of all articles for the decade (see app. C, table 6). Ioffe himself proudly noted the fact that while in 1931 only ten scientific research institutes and four factory laboratories were represented in articles published by *ZTF*, by 1936 the numbers had grown to thirty-four institutes, eleven *vtuzy* (higher technological institutes), and eight factory laboratories.[68] In spite of the attempts of government and physicists to secure the geographic distribution of physics institutes throughout the USSR, Moscow and Leningrad remained the centers and the desired locations for research and teaching positions.

A second measure of the vitality of a scientific field is the development of subdiscipline and the ability of scientists to secure funding to continue research in that area. Modern scientific research has become increasingly expensive, particularly in such fields as nuclear physics where Van de Graaff generators and cyclotrons, let alone modern accelerators, must be built. The success of Soviet physicists in generating government funding for research in solid-state and nuclear physics indicates the extent to which government and science agreed about the need to support the scientific enterprise in its modern setting, the research institute.

In LFTI, physicists continued to focus their research on the physics of the solid state, adding to their knowledge of the mechanical and electrical properties of crystals, metals, insulation, and dielectrics, and now turning to semiconductors, photoelements, and ferroelectricity. In a commemorative volume published by the Academy of Sciences in 1938, the leading representatives of Soviet science chronicled the achievements of their respective fields. A. F. Ioffe outlined work on the mechanical and electrical properties of the solid state, focusing on six major areas of "great importance" to world science: the electrical properties of amorphic bodies; electrical breakdown of dielectrics, including attempts to produce thin-layer insulation; the electrical properties of ionic crystals, including high-voltage polarization; the mechanical properties of crystals; ferroelectricity (*segnetoelektrichestvo*); and electronic semiconductors.[69] The last two represented newer developments in solid-state physics.

I. V. Kurchatov and P. P. Kobeko conducted a series of investigations into the properties of Rochelle salts in 1927–1929, and came to the conclusion that in several of these dielectrical crystals a peculiar electrical analogue to ferromagnetism takes place, as a consequence of which the phenomenon received the name "ferroelectricity." In March 1930, Kobeko and Kurchatov published their first work on this subject in which they attempted to explain the anomaly of high-voltage polarization in ferromagnetic salts.[70] In subsequent work the authors improved the methodology for measuring the distribution of the potential in the salt, and showed that the effect is not connected with high-voltage polarization.

More important to the development of the research program of LFTI were investigations in the physics of semiconductors under A. N. Arsen'eva, B. M. Gokhberg, V. P. Zhuze, B. V. Kurchatov, D. N. Nasledov, A. V. and A. F. Ioffe, and several physicists at UkFTI including V. M. Tuchkevich, later director of LFTI.[71] The study of semiconductors was of great interest to Soviet physicists from both a theoretical and a practical point of view. The theoretical investigations involved the study of energy levels of electrons and their behavior in a crystal lattice under the influence of heat, light, and electronic and magnetic fields. Quantum mechanics helped physicists to understand the properties of semiconductors.[72] Such young physicists as M. P. Bronshtein, N. L. Pisarenko, and B. I. Davydov were the first to turn their attention to the theoretical basis of semiconductor physics, for example the theory of kinetic phenomena. Later, such established scholars as Frenkel' and Tamm joined these investigations.[73]

Semiconductors were important for industrial application as well, because of their rectification properties (in radiotechnology, for alternating current, in high-voltage electronics, and so on) and their photoelectric properties (the change of resistance during illumination and the appearance of electromotive forces at the boundary with metals, on the order of hundreds and thousands of volts). Physics research at LFTI centered on (1) the electronic structure of semiconductors and the contribution of impurities to their electrical properties; (2) photoelectric properties; (3) rectification effects; and (4) kinetic phenomena.[74]

Throughout the 1930s, LFTI remained the center of work on semiconductors in the USSR, organizing conferences and symposia and carrying out detailed experimental investigations. In all events, theoretical work outdistanced applications. In September of 1932, the institute hosted the first all-union conference on the theory of the solid (nonmetallic) state, which was attended by Fowler, Bragg, and Dirac. The researchers focused on recent advances in understanding the differences between semiconductors, metals, and insulators, and on their thermal and photoelectrical properties, based on quantum mechanical interpretations of the motion of electrons or holes.[75]

At another conference on semiconductors, in Odessa in May 1934, physicists acknowledged that applications lagged; the factories which were responsible for the manufacture of semiconductors were criticized.[76] But work in this area of physics was quite promising and would continue apace, as B. M. Gokhberg, then academic secretary of the Physical Association of the Scientific Section of Narkomtiazhprom, reported to a November 1935 physics planning session. The mandate of the institute would include research on "the central problem of the electrical properties of the solid state," primarily semiconductors, where the application of wave mechanics had been especially helpful in providing a theoretical basis for experimental efforts, as well as the determination of energy levels of electrons in these materials; the role of electropositive and -negative impurities; the role of positrons and of "positive holes"; and the internal photoeffect.[77] In all, in the period 1931–1941 LFTI called or cohosted six all-union conferences and seminars on semiconductors.

In 1932, when experiments on the photoeffect in rock salt had been completed, and a series of articles, written by K. D. Sinel'nikov, A. K. Val'ter, A. I. Leipunskii, and G. D. Latyshev, on solid rectifiers and photoelements appeared in *ZTF*, Sinel'nikov and the others had already begun experiments to split the lithium atom with protons.[78] The researchers published a brief note about their experiment in one of the first issues of *Phys. Zeit. SU*.

NUCLEAR PHYSICS[79]

Physicists in the USSR began intensive investigations of the atomic nucleus soon after Chadwick discovered the neutron in 1932. They were well-prepared for this effort in terms of institutions, personnel, research program, and funding. In spite of the best efforts of Soviet Russia's leading physicists, however, the program in nuclear physics, like those in the West, often suffered from budgetary constraints leading to bottlenecks in the construction of the expensive equipment needed to break into the secret of the nucleus—cyclotrons, tesla transformers, and Van de Graaff accelerators.

Unlike their colleagues abroad, Soviets physicists became increasingly isolated from the international scientific community. Under the influence of autarkic policies, Soviet scholars were rarely permitted to travel in Europe or the United States and often unable to invite Western nuclear physicists to the USSR. They could not take full advantage of the wide-ranging international collaboration and close personal relationships which dominated the development of nuclear physics in the West under Rutherford, Fermi, Lawrence, and Bohr.[80] Even so, by the eve of the Nazi invasion in 1941, Soviet physicists had developed a research program advanced enough for them to begin work on an atomic bomb when called upon by their government to do so.[81]

In 1932, after his independent scientific career had spanned a quarter-century, A. F. Ioffe identified a new and promising area for expansion of the LFTI research program. At his urging, research groups at LFTI, UkFTI, and several other institutes began to probe into the nucleus, building upon the work of British, American, and German physicists. The contribution of Soviet physicists to this field before 1930 consisted largely of the work of D. V. Skobel'tsyn, who conducted experiments with a Wilson chamber at Leningrad Polytechnical Institute, and from 1924–1929 worked at LFTI investigating the interaction of photons with matter.[82]

The Rockefeller Foundation provided support to a core group of budding nuclear physicists: K. D. Sinel'nikov, George Gamov, and I. V. Obreimov joined Kapitsa in Cambridge at Ioffe's orders and worked under Rutherford, Cockcroft, and Walton in the Cavendish Laboratory, the major center of atomic physics in the 1920s, on the development of high-voltage apparatuses for proton acceleration. They provided the critical mass of scholars prepared to follow through on Ioffe's suggestion. These international contacts were vital to the nascent field of nuclear physics in the USSR, and the influence of research abroad carried over into the 1930s. Sinel'nikov returned to the USSR to organize a nuclear group at UkFTI in Kharkov which began to meet in the first half of 1931. Gamov was also a leading figure in bringing advances in the physics of the atomic nucleus to the attention of his Soviet colleagues, publishing a series of articles in *UFN* between 1930 and 1934 based on his work in Rutherford's laboratory.[83]

In 1932, the academic council of LFTI created a nuclear group. The group, with Ioffe as its head, included D. V. Skobel'tsyn, I. V. Kurchatov, D. D. Ivanenko, M. P. Bronshtein, and Gamov.[84] Ivanenko was responsible for organizing a regular seminar series, and Kurchatov, Ioffe's "deputy" and later head of the atomic bomb project, was in charge of organizing the first year of research.[85] The nuclear group met weekly beginning in November 1932, and invited physicists from other institutes. Researchers discussed current experimental and theoretical literature, relativistic quantum mechanics, and cosmic rays, in seminars attracting between thirty and forty scholars. Gamov opened the series with three talks on the atomic nucleus; Skobel'tsyn followed with two papers on cosmic rays. Between December 1932 and April 1933, Bronshtein, Ivanenko, Skobel'tsyn, Ioffe, and Kurchatov gave talks on various aspects of nuclear physics which reflected awareness of the major questions and current literature in this area, including the work of Lawrence and Henderson, Blackett, Fermi, and the Rutherford group.[86]

Soviet physicists recognized the need not only to follow developments in the international arena, but also to take initiative on a national scale in the Soviet Union. They did so through organizations created within the Academy of Sciences, LFTI, and UkFTI, and by holding periodic national conferences. A special commission for the study of the atomic nucleus, chaired by Ioffe, was organized within the Academy and included S. E. Frish, L. V. Mysovskii, and A. I. Leipunskii; it operated periodically in the mid-1930s. In 1938 the Academy created another commission on the nucleus under S. I. Vavilov.[87] Through these commissions physicists pressed the government for new equipment and research travel (which would not be forthcoming) and organized national nuclear physics conferences.

Soviet physicists moved to organize their first conference on the heels of the inaugural international conference on nuclear physics in Rome in October of 1931. In mid-1932 Ioffe wrote the directors of UkFTI concerning preparations for the meeting, which he hoped to hold in April of the next year. The conference would establish close contact between all institutions of the Soviet Union working in this area. The conference organizing committee established five major areas for the program: (1) methods of artificial splitting of atoms; (2) theory of nuclear structure and questions of relativistic quantum mechanics; (3) the neutron; (4) excitation and irradiation of the nucleus by neutrons, photons, and resonance; and (5) cosmic radiation.[88]

The first conference was held late in September 1933 under LFTI's sponsorship with funding from Narkomtiazhprom. In addition to the leading Soviet physicists, such foreign scholars as Frederic Joliot-Curie, Francis Perrin, P. A. M. Dirac, Franco Rasetti, Walther Bothe, and Victor Weisskopf attended. Intensive debate among the sixty participants revolved around the implications for nuclear theory of recent developments—the discovery of the neutron, its mass, questions regarding the basic constituents of the nucleus; and conservation of energy and beta decay. Dirac spoke about the theory of the positron and his theories of "holes" and second quantization (1929).[89]

In the late 1930s, two more all-union and international conferences drew attention to the achievements of Soviet physicists. The second all-union conference on the atomic nucleus, held in Moscow late in September 1937, drew over 120 Soviet scholars, and several physicists from abroad including Wolfgang Pauli, Rudolph Peierls, a longtime associate of L. D. Landau, and Fritz Houtermans. The participants read twenty-eight papers (twenty-three by Soviet authors) on five main problems: (1) the penetration of matter by fast electrons and gamma rays; (2) cosmic rays; (3) beta decay; (4) the interaction of the nucleus with neutrons; and (5) the theory of nuclear structure. There were also discussions of high-voltage particle accelerators.[90]

Ioffe's keynote address reflected the forces at work under Stalin and the fact that the field of physics was not immune to the political currents of the day. He spoke about the achievements of Soviet science in creating a network of physico-technical institutes and increasing the number of nuclear physicists fourfold in four years. He described how advances in nuclear physics served to verify the validity of dialectical materialism.[91] He praised the emergence of proletarian scientists who had replaced the old intelligentsia, concluding that "this phenomenon of a truly worker-peasant intelligentsia, a Soviet intelligentsia, is a fact of extremely great significance."[92] However, rarely was a peasant nuclear physicist seen in his institute.

On a more somber note, Ioffe acknowledged the failure of Soviet physicists as yet to achieve "any kind of practical applications."[93] There would be applications soon enough in the late 1940s and 1950s. And certainly work in solid-state physics revealed the extent of Ioffe's overstatement. But this kind of self-criticism was required in Soviet science. A protocol which praised the significant growth of the discipline in terms of institutes, research programs, and such young scholars as the Alikhanov brothers, V. I. Veksler, and I. M. Frank, also drew attention to the failure to begin construction of a new, more powerful cyclotron.[94]

By the time of the third all-union conference, held in Moscow in October 1938, Soviet physicists had made some progress in their assault on the nucleus. The conference participants discussed four major issues: (1) cosmic rays; (2) the penetration of matter by high-energy particles; (3) the existence of other subatomic particles (the neutrino and positron); and (4) properties of heavy particles.[95] The physics group of the Academy of Sciences passed a resolution at the end of the conference which called for the concentration of research within institutes of the Academy and the creation of a commission under S. I. Vavilov. The commission would coordinate national research efforts, including the search for applications; the allocation of radium, which was in short supply; training programs; and publication.[96] The resolution had the effect of temporarily turning the attention of the Soviet government to nuclear physics research in Moscow, and away from Leningrad. This caused consternation among LFTI physicists, who feared that resources would be diverted from their research, curtailing their efforts to build a cyclotron.

Soviet physicists had tried since the mid-1930s to acquire particle accelerators of various sorts. At that time, there were two methods for generating fast particles other than electrons—ions, protons, and neutrons—to produce nuclear transformations. The first involved using such high-voltage accelerators as the Cockcroft-Walton method of constructing a voltage-multiplier alternating-current circuit which uses rectifiers, Van de Graaff electrostatic generators, impulse generators, and Tesla coils (a resonance transformer). The cyclotron, invented by Lawrence, was the most promising for low-voltage acceleration.[97] Soviet physicists pursued both paths, primarily at LFTI, UkFTI, and the Radium Institute of the Academy of Sciences (RIAN).

By the end of 1932, UkFTI had taken the lead in building discharge tubes and Tesla transformers at 1.7 MV, a voltage-multiplier machine.[98] They had sent a telegram to the Central Committee of the Communist party after the experiment in which they successfully split lithium with protons: "The UkFTI in Kharkov as a result of 'shock work' in honor of the fifteenth anniversary of the October Revolution has secured the first successes in the destruction of the nucleus of lithium. The high-voltage brigade destroyed the lithium nucleus, and work continues." This provoked a series of articles in *Pravda, Izvestiia,* and a host of Ukrainian newspapers and journals.

Neutron bombardment continued over the next few years. Sinel'nikov and Kurchatov published a series of articles in *Phys. Zeit. SU.* in 1934 on their joint efforts, and in 1935 several publications by A. K. Val'ter, Rozenkevich, and Sinel'nikov duplicated the well-known experiments of the Fermi group the previous year. Ultimately, Sinel'nikov focused his attention nearly exclusively on linear accelerators, while Leipunskii inherited work at UkFTI on the physics of the neutron, moving to fast neutron reactors in the postwar years.[99]

The nuclear physicists then advanced to a Van de Graaff electrostatic generator with a ten-meter sphere, built under the direction of Sinel'nikov and A. K. Val'ter. Local industry—the Kharkov Electromechanical Factory—played a major role in the construction of the generator. The Van de Graaff machine began operation in the late fall of 1935, and Van de Graaff himself was deeply impressed by the facility, which he visited at that time. Sinel'nikov reported on the activities of the electrostatic generator to the second all-union conference on nuclear physics. But UkFTI physicists expressed disappointment with early results, with one scholar, A. S. Papkov, noting that the UkFTI generator had failed to produce one observable disintegration in two years of operation. Following the work of Tuve and others on a Van de Graaff machine, Sinel'nikov hoped to find a way to increase the current of ions in the tube fed by the generator.[100] World War II interrupted the UkFTI research effort.

Three Leningrad institutes—the Physico-Technical Institute, the

Leningrad Electrophysical Institute, and the Physics Institute of Leningrad University—combined resources to build a 600 kV impulse generator at the university.[101] LEFI and UkFTI also constructed Van de Graaff machines on the scale of those in the United States, at 2 MeV and 6 MeV respectively.[102]

Physicists at LFTI began work on an electrostatic generator in the late 1930s. But their efforts on this project were waylaid by the expense of hundreds of tons of metal, construction of a five-story building to house generators, financial shortfalls, and government inaction. The academic council of the institute, citing advances in nuclear technology in the West, approached Narkompros for increased funding. It took pains to show that LFTI not only copied Western efforts, building upon the work of the Cambridge and Fermi groups (published in *Ricerca Scientifica* in the spring of 1934) but simultaneously developed original technologies.[103] By that year, the atomic group had built a Cockcroft-Walton discharge tube at 500 kV to study the boron nucleus with the bombardment of neutrons; worked on a small cyclotron; and, like the Cambridge and Fermi groups, had observed nuclear processes with a Geiger-Mueller counter, an ionization chamber, a mass spectrograph, and a Wilson chamber.[104] In 1935 researchers commenced work on the equivalence of mass and energy in atomic processes with the mass spectroscope; constructed a fast-acting Wilson chamber (I. V. Kurchatov with G. Ia. Shenkin); engaged in slow neutron research (Kurchatov, Vibe, and Rusinov); and used beta particles and electrons to attack the nucleus (Skobel'tsyn with E. Stepanova, A. I. Alikhanov, Alikhan'ian, and Zhelepov).[105]

The continued failure to coax other particle accelerators to operate within acceptable parameters convinced the physicists that only a cyclotron would enable them to make contributions to nuclear physics on the level of the West. Alikhanov and Kurchatov assembled a "baby" cyclotron modeled after that of Lawrence and Livingston. As of 1934 it was the only functioning cyclotron outside of Lawrence's University of California facility. But the "baby cyclotron" did not work very well or for very long, and few experiments were conducted on it. The physicists seemed to be distracted by a cyclotron at the Radium Institute, which also operated infrequently and at low power.[106]

At an academic council meeting in September 1936, LFTI physicists voted to include "in the plan of the institute for 1937 construction of a Lawrence apparatus at 10 MeV."[107] Ioffe wrote Ordzhonikidze, commissar of heavy industry, in January 1937, urging him to support the expensive construction project. He pointed out that after a promising start in 1932 with 100,000 rubles from Narkomtiazhprom the institute's nuclear research had fallen on hard times. Nuclear research occupied first place in the major laboratories of Western Europe and the United States; and even "tiny" Denmark was building a cyclotron for Bohr.[108] But in the USSR the nuclear groups at UkFTI, LFTI, and Vavilov's Physics Institute of the Academy carried out research with minute quantities of radioactive preparations and underpowered cyclotrons. In Ioffe's opinion, conditions were worst of all at LFTI:

> The nuclear group of LFTI has none of these [materials] and conducts all of its work with rarely received preparations of emanations of radium from [the Radium Institute] in Leningrad. In the current conditions, the nuclear group of LFTI cannot therefore undertake any kind of first order problems. The technological basis which is available to it is comparable only to that which the small group of Polish physicists in Warsaw has at its disposal.[109]

The new cyclotron was vital to the interests of the USSR. Ioffe therefore requested 250,000 rubles for 1937 and 400,000 for 1938; a new five-story building; a high-frequency generator; materials for building powerful magnets and a transformer; one gram of radium; and two *komandirovki* to Lawrence's laboratory for six months, neither of which materialized. He asked Narkomtiazhprom to pressure the Bolshevik, Elektrosila, and other Leningrad factories to produce equipment in short order. Ioffe stressed that applications would not be limited to nuclear physics but that some day atomic power would be harnessed.[110] He instructed the academic council of LFTI to include in plans for 1939 and 1940 the fact that research on atomic reactions would have benefits for medicine and biology, in the same way that physicists in the United States, Japan, and Denmark secured funding for their research by stressing potential applications in the medical sciences.[111]

Because of government funding priorities in other areas, and the opposition of Moscow physicists who wished their nuclear program to be second to none, the Leningrad contingent experienced frustrating delays in building their cyclotron.[112] The centralization of planning and procurement did nothing to improve the ability of Leningrad factories to build the needed equipment, so the physicists used their own contacts to get results.

Kurchatov's experience at the Radium Institute and his repeated *komandirovki* to the Bolshevik, Elektrosila, and Krasnyi Vyborzhets factories in Leningrad saved the day. He prevailed upon factory management to accelerate work on the magnets and generators for the LFTI cyclotron.[113] At the end of 1937, builders arrived at the institute. Finally, a June 1939 resolution by the Council of Ministers approved increased funding, and work began in earnest. On September 22, 1939, the foundation of the cyclotron building was dedicated. In the fall of 1940, Kurchatov left the Radium Institute, where he had until that time directed the physics department, to focus all of his energies on LFTI. By the spring of the following year, the building was practically completed, a 20 kV generator built and designed at the polytechnical institute had arrived, and Elektrosila had delivered the electromagnet. On June 1, 1941, testing of the vacuum chamber began. In the growing excitement a journalist filed a story on June 22 in *Pravda* describing the construction of "a building which looks like a planetarium. . . . This is the first powerful cyclotron laboratory in the Soviet Union for splitting the atomic nucleus."[114]

But the cyclotron never operated. The German armies invaded the Soviet Union that day, forcing the physicists to abandon work and hurriedly evacuate much of the institute to Kazan by train. Soon Leningrad was surrounded by Nazi troops, and in the nine-hundred-day blockade which followed, two-thirds of the population died of cold or famine. While the government was unable to get food and supplies into Leningrad for nearly three years, special forces succeeded in dismantling the cyclotron and shipping it to the newly established "Laboratory Number 2" where Kurchatov and others worked on the atomic bomb.

CONCLUSION

The Great Break set off two decades of unimaginable suffering for the Soviet people. Forced collectivization, rapid industrialization, the purges, and World War II caused the deaths of millions. Rapid social and political change left no one untouched. The destruction of whole strata of society, of institutions, of individuals occurred at such a great cost that it is inconceivable today that anyone would argue it was worth the sacrifice.

Stalin's revolution from above sought to make the country a modern industrial power, and in this it succeeded. The government abandoned all reliance on market forces and turned to a command economy—and coercion where necessary. Investment and social policy reflected this state of affairs: collectivization was a tool not to modernize agriculture but to bring the peasantry under state control and ensure the requisition of enough food for the cities; the consumer sector was largely ignored; and science, technology, and education were put to use fully toward these policies.

Since the first days of the Revolution, Soviet leaders and scientists alike had recognized the important role that science and technology could play in the pursuit of national programs. Throughout the 1920s the scientist and engineer had maintained a large degree of autonomy in founding new research centers and in formulating and carrying out research programs. With the introduction of Stalinist science policy, physicists now met direct demands to make their research more accountable to the needs of the state.

Stalinist science policy—the requirement that all research activities be planned, that research be subordinated where possible to state economic programs, that applied research assume primacy over fundamental, and that policymaking be centralized within one bureaucracy, the Scientific Section of Narkomtiazhprom—led to a reevaluation of the relationship between the specialist and the policymaker, and his subjugation to Party and government organs of administration, planning, and control.

But as the case of LFTI indicates, the physicist took advantage of capable leadership and the increased financial backing which came his way during the industrialization effort. He was therefore able at once to undertake research of interest to the state, to expand the research basis and network through internal reorganizations and the establishment of new institutes, and to embark on new programs in nuclear and solid-state physics. How the scientist managed to face down the pressures of cultural revolution and hostile philosophical scrutiny are the subjects of the next three chapters.

6

Cultural Revolution and the Natural Sciences

The Stalinist revolution from above was accompanied by cultural revolution from below. Cultural revolution involved the penetration of increasingly militant young Party members into positions of responsibility and power in virtually all educational, scientific, economic, and social institutions of Soviet society. Concurrently, it involved class war against those who were deemed to be representatives of the old order: former Tsarist factory managers, schoolteachers, leftist intellectuals in the Party, and so-called bourgeois specialists in science and technology.

In the sciences, cultural revolution was based on the notion that "proletarian science" existed as distinct from "bourgeois science." Scholars of working-class origins had to master all fields of science without delay, bringing with them new organizational forms and Marxist methodology, and supplanting the bourgeois specialist. The advancement of worker-scientists occurred throughout institutes of Narkompros, Narkomtiazhprom, other commissariats, and the Academy of Sciences by several paths: accelerated graduate training, Marxist study circles in dialectical and historical materialism, and through the efforts of such Communist scientific organizations as the Institute of Red Professoriat.

How did representatives of the exact sciences confront the pressures of cultural revolution? How successful was the Party in infiltrating scientific research institutes, training working-class and Communist physicists, and creating a "proletarian science" in terms of methodology and worldview? What were the mechanisms of cultural revolution? And what was its impact at the level of the institute and the individual scholar?

Physicists were able to counter some of the more damaging effects of cultural revolution that other disciplines suffered. Unlike such professional groups as rural scholars, ecologists, architects, and engineers, who were divided along the lines of social profile, professional orientation, institutional affiliation, and political commitments, physicists shared a number of attributes and aspirations.[1] Despite their different fields of interest and diverse social backgrounds, they welcomed the role played by their leadership in tempering the assault of cultural revolution. They recognized the importance of formal training and research in creating a modern physics discipline. And they were shocked by the absurdity of many of the claims of the worker-physicists regarding the "proletarianization" of science. Indeed, historical materialist doctrines of class struggle made less sense in the exact than in the social sciences, allowing physicists to develop a common approach to ideological encroachment.

Perhaps more important, inasmuch as physics research was not far removed from the practical problems of "socialist reconstruction," it may have been simpler for physicists to pay lip service to proletarian science than for many other professional groups, especially concerning "the unity of theory and practice." Their research was needed in metallurgy, construction, communications, and electrification. Once they had demonstrated that educational policies advanced during the cultural revolution could in no way produce physicists in sufficient numbers, let alone train high-quality worker-physicists who understood the complexities of quantum mechanics, semiconductor, atomic, or nuclear physics, the Party scaled back its attack on the scientist.

Nevertheless, cultural revolution put in place processes which changed the class makeup and intellectual orientation of physics in the long run. Some of the changes were the result of unavoidable demographic changes: the aging of the old intelligentsia and training of younger specialists entirely within the Soviet context. But others—heightened ideological awareness, changes in curricula, and the creation of a distinct Soviet, even "proletarian," science— were a direct outcome of cultural revolution.

REVOLUTION FROM ABOVE: CULTURAL
REVOLUTION AND PROLETARIAN SCIENCE

Proletkult organizations had all but ceased to exist in Soviet Russia by the time of the cultural revolution, and had not been active in the sciences since the early 1920s. But many of the notions of "proletarian science" advanced at this time were rooted in Proletkultist understandings. According to the Proletkultists, proletarian science differed from bourgeois science in terms of origin, methodology, scope, and approach. Proletarian science was characterized by a close tie with practice, in this case technology and the industrial process; bourgeois science suffered from empty intellectualizing. Under capitalism, science would be exploitative, helping the owners of the means of production to solidify their economic and political control; under socialism, it would serve the masses, and assist in the rationalization of production. Soon, too, the division between mental and physical labor would disappear. Science would be "a handmaiden to technology, eschewing broader social or political questions."[2]

In terms of methodology, there was great disagreement among the advocates of proletarian science. The physician, philosopher, and Left Bolshevik A. A. Bogdanov advanced a new epistemology based on an eclectic combination of Marxism, Darwinism, Machism, and Ostwaldian energetism to develop rigorously materialist notions of the relationship between subject and object. This was a heavily experiential philosophy of science based on an empiricism grounded in the labor process. That is, the worker best knew "science" and technology through hands-on experience with the machine. The Proletkultists believed that the worker must possess science, purge it of its bourgeois inheritance, and create entirely proletarian scientific institutions: from encyclopedias to universities and research institutes.

Others, including the so-called Deborinites and Mechanists, did not go as far in rejecting the products of capitalism but called for the creation of a nonexploitative, utilitarian, and practical science. These groups differed on how to apply Marxist theory—in this case Soviet dialectical materialism—to the sciences, a debate covered in the next chapter. But whether grounded in vulgar materialist, naive realist, or positivist epistemology, all supporters of proletarian science envisaged the construction of a new society with the working class leading the way, in "full possession" of science. This would require the replacement of bourgeois specialists by trained worker-scientists.

Until the end of the 1920s, the effort to ensure the education of communist worker scientists had failed. In spite of relaxed admissions requirements, university officials still managed to favor the qualified white-collar student over the less-qualified worker or peasant. Scientists were adroit in avoiding interference in the organization and administration of their institutes. The concept of "proletarian science" had little influence in scientific management beyond encouraging scientists to pay homage to some of the tenets of "government construction" and "socialist reconstruction."

Proletkult was far more influential in helping to set the direction of government policy with respect to higher and mass education in such institutions as *rabfaky*, special schools set up to matriculate workers with minimal training. The *rabfaky* graduated 24,500 students in the four academic years from 1925–26 through 1928–29. By the 1930–31 academic year there were almost 90,000 students in *rabfaky* with the majority under the jurisdiction of Vesenkha (37.5 percent), Narkompros (33.6 percent), and Narkomzem (12.5 percent).[3] The number of *rabfaky* reached a peak of some 600 in 1932–33 and declined throughout the 1930s; they were abolished in the early 1940s. The Party also set up an institute for raising qualifications of research and technical personnel under Vesenkha, which offered some seventy courses to over 3,000 students by 1930.[4] These programs produced few first-rate scholars and turned out nearly exclusively social scientists. The working-class natural scientist was a rare commodity.

Cultural revolution stimulated the effort to accelerate the training of working-class Marxist scholars. This in turn triggered a debate between those individuals who accepted a Leninist view of cultural revolution and those who supported the Stalinist position, over how best to create cadres of the proletarian specialist who would, in the words of Stalin, "master science." This had attendant implications for the concept of "proletarian science." Vesenkha and, later, Narkomtiazhprom were the locus of support for "proletarian science," cultural revolution, and Stalinist policies toward science. When Glavnauka's "Leninist" policies with respect to the administration of scientific research and cultural revolution lost out to Vesenkha, its plans for expansion came to naught, and a majority of its research institutes were transferred to the jurisdiction of Narkomtiazhprom or the Academy of Sciences.

The Leninist view of cultural revolution was "a gradual and nonmilitant raising of cultural standards, achieved without direct confrontation with the old intelligentsia and involving above all the expansion of mass education and the spread of basic literacy."[5] In terms of science, this meant incorporating the achievements of bourgeois science into Soviet science, and attracting the "bourgeois specialist" to "socialist reconstruction"—through coercion only as a last resort. The major institutional center of this point of view in the late 1920s was Narkompros. A. V. Lunacharskii, the commissar of enlightenment, A. I. Rykov, former chairman of the Sovnarkom who had long called for the use of Western science and technology in Soviet economic development, N. I. Bukharin, and others who supported the creation of a new Soviet intelligentsia through Narkompros were seen as opponents of "cultural revolution" and supporters of "bourgeois counterrevolutionary" interests.[6]

Bukharin's position on this matter is important inasmuch as he was director of the Scientific Sector of Narkomtiazhprom, a major public speaker on the issue, and one whose views, which were discredited by Stalin and his supporters, had been adopted by 1934.[7] He rejected the creation of "proletarian science" as premature. Initially Bukharin had supported Proletkult, but he came to believe strongly in the importance of taking what bourgeois culture had to offer to Soviet science, technology, and industry. He praised capitalist science, its organization of serial and mass production, the novelty and power of its technology and military. Like many other old Bolsheviks, he singled out the United States as the example to follow.[8]

On two counts, however, Bukharin rejected the unthinking adoption of Western scientific practices in the USSR. Bukharin believed that science in the West embodied the notion of production for production's sake, and did not serve social needs as it would in the Soviet Union. It led to the exploitation of the worker. In addition, scientific research would lead to the concentration of industry in the city and exacerbate the contradiction between countryside and urban centers, whereas in the USSR it would help reestablish *smychka* [symbiosis]."[9] Bukharin recognized that the capitalist West was the ideological, economic, and military opponent of the Soviet Union. The Soviet Union had become an increasingly urban society with an economy grounded in large-scale, science-based production, but without cultural revolution the USSR would remain inferior to the West competitively inasmuch as its citizens could not understand the highest culture, modern technology. Cultural revolution would also secure the cooperation of the peasantry with the government, and help to combat the increasing bureaucratization of Soviet society.[10]

Repeatedly invoking the name and words of Lenin, Bukharin criticized the Proletkult movement for wanting to "soar up to the proletarian heavens" immediately after the Revolution. The Proletkultists had been carried away, "debating furiously" the question of the creation of proletarian science in conditions that approximated an experimental laboratory.[11] Soviet science had a firm institutional basis, and was no longer confined to scholars' studies but had "broadened the circle of its work . . . in new directions [which] merge theory and practice, science and life." A scientific revolution had occurred, and in many regions of the sciences, especially the social sciences, Marxism had already established its methodological hegemony.[12] Soviet science could therefore now harvest the fruits of bourgeois culture—mathematics, biology, physics—and leave behind aspects harmful to the proletariat—religion, philosophical idealism, and bourgeois social science.[13]

It remained to follow and learn from Western developments to avoid reinventing the wheel, to raise literacy among the masses, and to accelerate *vydvizhenie*. In Bukharin's view, *vydvizhenie* had achieved some successes in the Red Army and among technologists, but the need for more qualified engineers, especially among middle-level technologists, was pressing, since quite often Party workers were "not familiar with a whole series of concrete practical questions which demand special knowledge."[14] Over the next few years Bukharin repeated his call for moderation in cultural revolution. Socialist reconstruction required both taking from capitalism what bourgeois culture had to offer the modernization of the productive forces, and raising the cultural level of the masses.[15]

I. V. Stalin, V. V. Kuibyshev, head of Vesenkha, V. Ia. Chubar, head of the government of the Ukraine and soon a full member of the Politburo, and others opposed the Leninist view of cultural revolution.[16] They declared that new cadres were needed to wrest the control of important economic and scientific institutions from the hands of the old intelligentsia, and that empty intellectualizing needed to be replaced by the practice of socialist science. In the hands of the proletariat, they declared, "technology decides everything!" They impatiently called for science and technology to join with the practice of production in socialist reconstruction, and called for the masses to master science.

There were several practical outcomes of the victory of Stalin and his supporters. One was the introduction of Stalinist policies toward science. Another was the transfer of educational responsibilities to Vesenkha from Narkompros by 1930, when all higher technical education "had been transferred to economic organs."[17] Yet another was the attack on specialists signaled by the Promparty and Shakhty affairs.[18]

While cultural revolution was initiated from above, it also involved "a response on the part of the leadership to pressures within the Communist movement and society as a whole." In this light, the Shakhty and Promparty affairs "can be seen as a mobilization strategy designed to create an atmosphere of crisis and to justify the regime's demands for sacrifice and extraordinary efforts in the cause of industrialization," at the same time as the increasingly proletarian Communist party demanded class war against bourgeois specialists and privilege from below.[19]

REVOLUTION FROM BELOW: CLASS WAR,
VYDVIZHENIE, AND PHYSICISTS

Cultural revolution had an immediate impact on the administration and conduct of scientific research in the Soviet Union. It subjected such representatives of the old intelligentsia as physicists to "class war," and resulted in attempts to take over the administration of scientific institutes, primarily through the penetration of Communist cadres into the ranks of scientists through *vydvizhenie,* cooptation, and coercion. *Vydvizhenie* was the advancement of workers into positions of administrative responsibility in economic enterprises, and enhancement of their access to higher-educational and scientific research institutes on the basis of class origin and Party affiliation rather than merit, qualifications, or other traditional reasons for advancement. The call to class war and advancement of workers was heard from 1928 until 1934, and especially between 1929 and 1931. The process was often coercive and always involuntary from the perspective of the institutions, and included organizations ranging from the highest scientific body in the USSR, the Academy of Sciences, to smaller, lesser-known ones.

The summons for a Marxist cleansing of the R and D apparatus apparently was first heard at the Fifteenth Party Congress in December 1927. But rather than purges or *vydvizhenie,* the Communist party had in mind merely an accounting of activities, Party membership, and other relevant information from research institutes and scientific workers throughout the USSR. The Party charged the Central Control Commission of the Workers' and Peasants' Inspectorate with the "verification of the utilization of the scientific-technological forces in the country." The Central Committee of the Party simultaneously ordered committees established in Moscow and Leningrad to investigate the activities of the personnel of scientific institutions. Some two years later the inspectorate began annual inquiries into the status of Communist personnel in research institutes.

What the Party discovered caused great ;concern but was not altogether surprising. The Party was underrepresented throughout the scientific realm, especially in institutions of the exact sciences. The Leningrad regional commission investigated 717 workers and scientific workers in the State Radium and Optical institutes and the Leningrad Physico-Technical Institute.[20] Leningrad was the center of scientific activity in Russia, with the Academy of Sciences, LFTI, the Optical Institute, and numerous other institutes, but in 1929, of over five thousand scientific workers in the city, only thirty-nine—less than one percent—were Party members.[21]

The Workers' and Peasants' Inspectorate was critical of the administration of scientific research institutes. As a result of its investigation, it had concluded that most research institutes had been created "spontaneously" on the initiative of individual scholars, and that "several similar institutions" existed within one and the same physical plant but were found under different bureaucracies, an apparent reference to LFTI and its laboratory. The commission drew attention to the centralization of scientific forces in Moscow and Leningrad, to the detriment of the provinces.[22] All in all, this meant the Party had been inadequately involved in the administration of scientific research, and only the planning of science and the penetration of Party members into positions of responsibility could reverse these trends.

Since the first days of the Revolution, Party officials had called for the reorganization of scientific activity along "socialist" lines; this meant planning, collectivism, and avoidance of parallelism or "duplication" of effort. As early as the organizational meeting of the Russian Association of Physicists in Petrograd in 1919, a representative of Narkompros had encouraged physicists to consider planning of their research activities.[23] Scientists had rejected these suggestions out of hand, with the president of the Academy of Sciences dismissing planning as "Fordism" and "a great mistake and bewilderment."[24]

Two factors frustrated attempts to introduce all-union planning of science. The first was the rapid, almost unregulated growth of the number of institutes and scientists during the NEP. The second was the small number of scientists who supported the notion of planning, and perhaps the smaller number still who held Party membership, especially among natural and exact scientists. Nearly all of the leading physical and life scientists of the 1920s had received their education in Tsarist institutions and often identified themselves with the prerevolutionary intelligentsia, rather than with socialist ideals. Such organizations as the Institute of Red Professoriat, the Communist Academy of Sciences, and the Russian Association of Scientific Research Institutes of the Social Sciences (RANION) were established in the early 1920s in an effort to train specialists with the proper worldview, and an ever-growing number of Soviet *rabfaky, vuzy, vtuzy,* and industrial technicums had opened to train cadres entirely within the Soviet period.[25] On the eve of the first five-year plans, however, there were very few Communist specialists and a majority of scholars in virtually all fields except the social sciences had been trained in the Tsarist period. The "accommodation" of the bourgeois specialist to Soviet power went at a snail's pace, especially in crucial sectors of industry and the exact sciences.[26]

Among scientific workers, the percentage of Party members in the RSFSR, where most of Soviet Russia's scientists were stationed, seems not to have exceeded 5 percent (see app. A, table 6). On an all-union level, according to another source, as of October 1928 there were 24,500 scientific workers in the USSR of whom 5,500 worked in applied science, 3,127 in medicine, 1,038 in chemistry, 977 in physics, and 1,566 in the humanities; almost one-third had begun their active careers under the Tsarist regime (7,346), and 7 to 8 percent were Communist party members.[27] In an effort to accelerate the introduction of planning and to tip demographic factors (age, sex, nationality, and Party membership) in favor of a new intelligentsia, the Central Committee on June 26, 1929, passed a resolution, "On Scientific Cadres," calling on economic and Party organs to find ways to increase Party membership and working-class origin in the selection of scientific workers. Hereafter, as a rule, scientific institutions were infiltrated by Communist scientific cadres of working-class origin, primarily through *aspirantura* and cooptation, and also through the establishment of Marxist methodology study circles.

The Party was particularly unsuccessful in its attempts to bring more Communists into scientific research institutes and in training new communist scientific cadres in the natural and exact sciences, particularly as world-class scholars and especially as physicists. Most leading Soviet physicists joined the Party after some delay late in their careers. A. F. Ioffe, for example, became a Party member only in 1942; I. V. Kurchatov, in 1948, when already hard at work on the atomic bomb project; and B. P. Konstantinov entered the Party in 1958, only when appointed director of the Leningrad Physico-Technical Institute. Other scholars—Landau, Tamm, and Kapitsa—merely avoided the pitfalls of Party membership. Of 25,286 "scientific workers" in the USSR surveyed in 1930, only 2,007, or 7.9 percent, claimed membership in the party, and of these only 44 were physicists, with but 3 in Leningrad and 15 in Moscow. In fact, over two-thirds of Communist scientific workers were social scientists and only 8 percent worked in the exact sciences. Moreover, nearly two-thirds of those claiming Party membership were at the bottom of the scientific hierarchy (see table 9).

The Party attempted to accelerate the advancement of trained Communist scientists through the activities of the Institute of Red Professoriat (IKP) and the Communist Academy of Science, which later enveloped the IKP. These institutions sought to raise the qualifications of workers through special brigades which lectured workers on physics at such factories as Dinamo, Moselektrik, and Elektrozavod, and on chemistry at Kauchuk and Dorogomilov. The physics brigades carried out propaganda work on relativity theory, quantum mechanics, ideological tendencies in modern physics, and the views of Marx, Engels, and Lenin on the philosophy of physics.[28] At a meeting in May 1931, in recognition of the poor state of affairs of these mass propaganda efforts, all societies and

Table 9.   Communist "Scientific Workers," by Discipline
and Experience, 1930

	Level of Professional Achievement					
	World- Class	Basic	Begin- ning	Part- Time	Total (%)	
Scientific Discipline						
Exact sciences	4	45	111	1	161	(8.0)
(%)	(2.5)	(28.0)	(68.9)	(0.6)	(100)	—
Applied sciences	—	31	168	5	204	(10.2)
(%)	—	(15.2)	(82.3)	(2.5)	(100)	—
Medical sciences	1	52	215	—	268	(13.4)
(%)	(0.4)	(19.4)	(80.2)	—	(100)	—
Social sciences	27	433	884	30	1374	(68.7)
(%)	(2.0)	(31.5)	(64.3)	(2.2)	(100)	—
Totals	32	561	1378	36	2007	(100.0)
(%)	(1.6)	(28.0)	(68.6)	(1.8)	(100)	—

Source: G. Krovitskii, B. Revskii, *Nauchnye kadry VKP (b)* (Moscow, 1930), pp. 16, 51–83. "Basic" includes professors, docents, and senior scientific workers; "beginning" connotes technical personnel, junior scientific workers, and advanced *aspirants*.

associations of the Institute of Red Professoriat were instructed to work more closely with various factories and associations; "Red" physicists were indicated in particular for Dinamo and Elektrozavod.[29]

IKP had several other activities. To raise the political awareness of scientists, it cooperated with several research institutes, including the Physics Institute of Moscow State University but few other prominent scientific organizations.[30] It even had a contract with Soiuzkino to produce popular-scientific films in conjunction with the institutes of its natural science division.[31] And it trained future Red specialists.

The IKP and the Communist Academy established graduate education, *aspirantura*, in a series of natural and social scientific fields, with special emphasis on the retraining of young Party members. In physics the coursework for first-year students covered such topics as the mechanical worldview and classical physics, laws of matter and energy in motion, the structure of matter in classical theory and contemporary physics based on an elucidation of quantum and wave theory, the problem of causality, and the crisis in contemporary physics and idealist trends in light of Leninist analysis. The second year was far more ideologically motivated, including such seminars as "The Dialectic as Logic and Epistemology," with discussion of idealist and vulgar materialist understandings of the relation between being and consciousness, criticism of "menshevizing idealism," and an attack on the "creeping" empiricism of de Broglie, Heisenberg, Dingler, and Schlick.[32]

But even Communist strongholds failed to produce good results in terms of numbers of students or their class origin. Three-quarters of *aspirants* who graduated and 50 percent of students admitted to IKP in 1931 were white-collar. Indeed, the natural science division of IKP appeared to be a stronghold of Jewish intellectuals in 1929–30, with 42 percent of its staff Jewish and one-third Russian, and with only 8.2 percent worker, 4.1 percent peasant, and 87.7 percent white-collar. Of twelve students in the division in 1928, seven were Jewish, two were Russian, and all but one were white-collar, although all were Party members.[33]

The IKP attempted to enlist physicists' help in propagandizing the exact sciences and in attracting workers, but these efforts failed. Only eleven mathematics or physics students entered IKP in 1930–31, while seventy-one chose medicine and twenty-three biology or agronomy. Proposed talks by Ioffe, Lazarev, Tamm, and Timiriazev did little to generate student interest in the exact sciences. The fact that the budget of IKP was cut at the beginning of *vydvizhenie* did little to help.[34]

Aspirantura was no more successful in natural scientific institutes. Such facilities as the polytechnical institute in Leningrad had continued to train students of white-collar background, while the *rabfaky* and other organizations attended to the worker. During *vydvizhenie*, Soviet-style equal opportunity for workers and peasants, candidates for admission had a better chance of being selected if they had Party affiliation and working-class roots. These were the *vydvizhentsy*.

Between 1929 and 1931 alone, the number of scientific workers in the USSR of working-class origin increased from 16.5 percent to 30.4 percent, while the number belonging to the party grew from 49.5 percent to 64.3 percent. But after initial successes, the pace of *vydvizhenie* slowed noticeably over the next two years, and the number of communist *vydvizhentsy* declined precipitously.[35] By 1934 within Narkomtiazhprom there were only 387 *aspirants*, of whom only 74 worked in physics institutes, a mere fraction of the thousands of working physicists and researchers. In essence, plans to expand *aspirantura* to ensure Communist scientific cadres on a national level were scaled back as *vydvizhenie* ran its course.

Why did *vydvizhenie* fail in the physical sciences in the short run? There are several reasons. First, the *vydvizhentsy* often lacked skills, it was difficult to train them, and, in fact, there were very few "worker-physicists." Tales of the incompetence of *vydvizhentsy* are legion for all regions of the Soviet economy. Second, for the entire USSR in 1934 there were roughly 4,200 *aspirants*, hardly enough to make a dent in the Party membership of the scientific cadres, which numbered in the tens of thousands. On January 1, 1932, there had been nearly 6,000 *aspirants* in the USSR, of whom nearly half were Party members, another 15.5 percent were in the Komsomol, and approximately 31 percent were of working-class origin. For Narkomtiazhprom the numbers were 1,116 *aspirants* with 682, or 61.1 percent, Party members; 136, or 12.2 percent, Komsomol; and over half working-class.[36] Yet excluding *aspirants* (who in 1932 would almost double the absolute number of Party members in scientific research institutes), the percentage of scientific workers in institutes of Narkomtiazhprom who belonged to the Party grew only from 2.9 percent to 4.3 percent between 1929 and 1932.[37] And of all professional scientific groups in the USSR, physicists had the lowest percentage of Party members.[38]

Third, there was a strong inclination against Party membership among the older, more established scholars, making it difficult for the Party to make inroads, especially as it grew increasingly unwilling to replace scientific specialists with workers with inadequate training. For those who began work in 1925 or earlier, fewer than ten percent belonged to the Party; one-quarter of those who entered scientific research institutes between 1926 and 1929 were Party members. In 1930 and 1931 at the height of the cultural revolution, however, the percentages were seventy and nearly sixty respectively.[39] While there was greater success in 1934 and 1935 in bringing a higher percentage of Party members (some forty percent) and researchers of working-class origin (some twenty-five percent) into *aspirantura*, the decline in absolute numbers by an order of magnitude, from the thousands to the hundreds, lessened the overall impact.[40]

Finally, *vydvizhenie* encountered opposition from leading scholars, even at the height of the movement. A. F. Ioffe urged caution in carrying out or accelerating *vydvizhenie*, from both a quantitative and a qualitative point of view. He realized that the staffing of new scientific institutes and industrial research centers during the First Five-Year Plan was a question of paramount importance. He allowed that the methods of *udarnichestvo* (shock work) and *sotssorevnovanie* (socialist competition) would invigorate the "all-union physical organism." But he had great concern about the potential of *vydvizhenie* to turn out scientists fast enough, since "a huge army of physicists" had to be trained and made available for new regions of the country and economy.[41]

More significant, Ioffe doubted the abilities of the *vydvizhentsy* from a qualitative standpoint. "We need," Ioffe argued, "sociopolitically conscious and steadfast workers," to be chosen not merely on the basis of class consciousness, but also because of

> their natural gift for scientific-research creativity. The fact of the matter is that . . . creativity is art. And in the same way that a well-disciplined musician who does not have natural talent will never be a true creator, a composer, so too the person who is devoid of talent, even though he works conscientiously in the area of science, will not become that creator who creates new socialist technology in the Union. It is necessary, consequently, to find criteria for the selection of talent.[42]

In a speech before the radicalized professoriat of Moscow State University he added, "If each scientific worker possesses creative talent, powers of observation, fantasy, and the ability to discover and invent . . . , that's all that matters.[43] He warned that it would be difficult to identify and train these individuals, whose number might not equal the needs of Soviet Russia. But Boris Gessen, the leading Marxist physicist at the university, disagreed with Ioffe, saying it was precisely such a mass of Communist scholars that was needed to satisfy the manpower needs of "socialist reconstruction," change the face of the sciences, and promote an interchange of ideas between the older order and "the young proletarian *aktiv*."[44]

The effort to train more specialists achieved several important successes, but only over time, changing the social composition of specialists forever. From 1921 when the percentage of members of the Communist party of the Soviet Union of working-class origin had reached its lowest point (41 percent), the proportion of working-class Party members grew almost yearly, reaching 65.3 percent in 1930.[45] The same trends held for scientists. According to a November 1933 survey of 82,689 Soviet specialists, almost 19,000, or 22.9 percent, were now Party members, and 32,000, or 38.7 percent, boasted working-class origins. Party membership was higher among those of working-class origins, reaching some 43.4 percent of the total. But again, when controlling for those who had higher education, the number of Party members dropped to 1,836, or barely 2.2 percent. Party membership was highest among those without higher education.[46]

The reasons for the failure of *aspirantura* are highlighted by the cases of the Leningrad Physico-Technical Institute and the Academy of Sciences, arguably two of the most important scientific establishments in the USSR at the time. Since the first days of the Revolution, the Academy had stood aloof from the political and economic turmoil of Soviet Russia. Under pressure from an increasingly militant Communist party and charges of elitism, in 1927 the Academy revised its charter. At the same time, it increased membership from forty-five to seventy, allowing scholars with Marxist inclinations and underrepresented disciplines—the social sciences—the opportunity to become members. But this was not enough for the government, which voted in April 1928, without consulting the Academy, both to raise membership to eighty-five and to give to organizations outside of the Academy the right to nominate candidates.[47]

When the Academy voted on membership, however, it blackballed several Marxist scholars, most notable among them the philosopher A. M. Deborin, while electing twenty-eight scientists who were already corresponding members. This infuriated such Marxist strongholds in the scientific world as the Society for Militant Materialists, Moscow State University, VARNITSO, and the Institute of Red Professoriat, the latter of which called for a purge or disbanding of the Academy. The Academy leadership recognized that the time had come to become a Soviet institution. Ioffe, an Academician since 1920, and in his career twice vice-president of the Academy, worried that the campaign against the Academy in the press would require scientists "to cook in the kasha" of the Great Break.[48]

Over the next two years, thirty more members were elected, including a number of Marxists. There was criticism of the state of affairs regarding physics. Boris Gessen, active at Moscow State University and the Institute of Red Professoriat, complained that the Academy remained, as it was before the Revolution, a center for geophysics and abstraction. There were no physics laboratories, nor physicists, he claimed, let alone Marxist physicists,[49] in spite of the fact that D. S. Rozhdestvenskii, S. I. Vavilov, and L. I. Mandel'shtam were newly elected.

The election of new members marked the turning point in the life of the Academy. It was followed by a purge of Academy workers, the reorganization of branches and activities, and an increase in the number of scientific institutes; the introduction of graduate education to advance Marxist students; and the penetration of the Communist party into the administration. The Party had put the Academy at the service of socialist reconstruction.[50]

Yet periodic purges of "undesirable" elements and unqualified students, and uneven admissions standards and policies limited the impact of *vydvizhenie* through graduate education within the Academy. The number of *aspirants* increased between 1930 and 1933. A purge then eliminated fully 76, or one-fifth, of them.[51] Throughout 1933 a purge (*chistka*) continued, although 45 managed to finish successfully. The inability of the Party to promote workers in the exact sciences persisted: there were 4 graduates in chemistry, 3 in physico-mathematical sciences, 14 in biology, and 22 in the social sciences.[52] One year later the situation was somewhat better. In the physico-mathematical sciences 7 *aspirants* defended, 12 in chemistry, 15 in the social sciences and 10 in biology. But in 1934 the insufficient number of highly qualified candidates had begun to trouble Academy officials. Some 32 were transferred into *vuzy*, so that only 112 remained. Their "weak preparation" made defense of final work impossible before the summer of 1935.[53]

The Communist takeover of the Academy of Sciences was important on two counts. First, it served notice to the rest of the academic world that the aloofness of scholars from the social and political forces at work in the USSR would no longer be tolerated. Second, the mechanics of the takeover may have been a model for the assault on other scientific institutions.

In other institutes and higher schools, "Party leadership" also lagged considerably behind plans. In Leningrad Polytechnical Institute as of November 1, 1929, only one of eighty-three professors was a Communist, and only three had received education fully within the Soviet system. On the eve of the First Five-Year Plan, for all Leningrad *vuzy* and scientific institutes only one-quarter of all scientific-technical cadres were Communist, and less than one in fourteen had working-class origins. As a result, the Party sent N. I. Podvoiskii as chairman of a commission "to speed up" the training of communist specialists in Leningrad *vuzy*. Eleven teachers from the Institute of Red Professoriat arrived in April 1929 to assist in the training of cadres; they worked in the Institute of Students-Vydvizhentsy, created in 1926, to improve Party and social makeup.[54] As of May 1930, while purges had been carried out in organizations of Narkomtrud, Narkomzem, Narkomfin, and other commissariats, however, they had just begun in such institutions of Narkompros as LFTI.[55]

There were few Party members within LFTI before World War II, and *vydvizhenie* failed to achieve desired results. Of ninety-six staff members of the Ioffe Institute in July 1926, only one of the leading administrative and scientific personnel, V. N. Glazanov, was a Party member. One unidentified scientific worker also belonged to the Party. None of the leading physicists—Ioffe, Chernyshev, Obreimov, or Frenkel', to name a few—was a Communist. The Party did boast members among technical personnel and staff. Five Party members and two candidate members, eight Komsomol members, and one candidate worked in labs or workshops of the institute. Because the Bolshevik party's sources of strength were in the working class, most LFTI Party members were workers, the majority of whom entered the Party in 1922 or later (see app. B, tables 6 and 7).

In order to bring more communists into scientific research institutes (as opposed to "bringing science to the masses"), the institute adopted a program in which *aspirantura* for workers played a large role. In February of 1931 the first worker-*aspirant*s arrived at LFTI. This event, in line with the cultural revolution, acquired, in the words of an observer, "all the more political significance at this moment of the rebuilding of the methods of scientific work."[56] At first Ioffe himself was hopeful that *vydvizhenie* might work, although he worried about the shortage of qualified individuals.[57] In a preliminary investigation, a Komsomol brigade from the Academy of Sciences created to inspect the work of *aspirants* in the summer of 1932 praised the institute's program. The brigade was impressed with the yearly and long-term (five-year) plans which provided scientific themes upon which *aspirants* concentrated. While plans did not exist for individual *aspirants*, their efforts were "closely linked" to the collective activity of the general research program of the institute. The worker was involved in laboratory practica and theoretical and methodological instruction, and had to give weekly presentations on his work. The Komsomol brigade concluded that this method should be applied "within the Academy of Sciences."[58]

Yet *vydvizhenie* at LFTI was a dismal failure. The group of students consisted of inexperienced workers of rural origin and some urban residents with elementary education, most of whom could not even work with fractions. After five months of mathematical training and introductory chemistry or physics, the institute's directors painted the best possible picture, arguing that there had been some success and that the "skeptical views of several scientists on the rapid training of cadres from workers have been proven incorrect." The greatest successes were associated with workers whose tenure in productive capacity had been the longest, that is, among those who undertook laboratory work in their specialty.[59] But the tempo of training of the workers was rather slow—their teachers at the institute and the polytechnical institute's physico-mechanical department seemed to ignore them. There were also a number of problems which arose in regard to, students' "weak Party ties" and lack of the "proper" social origin. Some even had "the wrong frame of mind from the 'sociopolitical' point of view." The Party urged more time and effort, including entrance interviews to select better candidates and to form a larger group.[60]

More revealing of the failure of *vydvizhenie* is the fact that projections for its expansion were never met. According to the draft five-year plan for LFTI, over a three-year period as part of its general growth, Party membership would quintuple between 1930 and

1932 alone. *Aspirantura* would be the major source of this increase. The plan included 55 *aspirants* in 1931, of whom 22 would be workers and 45 Party or Komsomol members; and in 1932 there would be 150 *aspirants* with over 40 of working-class origin and a commensurately large number of Party members. Between 1930 and 1932 the percentage of scientists of working-class origin would grow from 8.6 percent to 17.1 percent, declining slightly to 14.9 percent; the percentage of Party and Komsomol members would grow from 5.1 percent to 14.0 percent to 23.8 percent (see app. B, table 8).

The number of worker/*aspirants* never came close to the projections of the five-year plan. There were only eight at the institute in 1932, at the same time as there were twenty-five at the Leningrad Electrophysical Institute and twenty-three at the Optical Institute.[61] In 1934 *aspirantura* had expanded to include four specializations: solid-state physics (semiconductors, photoelements, rectifiers, larger crystals, and amorphic bodies); nuclear physics (artificial splitting of the nucleus, X- and cosmic rays); electrical emission (photoelements and diffraction); and theoretical physics.[62] But the number of communists in these specialties was small: eleven of a total of twelve (see app. B, table 9).

In sum, the plan forecast annual increases—through *vydvizhenie*, cooptation, and new hiring—in the quantity of scientific workers who were Party or Komsomol members, discipline by discipline. However, the Party was unsuccessful in infiltrating the institute. There were very few *aspirants* at LFTI to augment Party membership, and fewer still Party physicists who could hold their own with the directors in matters of science and technology policy. As of 1990, in natural science institutes of the Academy of Sciences only 13 to 18 percent of all scientists are Party members, although 60 to 70 percent of laboratory heads belong to the Party, and almost 100 percent of institute directors.[63]

When *vydvizhenie* came to an end in 1934, the authorities restored a great measure of authority to scientists: they granted institutions the right to award advanced degrees, both with and without dissertation defense, a right they had lost at the beginning of the Revolution.[64] LFTI was no exception and began to bestow upon its leading physicists academic titles for the first time in years. The list of those leading physicists who finally received degrees includes most of the first generation of Soviet physicists and two Nobel Prize winners. In March 1934 an Academy commission for the qualification of scholars set up under V. F. Mitkevich, S. I. Vavilov, and D. S. Rozhdestvenskii began to award doctors of physics degrees *honoris causa* (without defense), and among the first selected V. K. Arkad'ev, V. V. Shuleikin, P. L. Kapitsa, N. K. Shchodro, G. S. Landsberg, Ia. I. Frenkel', V. A. Fock, Iu. A. Krutkov, I. E. Tamm, P. I. Lukirskii, I. V. Obreimov, D. A. Rozhanskii, D. L. Talmud, P. M. Nikiforov, and several others. Most were physicists at LFTI.[65] In November the commission, newly constituted under Ioffe's chairmanship, awarded a degree to B. P. Gerasimovich; in December to N. D. Papaleksi, A. F. Val'ter, N. N. Andreev, A. I. Tudorskii, V. S. Ignatovskii, and P. A. Rebinder; and in January 1935 to S. N. Rzhevkin and I. M. Mysovskii.[66]

STUDY CIRCLES IN MARXIST PHILOSOPHY

In addition to *vydvizhenie* and *aspirantura*, there were two other mechanisms through which the Party infiltrated scientific organizations and *vuzy* during the First Five-Year Plan. One was conversion and cooptation, the other was coercion and terror, which will be discussed in the final chapter. Conversion and cooptation involved setting up study circles on the history of the Party, dialectical materialist philosophy, and new Marxian methodologies, and the development of curricula for the dialectical materialist teaching of physics. The goal was to allow any Soviet citizen, and not only first-class scholars, academicians, professors, or *aspirants*, but anyone interested in self-education—from workers to peasants—to become proficient scientists, and to provide guidance in the development of a Marxist philosophy of the natural sciences.[67]

Initially, especially when older, more established scholars dominated the sciences, study groups and circles rather than curriculum changes were the main vehicle for the proletarianization of science in research institutes, laboratories, and universities. There were two kinds of study circles, independent all-union organizations and circles which were founded within scientific research institutes. The spread of both kinds accompanied *vydvizhenie*, but the all-union variant declined in activity when the momentum of cultural revolution was spent. They were then incorporated in such existing organizations as Moscow State University. Until recently the dialectical materialist circles in scientific institutes have continued to operate, as "methodological seminars," under the watchful eye of the Party, although they had lost their ideological vigilance by the early 1960s in institutes of the exact sciences.

One of the first study circles was established at Moscow State University in 1924. It seems to have been connected with the newly founded Institute of Red Professoriat. Later, in the spring of 1927, the university Party cell organized a special commission with representatives from the administration, the Party, and scientific workers, which called for the establishment of circles for the study of dialectical materialism for all *aspirants*. Some ninety-seven students originally took part.[68]

The Circle of Physicist-Mathematician-Materialists, which was founded in December 1927 within the division of natural and exact sciences of the Communist Academy of Sciences, grew out of the university circle. It sought to apply Marxist methodology to mathematics and physics to ensure that scientific workers, all too many of whom had Tsarist training, were not only "natural [*stikhiinyi*]" materialists in their understanding of modern physics but conscious dialectical materialists in all of their work. Participants in the circle read Engels' *Anti-Dühring* and *Dialectics of Nature* for the theoretical background necessary to comprehend fully the kinetic theory of matter, and then Lenin's *Materialism and Empirio-criticism* for clarification of the electronic theory of matter. Such reading was intended to overcome nearly two decades of idealism in the "bourgeois" physics of Eddington, Jeans, and Nernst, the mathematics of Hilbert and Weyl, and so on.[69]

In the first six months of its activity, the circle heard papers by Z. A. Tseitlin, on the relationship between universal gravitation and electromagnetism,[70] V. G. Fesenkov on solar evolution, and several others on mathematics and probability. While some of these talks had the appropriate level of sophistication, however, others were too popular for the audience of advanced scholars; they were not converted immediately to Soviet Marxism. The circle's leaders hoped to publish the papers, revised with this failing in mind, to overcome this problem.[71]

The Physicist-Mathematician-Materialists circle quickly acquired a reputation for acumen in the philosophy of science. A. A. Maksimov, then head of the natural science department at the Institute of Red Professoriat, was asked to designate leaders for dialectical materialism study circles at such established scientific research centers as the Karpov Chemical Institute. The Agitprop and Party committee of the Karpov Institute suggested "it would be desirable if the director of the circle were well oriented in questions of chemistry and physics."[72]

Most of the activities of the Physicist-Mathematician-Materialists circle took place in and around Moscow. The circle grew to have over one hundred active members, most of whom were connected with the university as faculty, researchers, graduate students, or *vydvizhentsy*. There were twenty-seven Communists including B. M. Gessen, A. K. Timiriazev, Ernst Kol'man, and A. A. Maksimov. None of the leading physicists—L. I. Mandel'-shtam, the leading Moscow theoretician, I. E. Tamm, future Nobel Prize winner, M. A. Leontovich, who later worked on thermonuclear synthesis, and the specialist in astrophysics V. L. Ginzburg—belonged to the Party at that time; several joined later. Soon the circle had grown to include fifty-three "Communist physicists," but most of these were university *aspirant*s or *"komsomol'tsy-fiziki* [Communist youth league physicists]." The circle had thirty-one institutional members, five of whom were associated with the university.[73] Efforts to set up filials in Baku, Kazan, Kharkov, Kiev, and even Leningrad, the traditional center of Russian science, fell short of expectations.

In Leningrad a society of "mathematician-materialists" was established in 1928, and quickly grew to fourteen members, including five in the Party or Komsomol. This society became a section of the Society of Militant Materialist Dialecticians in 1929, and then the Society of Mathematician Materialists in 1930 with fifty-three members. The problem was that the membership included mostly philosophers and no natural scientists.[74] The Leningrad division of the Communist Academy of Sciences (LOKA), created "to organize and plan scientific research, to fight bourgeois idealism, to train cadres, and to provide methodological guidance for *vuzy* and scientific research institutes of Leningrad," was more successful in assisting study groups to advance the "struggle" for a dialectical materialist *Weltanschauung* among scientific workers.[75]

But at a May 1931 meeting, Maksimov and others noted problems with the work of the LOKA natural science section, removed
the director, Uranovskii, replaced him with one "Iakson," and
ordered two Moscow members, Egorshin and Iakobson, to proceed to LFTI for organizational and propaganda efforts.[76]

Another study circle, a workshop established in the summer of
1929 under the auspices of the Communist Academy to enlighten
teachers of physics about the wonders of dialectical materialism,
met with interesting results. During an intensive three-week session, under such representatives of early dialectical materialism as
B. M. Gessen, K. F. Teodorchik, and several leading physicists, the
participants took nine courses on such topics as relativity and
quantum theory, atomic structure, biological physics, and cosmology. Even though there was little promise of financial support and
most attendees paid their own way, one hundred aspiring dialectical materialists took part. Of these, eighty were secondary school
teachers. While seventy percent had higher education, only fifteen
represented physics as a speciality and only five were Party
members. The workshop leaders therefore resolved to organize
teaching expeditions to several of the major physics institutes in
and around Moscow, and proposed to publish the most interesting
papers: Gessen's "Problems of the Methodology of Physics,"
Tamm's "The Physical Foundations of Relativity Theory," S. I.
Vavilov's "Fundamentals of Quantum Theory," Landsberg's
"Atomic Theory," and Lazarev's "Lecture on Biological Physics."[77]

The creation of circles in scientific research institutes followed
quickly on the heels of the creation of the Circle of Physicists-
Mathematician-Materialists. Some 300 scientific workers took part
in circles in the Academy of Sciences to develop "Marxist scientific
methodology"; at LFTI six groups with 100 participants met regularly to discuss dialectical materialism. Throughout the USSR,
physico-technical institutes developed programs in the area of
Marxist-Leninist methodology and philosophy of natural science.[78]
By 1932, study groups existed in institutes in Moscow, Leningrad,
Kharkov, Bukhara, Tashkent, and Gorki, and the natural science
division of the Institute of Red Professoriat had sections in the
North Caucasus, Tashkent, Azerbaidzhan, Sverdlovsk, Nizhni
Novgorod, Saratov, and Rostov. But there were mostly historians
of the Party, philosophers, and economists, but no physicists.[79] As

of January 1, 1933, in Leningrad alone, there were over 500 circles on Party history, and 168 on historical and dialectical materialism.

For some reason, by the end of the following year only twenty-eight groups for 630 participants remained in Moscow, and fifteen groups for 456 specialists in Leningrad. The onerous requirements of a year-long, 220-hour course on dialectical materialism, historical materialism, and the dialectics of nature may have turned some scholars away.[80] But several thousand scientific and technical workers ultimately participated in the circles.[81] Judging by the increasing Communist party membership among scientists, the circles were successful. As of January 1, 1933, of 14,772 scientific workers in Leningrad, 1,761, or roughly twelve percent, and a fifty percent increase from five years earlier, were Communist, although few of them represented the exact or technical sciences.[82]

The LFTI circle was successful in securing the participation of the leading representatives of Soviet physics, although its reach does not seem to have extended to mid-level researchers or worker-physicists. The circle was founded in 1928 or 1929 under Rubanovskii, a Hegelian and 1928 graduate of the Institute of Red Professoriat who killed himself in 1934 because of unyielding pressure on the Deborinites and incessant baiting by the mechanist A. K. Timiriazev[83] and M. L. Shirvindt, a professor of philosophy, who accused him of being a "menshevizing idealist." The circle had "general-institutional" significance, although it belonged formally to the physico-mechanical section of LFTI. According to the draft five-year plan of the institute, the workers of the philosophical circle

> should master the new physics and, being qualified Marxists, fertilize the work of other departments. . . . The most urgent problems of the methodology of physics are the composition of the Marxist history of physics (according to the task of the Central Committee of the VKP(b)), and also philosophical analysis and criticism of wave and quantum mechanics. A more specialized but also extremely pressing problem is the Marxist methodology of measurements. The reading of lectures, the conduct of seminars . . . has great significance for Marxist training of workers of the Institute.[84]

As the philosopher I. D. Rozhanskii recalled, Shirvindt conducted the philosophical seminar in a lively fashion, as a result of which it acquired a quasicomic character.

This was the time of the predominance of that trend in Soviet phi-losophy which received the name "menshevizing idealism." Shir-vindt conducted the seminar fully in the spirit of this trend, trying to assimilate in the consciousness of physicists the bases of dialectical logic, which he set forth rigorously according to Hegel. Sometimes he was thirty to forty minutes late, or didn't show up at all, in which case his assistant, Tat'iana Nikolaevna Gornshtein, a young, attrac-tive woman, who was captivated by her conviction and sincere sur-prise on account of the fact that physicists remained immune to the abstract hairsplitting of the Hegelian dialectic, opened the seminar. I remember only the general meaning of a series of presentations of my father and Iakov Il'ich [Frenkel']. From the point of view of the leaders of the seminar, these presentations were at times "hereti-cal," although of a different kind. My father openly defended Mach, with a soft smile that was characteristic of him, and grounded his philosophical views . . . on some kind of variety of physical positiv-ism. Iakov Il'ich conducted himself entirely differently. He often be-came excited, passionately protesting against attempts to fasten spe-culative schemes which were developed by philosopher-idealists at the beginning of the nineteenth century to physics. He caustically ridiculed the scientific absurdities which were contained in Hegel's *Philosophy of Nature.*[85]

In 1934 the academic council of the institute voted to create a section on the history of physics and invited S. F. Vasil'ev to direct its work, and also established a methodological circle under I. P. Selinov.[86] Throughout the 1930s, Vasil'ev conducted a seminar at LFTI on the history of materialism, and he published a book of essays on the subject, covering the philosophy of science from Kepler and Galileo to Descartes, the rise of the mechanical world-view, and the nineteenth-century crisis leading to the work of Marx and Engels and the creation of dialectical materialism.[87] To-ward the end of the decade, Rusinov, chairman of the local Party committee, Kuprienko, its secretary, Fedorchenko, secretary of the institute's Komsomol organization, and L. A. Artsimovich orga-nized and conducted a seminar on dialectical materialism at the institute. Talks scheduled for 1938 included those by Tamm on Dirac's theory, Fock on the principle of complementarity, E. Kol'-man on universal constants in contemporary physics, K. V. Nikol'-skii on the mathematical method in contemporary physics, and an as yet untitled talk by S. I. Vavilov.[88]

Changes in curriculum which accompanied the transfer to Vesenkha from Narkompros of many secondary and higher-educational institutions in the first years of the five-year plans represented the most blatant attempt to proletarianize the sciences. A concomitant goal was to train an increasing number of graduates with the proper social origin and worldview. In the first years of *vydvizhenie*, changes in curriculum amounted to little more than criticism of textbooks for failure, among other things, to include the concepts of class struggle and socialist reconstruction in their exercises, or to mention the relevance of Lenin's *Materialism and Empiriocriticism* to modern physics.[89]

The revised version of A. F. Ioffe's *Kurs fiziki* (1927), originally published in 1919, fared poorly in this light. Ioffe was accused of "ignoring practice" in his book, especially in his treatment of curvilinear, rotational, and rectilinear motion; of contradicting the dialectic by asserting that all systems eventually reach some equilibrium state; and of making such factual errors as abandoning the ether.[90]

V. P. Egorshin, one of the early graduates of the Institute of Red Professoriat, was a vocal advocate of the need for new texts. While on lengthy *komandirovka* in the provinces he lectured extensively to workers in various settings on his specialty, the history of astronomy, from the historical materialist point of view; at Dom uchenykh in Rostov he spoke on dialectical materialism; and in Novocherkassk on Marxist methodology. He taught in several capacities and worked in the office for the history and philosophy of science in the physics department at Moscow University. This contact with the *vydvizhenie* convinced him of the need to have texts which conveyed the appropriate sense of "proletarian science."[91]

The proper physics textbook would elucidate the Marxist-Leninist view of science; relate critically to contemporary literature in view of the ongoing crisis of bourgeois science most evident in the quantum mechanical concepts of indeterminacy and acausality; and help science serve socialist reconstruction. Such a book, according to Egorshin, would offer a series of concrete examples to demonstrate a connection between theory and practice, thereby avoiding the abstract nature of most texts. It would present the dialectical relationship between the three states of matter—gaseous, liquid, and solid—and their qualitative and quantitative differences. It would be "politically acute," pointing the way for technology to "reach and surpass the West" during the First Five-

Year Plan. It would stress the class struggle and *partiinost'* (party-ness) when referring to theoretical physics. Finally, as a new Soviet textbook, it would be written in a style, language, and methodology of political significance not for the intelligentsia but for workers and peasants. This required that the physics of mechanics, heat, sound, light, and electromagnetism be interpreted and taught, as Engels had observed, as definite forms of matter in motion.[92]

Vesenkha and Narkompros eventually adopted the perspective suggested by Egorshin and others. By the end of the 1930s, course requirements, from reading lists to hours per subject, had become ossified in all-union prescriptions that invoked such classics as Beria's *K voprosu istorii bol'shevistskikh organizatssi v Zakavkaze* (*The History of Bolshevik Organizations in the Caucasus*) under the rubric of "Basic Laws of the Materialist Dialectic."[93] Having exposed all aspiring Soviet specialists from the early 1930s onward to dialectical and historical materialism, the Soviet Union could not fail ultimately to produce a new generation of physicists, chemists, and biologists who were deeply aware of the worldview expected of them.

Soviet physicists met the pressures of cultural revolution head-on in one final important way: through the publication of articles and monographs geared to a less sophisticated audience of beginning students and workers. These publications sought to highlight the achievements of Soviet physics and perhaps to head off criticism that the new physics was imbued with "bourgeois idealism." One medium of popularization of the "new physics" was the radio. The theoretician M. P. Bronshtein, who worked at the Ioffe Institute until he perished in the purges in 1937, gave a series of popular talks on Leningrad radio on such topics as "contemporary theoretical physics and prospects of its development."[94]

A more important vehicle was popular-scientific publication. Scientists instituted the series "Popular Scientific Library" and "Biographical Library" to bring science to the "masses." Another publication series, "Classics of Science [Klassiki estestvoznaniia]," under the editorship of I. I. Agol, M. Ia. Vygotskii, B. M. Gessen, S. I. Vavilov, and A. A. Maksimov, was devoted to publishing new editions of classic science texts, to assist in discussions of Marxist philosophy of science. A short-lived but important series was "Problems of Contemporary Physics [Problemy noveishei fiziki]," edited by Ioffe, S. F. Vasil'ev, and D. Z. Budnitskii (all from LFTI), which focused primarily on reporting the research results of the Association of Physico-Technical Institutes. Finally, the series "Our Scientific-Technical Achievements," which Ioffe, A. E. Fersman, V. M. Sverdlov, and Iu. I. Dmitriev-Krymskii edited, produced such books as *Nad chem rabotaiut sovetskie fiziki* (1930), written by Ioffe and A. K. Val'ter, which were clearly produced to show the *vydvizhentsy* how the work of Soviet physicists reflected the organic unity of theory and practice.

In spite of the attempts to meet the scientific interests of the masses, scientific and technical publications came under attack for their failure to express clearly the social and cultural needs of cultural revolution. They were therefore reorganized into the Association of Scientific-Technical Publishing Houses, under the watchful eye of the Sector of Technical Propaganda of Narkomtiazhprom.[95] The new house produced such standard "popular-scientific" treatments of recent developments in physics as K. D. Sinel'nikov and A. K. Val'ter's *Fizika dielektrikov* (1932) and Val'ter's *Ataka atomnogo iadra* (1932) and its update, *Atomnoe iadro* (1935). So in scientific publication, too, cultural revolution had run its course.

CONCLUSIONS

The Communist party scaled back *vydvizhenie* by 1934; the insistence upon working-class or peasant social origins and Party membership as prerequisites for admission to universities or institutes, or a position as factory manager or engineer, had failed to produce qualified proletarian specialists. In industry the *vydvizhentsy* destroyed sophisticated equipment, ignored advice, and abused the engineer.[96] In science they never got the chance to take control of the institute; too few had been trained, and fewer still mastered modern science and technology. Research institutes were forced to bring more and more Communist workers within their walls. But with the exception of institutes of the social sciences, the *vydvizhentsy* failed to supplant the so-called Tsarist expert. In 1935, in a surprising *volte face*, the Party actually reinstated the award of advanced degrees for the first time since the Revolution.

Physicists blunted the impact of cultural revolution on their discipline. There was a broad degree of consensus on the dangers of *vydvizhenie* and the absurdity of "proletarian science." Physicists recognized that it was impossible to train capable researchers in less than six or eight years. Some worried out loud that cultural revolution might make the Soviet Union fall further behind the science and technology of Europe and the United States. "We lag even further behind the leading capitalist countries and we have only begun to work, so it is not surprising that the vast majority of important discoveries are produced not by us but abroad," M. P. Bronshtein warned. "In the region of physics we are still to a large measure the students of West European researchers."[97] There was also agreement about the fact that "proletarian science" had little meaning in physics. Physicists participated in Marxist study circles and paid lip service to dialectical materialism, but they had little use for the concepts of the unity of theory and practice or class struggle in quantum mechanics and relativity theory.

And yet, the impact of cultural revolution was considerable on several counts. First, it lead to the formulation of standardized natural sciences curricula which included political indoctrination and stressed rote learning. To this day physicists complain about how this system stifled individual initiative and flights of fancy. Second, in the long run, in spite of their failure from 1929 to 1934 to bring about change in the social makeup and philosophical orientation of scientists, Marxist study circles and curriculum requirements altered the worldview of young physicists, while demographic trends ensured the increasing domination of the sciences by individuals trained entirely within the Soviet context—especially as older scholars died off. And third, cultural revolution meant the victory of the Party over the specialist. The natural scientist had reached an accommodation with the organs of government— Vesenkha, Narkompros, and so on over the funding and direction of their research. But the Party now insisted that it would be the final arbiter in matters of science policy.

Cultural revolution had one more important effect on the physics enterprise. It provoked disputes on the ideological front between ideologues and scientists, and between Leningrad and Moscow physicists.

7

Theoretical Physics: *Dramatis Personae*

Political and cultural revolution led to changes in the physics discipline in Soviet Russia in terms of its organization, government-science relations, and the social makeup of physicists. The revolution in worldview had profound impact on the philosophical concerns and research orientation of scientists. The scientific revolution brought about by relativity theory and quantum mechanics required the overturning of classical conceptions of space, time, matter, and energy, introducing ones based on complex understandings of the discrete and continuous properties of matter and energy, the importance of the observer and measurement in subatomic processes, and changes in beliefs about atomic and nuclear structure with attendant revision of theories of eletromagnetic fields, action at a distance, and the conservation of energy.

A crucial aspect distinguishes this revolution in Soviet physics from that in Western physics, namely the requirement that Soviet researchers adapt scientific concepts to Marxian ones as set forth in the official doctrine of the state, dialectical materialism. Through its ideological henchmen, the Party sought the power to inform physicists which features of relativity theory and quantum mechanics were acceptable, and which were "idealist" and antithetical to the Soviet regime. Physicists had to learn to tailor their theoretical pronouncements to avoid the appearance of idealism. In the 1930s and 1940s, in an environment of hostility and suspicion engendered by the Stalinist revolution from above, they often failed to convince Stalinist philosophers that they were at one with the Party on this matter.

Political revolution involved encroachment on physicists' independence in the determination of the research agenda. Cultural revolution served notice that the specialists' hegemony in science administration, hiring of staff, and curriculum planning was at an end, and through its advocacy of proletarian science set the stage for intrusion in the methodological sphere. The revolution in the philosophy of science cut more to the heart of the discipline, and of theoretical physics in particular, since it had implications for the way physicists did their work, how they determined which topics merited study, and even how they interpreted results.

Scientists did not at first anticipate how serious the scrutiny would become. But seemingly harmless philosophical discussions became politicized by a seige mentality brought about by the succession struggle and the defeat of Trotskii; cultural revolution and the attack on "bourgeois" specialists; and the cutthroat tempo of industrialization.

Philosophical battles were therefore among the most crucial that physicists fought in the 1930s. Scholars at the Leningrad Physico-Technical Institute had struggled with some success to maintain their autonomy against *vydvizhenie* from below and the institution of Stalinist politics of science from above; they had conceded to pressures to adopt long-range planning for such areas of their discipline which were not strictly compatible with the dictates of hard-and-fast targets as theoretical physics; and they had accepted Marxist indoctrination for cadres, although not willingly or without struggle. Now they encountered pressure from both fellow physicists and Marxist ideologues to abandon certain allegedly "idealist" philosophical concepts and conform to the "dialectical materialist" point of view and the dictates of "proletarian science" in the area of methodology.

The debates over the philosophy of physics were the point at which physicists drew the line. They could not allow ideologues to dictate what was "materialist" physics and what was "idealist," nor could they allow them to determine the philosophical content of new theories. Specialists from Marxist scientific institutions commenced the assault on the philosophy of physics. Later, from within the physics community itself, and in particular from a group of vindictive and grotesquely self-righteous Moscow scholars, the entire discipline fell prey to dangerous philosophical gamesmanship. Taken together with the strong emphasis on applied over fundamental research, this could have had a devastating impact on theoretical physics.

But the participants' stable institutional bases, in particular active theoretical physics departments in scientific research centers, and angry and steadfast rejection of ideological intrusion saved physics from the "proletarianization" of science which devastated genetics under Lysenko. This chapter sets the stage for discussion of the philosophical disputes of the late 1920s and 1930s with a description of the institutional and individual actors, from the physics community to Marxist philosophers.

THEORETICAL PHYSICS AT THE
LENINGRAD PHYSICO-TECHNICAL INSTITUTE

By the eve of the cultural revolution, the department of theoretical physics at the Leningrad Physico-Technical Institute was one of the leading centers of fundamental research in the USSR. It had solid institutional backing, gifted leadership, and a broad research program. Along with the physics institutes at the Leningrad and Moscow universities, theoretical sections at the Institute of Physics and Biophysics, and the Ukrainian Physico-Technical Institute, it was a major center for the development of relativity theory and quantum mechanics, as well as of the electronic theory of solids, quantum theory of the solid state, and theories of molecular and nuclear structure. Its general purpose was to provide such theoretical grounding and interpretations for the work of other departments. The department's most prominent figures were V. R. Bursian, from 1918 to 1932 its head, Iu. A. Krutkov, and Ia. I. Frenkel', who was its director from 1932 to his death in 1952 and the advisor of such students as G. A. Grinberg, M. P. Bronshtein, L. V. Rozenkevich, N. N. Miroliubov, B. N. Finkel'shtein, G. Kh. Gorovits, and L. D. Landau. After Frenkel' joined LFTI in 1921, he attracted other theoreticians to the institute including V. F. Frederiks, F. A. Miller, and briefly, V. A. Fock and D. D. Ivanenko. Frenkel' was the most active of the Leningrad theoreticians, publishing over one hundred articles and twenty full-length monographs in his life.

Frenkel' and Frederiks, along with A. A. Friedman, were the leading early popularizers of relativity theory in Soviet Russia.[1] Frenkel', Tamm, and several other Soviet physicists had been stranded in the German-occupied Crimea during the war. Unlike most other Soviet scholars, they had access to current German periodicals, and were well prepared to introduce their Moscow and Leningrad colleagues to Einstein's work. Frenkel' published the first "nonpopular" guide to relativity theory in Russian, *Teoriia otnositel'nosti*, for students with a university mathematics background. After describing the historical bases of the theory, Frenkel' gave full endorsement to its physical results, which included the banishment of the "anthropomorphic ether." The ether had served as a kind of universal frame of reference for classical explanations of transmission of electromagnetic energy through space, and was particularly important in the late nineteenth century in explanations of the behavior of light with respect to the observer. Increasingly, however, failed attempts to determine its properties or detect its existence, notably those of Michelson and Morely, indicated that something was not right. Nevertheless, in order to maintain the classical status quo, physicists held on to the imponderable, undetectable medium.[2] The theory of relativity, which rejected absolute space and time and established the speed of light as an absolute magnitude, dispensed with the ether.

Frenkel' saw the classical "materialization of the ether" as a mathematical fiction. To his horror, such physicists as O. D. Khvol'son and the mechanist A. K. Timiriazev refused to abandon it, continuing to explain electromagnetic forces essentially in mechanical terms. Supporters of the ether cited the May 1920 speech at the University of Leiden in which Einstein admitted that in theory it would be possible to have an ether and the special theory of relativity.[3] But they had merely confused Einstein's willingness to grant the theoretical possibility of its existence with his support for the ether hypothesis. In fact, Dayton Miller's attempts to prove ether drift in a repetition of the Michelson-Morley experiment, which was widely published in Russia as in the West, did little to convince leading scientists of its existence; such physicists as S. I. Vavilov and T. P. Kravets pointed out the weaknesses and irregularities of the experiments.[4] Frenkel', for his part, publicized the death knell of the ether in semipopular journals on new advances in physics. In a bitter and sarcastic comment called "The Mysticism of the Universal Ether," Frenkel' traced the history of the concept from Huygens and Fresnel through Faraday and Maxwell to Lorentz. He argued that this unknowable, undetectable "material all-encompasser [*vsederzhitel'*]" had been transformed into a god of physics in the nineteenth century: all late-nineteenth-century physics had become the study of the physical properties of the ether and its contradictions. This "teleological" belief in the ether provoked a reaction in Frenkel' like that of Lenin to the "God-builders [*bogostroiteli*]" and "God-seekers [*bogoiskateli*]" in 1909; Frenkel' castigated the Mechanists as "ether-seekers [*eferiskateli*]."[5] Yet he recognized that much of Newtonian physics remained in force:

> The question may arise: in what is the value of the general theory of relativity if it practically leaves the classical *Weltanschauung* in force? The answer is quite simple: in the first place, relativity theory frees human thought from those logical contradictions which flow from the absolute character of acceleration in classical mechanics. Second, it leads to a more exact theory of gravitation in relation to which the theory of Newton is . . . only a first approximation. And last, and third, it connects in a unitary whole electromagnetic phenomena with gravity. We recall that the key to this theory is the identification of inertial and gravitational forms flowing from the principle of the relativity of accelerated motion, and leading to the fundamental expression for the impulse of work of the forces of gravity.[6]

These qualities convinced Frenkel' that the new four-dimensional notion of space-time enabled physics to escape "pseudophysical" constructs and to overcome the contradictions of the classical approach.

In *Elektrichestvo i materiia* (1925), published about the time he departed for German centers of theoretical physics, Frenkel' presented a popular contemporary description of the structure of matter, and the relationship between matter, mass, and "positive and negative" particles of energy (protons and electrons). This provided the way to explain the materialization of electricity and the electrification of ordinary matter. While at a loss to explain the nature of forces of molecular cohesion, he offered hypotheses on the electrical nature of electrolytes, gases, and solids, and set forth an electronic theory of metals based on the role of free electrons. In the last three chapters of his book, Frenkel' focused on atomic structure and recent developments in quantum mechanics.

But two aspects of his book which today seem innocuous drew Frenkel' into the sights of Marxian philosophers and physicists who had a reductionist, mechanical worldview. First, he relegated the ether to the "providence of shade," and second, he advanced a theory of point electrons where he proposed to consider the electrons not only physically but also geometrically indivisible; that is, to view them as material points, unextended force centers, and to look at mass as a primary quality, independent of charge. A similar "dynamic" view of the nature of elementary particles of matter was already long ago advanced by [two] very different philosophers: Leibniz and, especially, at the end of the eighteenth century, Boskovich.[7]

This latter point of view provoked hostile reaction for a number of reasons. First of all, it seemed to counter Lenin's (and Engels's) view that matter underwent fundamental change. Second, Frenkel' had openly referred to his "idealist" philosophical heritage of Leibniz and Boskovich. Third, his comment on force centers ran counter to several physicists' opinion of the impossibility of action at a distance without an intervening medium. In particular, as discussed below, V. F. Mitkevich harked back to "Faraday-Maxwell" force lines for mechanical explanations of electromagnetism.

V. K. Frederiks and A. A. Friedmann also wrote widely on relativity theory, in this case the general theory of relativity. Friedmann, at one time associated with the physico-mechanical department at Leningrad Polytechnic Institute and later with the Main Geophysical Observatory, is the better known of the two by virtue of his work "Über die Krümmung des Raumes," which first proposed equations describing a nonstatic universe.[8] He wrote little more on the subject owing to his general interest in mathematics and geophysics, as well as his premature death in 1925 from typhus of the stomach.[9] Frederiks, who had studied under Hilbert in Göttingen from 1914 to 1918, published in 1921 the first Russian description of the general theory of relativity, in which he noted the role of non-Euclidean geometry, time as a magnitude inseparable from spatial measurements, and the rejection of absolute space and time in the new approach to gravitation.[10]

In their first joint work, *Osnovy teorii otnositel'nosti,* Friedmann and Frederiks used tensor analysis to "set forth relativity theory sufficiently rigorously from a logical point of view." The book was published only after a two-year delay because of "special technical problems [e.g., paper shortages] . . . during which time many foreign books on tensor analysis appeared which are better than ours."[11] Frederiks and Friedmann viewed relativity theory as the beginning of the "axiomatization" of physics, which, as Joravsky

writes, involved an implicit challenge to the Marxist *Weltanschauung* when "they equated the progress of 'axiomatization' with
the growth of scepticism."[12]

The books by Frederiks, Friedmann, and Frenkel' were standard
fare among Soviet physicists. O. M. Todes recalled,

> When still in the first year of university, I decided to study the
> theory of relativity independently, and courageously laid my hands
> on the book of Frederiks and Friedmann. . . . Alas, . . . after a
> beautiful and melancholy-lyrical forward, it turned out that it con-
> tained only the mathematical apparatus of relativity theory—tensors
> and operations with them. The abstractness and difficulty of this
> mathematical apparatus, especially for a first-year student who was
> familiar only with the foundations of differential calculus, could
> have repulsed any desire for work in theoretical physics for my en-
> tire life. However, while I still continued to make my way through
> the mathematical debris of tensor calculation, the small book of Ia. I.
> Frenkel', *Teoriia otnositel'nosti'*, the 1923 edition, fell into my hands. I
> read it without pausing for breath, and the reading of this book fore-
> ordained my interest in and attraction to theoretical physics for all
> the following years. The physical meaning of the theory is regarded
> as of paramount importance in this book. The explanation of the
> revolution in basic physical concepts and properties of space, time,
> and matter by relativity theory is brought to the forefront. Nowhere
> does the author steer clear of the mathematical apparatus, and the
> physical meaning of the results is clarified beautifully with the help
> of original and far-carrying geometrical and physical analogies.[13]

In much the same way as they had greeted relativity theory,
Soviet physicists on the whole rapidly embraced quantum mechanics and debated its philosophical underpinnings, epistemological
implications, and physical content. They were actively engaged in
the development of quantum mechanics from the mid-1920s onward, either while abroad (Fock and Frenkel' in Germany; Landau
in Denmark; Gamov in England and Denmark), or in any one of a
number of seminars in Soviet institutes. Soon after the first articles
by Werner Heisenberg, Erwin Schrödinger, and Niels Bohr on
quantum mechanics appeared in *Nature, Die Naturwissenschaften,*
and *Zeitschrift für Physik,* Soviet physicists came out with a series of
papers in their own journals. *UFN* began to publish articles on
quantum mechanics in 1926, including translations by European
scholars.[14] *ZhRFKhO* published articles by V. A. Fock, L. D. Landau, T. P. Kravtsov, and P. S. Ehrenfest beginning in 1927. By the
time of the sixth congress of the Russian Association of Physicists

in 1928, Soviet theoreticians had turned their attention wholly to quantum mechanics and the experimental data which had accumulated in its support, and debated the interpretations of the Copenhagen and Göttingen positions with Western scientists.

The activities of three physicists associated with LFTI, Ia. I. Frenkel', V. A. Fock, and V. R. Bursian, in particular, stimulated the development of quantum mechanics in Soviet Russia. Bursian's seminar at the institute seems to have served as an informal "quantum commission," much like the atomic and molecular commissions before it. Bursian's career followed the pattern of many other Leningrad theoreticians, from his study abroad in Germany at Tübingen in 1906 and his return to Germany in the mid-1920s to study with Frenkel', Krutkov, and others, to his arrest and incarceration during the purges in October 1936.

Bursian was born into the family of a prominent Petersburg doctor in 1887; his father died during the blockade of Leningrad in 1942. Bursian entered the physico-mathematics department of Petrograd University in 1904, graduated in 1910, and worked in the department until 1918 when he was invited to join the physico-mechanical department, or *fizmekhfak*, at the polytechnical institute as head of its theoretical division. Bursian was connected with Ioffe's prerevolutionary seminar, and continued his association with Ioffe as the first academic secretary of LFTI and head of its theoretical department until 1932, when he became deputy director of the newly created scientific research physics institute of the university. He was chairman of the physics section of the Physico-Chemical Society and a member of the RAF and of the atomic commission at the State Optical Institute. In 1936 Bursian was arrested with V. K. Frederiks, Iu. A. Krutkov, and P. I. Lukirskii on the basis of false testimony, and died in 1945 in the camps. He was rehabilitated in 1956. In all, Bursian published over forty-five articles and two books, in regions of theoretical physics ranging from purely mathematical research to the nature of chemical forces of molecular bonding, to quantum mechanics.[15]

Bursian's quantum mechanics seminar involved the major Soviet figures in the history of quantum mechanics—Fock, Frederiks, Frenkel', as well as Krutkov, Khvol'son, and Kravets, who had come to Leningrad from Kharkov with D. A. Rozhanskii.

The seminar considered the work of de Broglie, Schrödinger, and Born, based on formal presentations by the participants and a general discussion of literature from *Zeitschrift für Physik, Proceedings of the Royal Society,* and other journals.[16] The first major Soviet work on quantum mechanics, *Osnovaniia novoi kvantovoi mekhaniki* (1927), is a result of papers prepared for Bursian's seminar. With the exception of Fock, the eight authors of the volume were representatives of the Ioffe school, and Fock, too, worked at LFTI from 1924 through 1936.

As Ioffe noted in his introductory essay to the book, Born, Heisenberg, de Broglie, Schrödinger, and Einstein had gone a long way toward resolving problems tied to the wave-particle duality of light, and had established a rigorous and formal, if not physical, mathematical apparatus to explain atomic phenomena. In view of the significance of quantum mechanics for physics and the general *Weltanschauung,* Ioffe urged that "the achievements be accessible to broad circles of both physicists and all others who are interested in it."[17] The introduction focused on the novelty of quantum mechanics, especially in its matrix form; on the correspondence principle; on the similarities, differences, and difficulties of the approaches of Schrödinger, Hamilton, Klein, and de Broglie; and on the problem of the development of relativistic quantum mechanics. This was followed by rather lengthy and detailed analyses by P. S. Tartakovskii, a theoretician who left Leningrad for Tomsk and the Siberian Physico-Technical Institute in the early 1930s, and by Bursian, Frederiks, Fock, and others.

Courses on quantum mechanics were soon introduced into university curricula. Frenkel' offered his first course on quantum mechanics in the late fall of 1926, when he had just returned from a sabbatical in Germany where he had met with the leading European theoreticians; his lectures were attended by students and academicians alike.[18] Fock finished Petrograd University in 1922, and taught theoretical physics primarily at Leningrad University, but also at the Ioffe Institute, the State Optical Institute (1928–1941), the Physics Institute of the Academy of Sciences (1934–1941, 1944–1953), and the Institute of Physical Problems (1954–1964). Fock, too, was arrested in 1937, but surprisingly was released after a few days. He was a major defender of the commensurability of quantum mechanics and dialectical materialism, especially in the late 1940s and 1950s, and he was a leader of the Copenhagen interpretation.[19]

Fock conducted the first systematic course on quantum mechanics in the physico-mechanical department of the polytechnical institute in the fall of 1928, also after returning from European *komandirovka*. Fock's courses set forth the basic ideas of quantum mechanics axiomatically. Not surprisingly, even the leading scientists attended these lectures to learn about recent developments. As O. M. Todes recalled,

> Providing an example to all [scientists] and students, Ioffe himself went to the entire first semester of lectures, and was accompanied by an entire retinue of [institute scientists]. Therefore, in opposition to the norm, Fock began the lecture only after the arrival of the "highest listeners."[20]

In the spring semester of 1929, Frenkel' offered a course which began with an historical analysis of the development of wave mechanics.[21] He was well versed in this story, and aware of the unresolved problems which faced quantum theoreticians, including how to take into account the forces of interaction among a changing number of particles and the need for a more complete picture of the analogy between energy waves and matter waves.[22]

The head of the theoretical department at Moscow State University, Leonid Isaakovich Mandel'shtam, began to lecture on quantum mechanics about this time, but under the watchful eye of Boris Gessen, the "Red" physicist at the university. Mandel'shtam (1879–1944) came from a middle-class Jewish family. After being expelled from Novorossiskii University in 1899 for participation in student demonstrations, he went to Strasbourg where he studied with Carl Ferdinand Braun, 1909 winner of the Nobel Prize in physics. P. N. Lebedev, B. B. Golitsyn, P. P. Lazarev, and other Russian physicists had studied there in their time. He stayed in Strasbourg, ultimately as a "titular" professor, until war broke out in 1914. He then returned to the Russian Empire, to work in the Kozitskii factory, Tiflis University, and in 1918 as an ordinary professor at Odessa University, where he helped organize Odessa Polytechnical Institute as head of its physics department. He invited N. D. Papaleksi and the future Nobel laureate I. E. Tamm to work with him. In 1925 he joined Moscow State University and became the central figure of the physics department, training in his lifetime such scholars as A. A. Andronov, M. A. Leontovich, A. A. Witt, and S. M. Rytov. He worked jointly at the Physics Institute of the Academy of Sciences after its founding in 1934.

A specialist in the physics of the radio, the propagation of light, and other phenomena of oscillation, Mandel'shtam possessed deep theoretical insights, as revealed by his five-volume collected works which included many of his lectures and were published posthumously. His lectures on statistical physics (1927–28), wave mechanics (1929–30, 1938–39), and relativity theory (1928–29, 1933–34) provoked a storm of disapproval from Stalinist physicists and philosophers in the late 1940s, leading to the censure of his work by a special commission of the Physics Institute of the Academy of Sciences in 1951. The lectures were republished in 1972, but without mention of why the fourth and fifth volumes of the previous edition had become a rarity.[23]

In an essay written in 1938, Fock outlined three stages in the development of quantum mechanics against which he judged the history and success of the Soviet program.[24] In the first stage (1924–1926), during the creation of the mathematical apparatus, Bohr, Heisenberg, de Broglie, and Schrödinger provided the foundations for the field. In phase two (1927–1928), physicists clarified its physical meaning and devised quantum mechanical systems with constant numbers of particles. Heisenberg, Pauli, Dirac, Bohr, Weyl, and Frenkel' applied quantum explanations to the behavior of electrons, molecules, and metals. In the third stage (1929–1932), scientists extrapolated quantum mechanics to systems with a changing number of particles, with Heisenberg and Pauli constructing quantum electrodynamics on the basis of the Dirac theory of radiation, upon which Fock, Fermi, and Podolsky expanded. But by 1932, with the discovery of the neutron by Chadwick and Joliot-Curie's discovery of artificial radioactivity, experimental physics had outstripped theoretical. This had slowed the development of the field because of problems in explaining quantum conditions in the nucleus.

Fock concluded that there were four areas where Soviet physicists made major, albeit later, contributions: general quantum mechanics; quantum theory of atomic structure; quantum theory of the solid state; and quantum theory of the nucleus and positron.

Landau, Ivanenko, Tamm, and Fock himself were responsible for work in the first area.[25]

Landau, recipient of the Nobel Prize in 1962 for his "pioneering work in the theory of condensed media," was one of the most beloved theoreticians in the Soviet Union. Sharp, caustic, even sarcastic, and witty, but not impolite, he attracted the most promising young scholars. He was the author of the famous "theoretical minimum," a one-on-one rite of passage with Landau to demonstrate breadth of theoretical knowledge, and co-author with E. M. Lifshits of the world-renowned seven-volume course of theoretical physics. In twenty-seven years only forty-three individuals passed the *"teorminimum"*—including seven future Academicians and another sixteen doctors of science, the Soviet equivalent of full professor.

Landau was born in Baku in 1908, attended Baku University, and then in 1924 moved to Leningrad University, the center at that time of Soviet theoretical physics. He graduated in 1927 and joined LFTI as a graduate student. He spent eighteen months in Cambridge, Copenhagen, Berlin, Göttingen, and Zurich, before moving to Kharkov in 1932 as head of the theoretical department of the Ukrainian Physico-Technical Institute. He then moved in 1937 to Kapitsa's Institute of Physical Problems where, after arrest and a year spent in jail—not so much for his Trotskiite sympathies as for the arbitrariness of the purges—he became head of the theoretical department until his death in 1968. His work covered such areas as phase transformations, superfluidity, and nuclear and solid-state physics.[26]

Regarding the quantum theory of atomic structure, Landau first solved the difficult problem of adiabatic inelastic collisions of atoms and molecules, while Fock offered a solution to the quantum problem of multiple bodies. Sir Rudolph Peierls, a contemporary of Landau's offered this praise for his work:

> In the early 1930s, when the quantum theory of solids was very much a proving ground for the new quantum mechanics, Landau was very interested in this field, to which he brought his characteristic depth of insight, and his insistence on fundamental understanding. His paper on the diamagnetism of free electrons is well known. He explained that the domain structure of ferromagnets was not entirely accidental, but that the configuration of lowest energy took, because of the magnetic field energy, the form of alternating layers

of opposite magnetization, which thickness was calculable but dependent on the shape of the body. . . . In addition to these papers, his influence was exerted through many discussions with colleagues, which often helped them to clarify and extend their views.[27]

Landau's work in the diamagnetism of free electrons was later developed abroad by Peierls and Edward Teller. Landau and Frenkel' were also active in the development of the quantum theory of superconductivity.[28]

Considering the major research areas of LFTI, it is not surprising that physicists focused on the quantum theory of the solid state. Frenkel' was the leading figure here, setting forth important research on the question of the theory of electroconductivity and the photoeffect in metals. In particular, Frenkel' applied "the equations of Fermi to the calculation of the surface electrical field of metals" and conducted research on the magnetic properties of metals at absolute zero. In the area of electronic theory of the solid state (metals, dielectrics, theory of crystals, and semiconductors), Frenkel' wrote more than seventy works.[29] Other physicists at the institute were active in the development of theoretical physics, especially concerning quantum mechanics and its application in the theory of gravity waves.[30]

Thus when the debates among Marxists over the epistemological basis of the new physics began to engage physicists, the scientists were prepared. They conducted seminars, published a substantial number of monographs and articles, and offered university courses to introduce their colleagues to new regions of theoretical physics. When discussing the philosophical implications of the new physics, they paid little attention to ongoing debates among their Marxist colleagues. They considered the changes in *Weltanschauung* required by relativity theory and quantum mechanics little cause for crisis, although some were hesitant to abandon classical conceptions of matter, energy, space, and time.

MARXIST PHILOSOPHERS AND
THEORETICAL PHYSICS

Beginning in the late 1920s, Soviet physicists began to encounter interference from Marxist philosophers. During the NEP, as in the

1904–1909 period of Party history, there was a tendency to keep philosophical debates separate from Party politics. In this environment most universities and higher-educational and research institutes maintained philosophical independence and paid little attention to Marxist scholars who deliberated over dialectical materialism at the Communist Academy of Sciences, *rabfaky*, and other special Marxist higher-educational and research institutes. This freedom extended to the then recently recovered national publishing industry, which issued a wide range of historical and philosophical works later to be considered "idealist." In fact, the NEP helped to bring about the complete recovery of scientific publication, which had been disrupted by the Revolution. Over twenty different tracts appeared on relativity (and later a larger number on quantum mechanics) including works by so-called idealist and Machist philosophers (Pavel Florenskii, A. A. Bogdanov, and his followers), and such scientists as Frenkel', Frederiks, and Friedmann, S. I. Vavilov, Tamm, Ioffe, and O. Iu. Smidt.[31]

The close examination of the work of natural scientists and philosophers by Marxist scholars began in the mid-1920s, perhaps in part because of the publication of one too many of these texts, and degenerated into a full-blown, bitter struggle among two groups of Marxists, the Mechanists and the Deborinites.[32] The Mechanists and Deborinites disagreed fundamentally over the proper interpretation of the works of Marx, Engels, and Lenin. There is a tension between the Hegelian or dialectical aspects and the naive realist or even vulgar materialist tendencies of Marxist philosophy of science. The former trend is well illustrated in Lenin's *Philosophical Notebooks* and Engels's *Dialectics of Nature*, the latter in Lenin's *Materialism and Empiriocriticism* and to a certain degree in Engels's *Anti-Dühring*. Articles and letters by Marx, Engels, Lenin, Plekhanov, and others on issues other than the philosophy of science also compelled their followers to engage in confused and detailed philosophical exegesis. Without going into analysis of the "fundamental texts" which served as reference points, if not bibles, for the Deborinites and the Mechanists, suffice it so say that the Mechanists were enamored of the vulgar materialist aspects of Marxism, whereas the Deborinites found the dialectical aspects much more compelling.

THE MECHANISTS

The Mechanists' stronghold was the Timiriazev State Scientific Research Institute (Gosudarstvennyi timiriazevskii nauchno-issle-dovatel'skii institut), from which they published five volumes of collected articles, "studies and popularization of the natural scientific bases of dialectical materialism," between 1926 and 1929. After their official defeat at the second all-union conference of Marxist-Leninist institutions in April 1929, they found a home—and solace—at the Physics Institute of Moscow State University. The Mechanists sought to bring scientists and philosophers together under one methodological roof in the Timiriazev Institute, to convince both groups that all processes in the external world—and not merely physical phenomena but perhaps even social dynamics—could be explained in terms of laws of classical mechanics.

The Mechanists desired to fight idealism, which they believed was hidden in the dialecticians' approach and lurked behind the bourgeois specialists' every word. As the editors of the institute's journal, *Dialektika v prirode* (*Dialectics in Nature*), explained,

> The study of the objective dialectic of the material world on the basis of the achievements of natural science should shed light on the internal connection of a whole series of regions and facts that have been independent up until now, and give it a methodological compass, directing the evidence by which to transform the common scientist from a narrow empiricist who is lost in the forest of a huge quantity of facts, into a conscious organizer of science and a creator of broad and generalized scientific constructions.[33]

Both scientists and philosophers lacked the proper tools and worldview. Scientists had too long attempted to philosophize without the necessary schooling; philosophers had long lacked the "concrete practical soil of science" for the development of materialist philosophy and had lapsed into idealism. The Mechanists had the solution to these problems: "The elaboration of the truly scientific method of the natural sciences, the elaboration of the materialist dialectic is only possible in the case when theoretical thought goes hand in hand with practical verification with the help of positive science. Our institute attempts to realize such a synthesis of theory and practice." The synthesis would come about through the study of the works of the leading Marxist scholars—Engels and Lenin in particular—on mathematics, mechanics, electrification, chemistry, and biology, and an examination of the organic and inorganic worlds to reveal their interconnections. This would demonstrate that mechanical processes inherent in both worlds reduce to the notion of matter in motion and to the concept that all qualitative differences are differences of quantity.[34]

What were the Mechanists' criticisms of the new physics, and in particular of the representatives of the Ioffe school? According to the philosopher A. Var'iash, three major trends which characterized twentieth-century physics required careful philosophical analysis: the rise of the statistical approach to atomic and molecular processes; the concept of relativity and the reworking of such basic physical categories as space, time, motion, and mass; and the apparent discrete nature of radiation. Marxism would strive to reform classical mechanics to reconcile it with concepts of continuity and discontinuity in contemporary physics.[35] However, it was the notion of relativity that caused the Mechanists the greatest difficulty, on both physical and philosophical grounds.

The Mechanists opposed the rejection of Newtonian mechanics required by relativity theory to explain the behavior of elementary particles and speeds approaching the velocity of light. This objection went so far as to require the maintenance of a mechanical ether to explain the transmission of electromagnetic energy through space. The Mechanists relied on Dayton Miller's experiments to support their position,[36] although mainstream Soviet physicists had repudiated them.[37] The Mechanists also rejected relativity theory on philosophical grounds, asserting that Einstein's theory led to philosophical relativism. They claimed that the trend toward divorcing science from philosophy prevalent among the Deborinites had gained strength because of this philosophical relativism.

More crucial was the fact that relativity theory was the bugaboo of A. K. Timiriazev. When it became clear that he would not succeed in narrow Marxist intellectual circles, let alone among scientists, in proving the "idealism" and philosophical relativism of Einstein's work, he attempted to widen the circle of discussion to include Party philosophers.

A. K. Timiriazev (1880–1955), son of the famous agronomist and biologist K. A. Timiriazev, a close associate of Maxim Gorky, was a professor of physics at Moscow State University and a Party member from 1921. Not an outstanding scientist, Timiriazev was distinguished by his unceasing devotion to classical physics. The author of *Vvedenie v teoreticheskuiu fiziku, Kineticheskaia teoriia materii,* and over twenty articles in *Pod znamenem marksizma* alone, Timiriazev maintained strict allegiance to classical Newtonian conceptions of space, time, matter, and motion in everything he wrote.[38]

Timiriazev had few friends since he attacked everyone who differed from him in the slightest, including such "allies" as A. A. Maksimov and Vladimir L'vov, two of the staunchest, most self-serving defenders of Stalinist philosophical orthodoxy. Those who would be shot or sent to the camps in the purges of physics— Gessen, Rumer, Bursian—quickly earned Timiriazev's appellation "enemy of the people." Indeed, perhaps only Stalin took a liking to Timiriazev. Who else would have searched the great leader's writings back to 1906 to find some evidence that Stalin had theoretical notions with bearing on modern physics?[39] Timiriazev ultimately turned the physics department at the university into a kind of armed ideological camp, forcing graduate students and faculty alike into his circle of militant mechanist physicists, heaping vituperative "kasha" on Academicians Mandel'shtam, Ioffe, Vavilov, and Tamm, and future Academicians M. A. Markov, M. A. Leontovich, and L. D. Landau. To this day, he is referred to without reverence as "the son of the monument [*syn pamiatnika*]," as if his father were his only valid claim to fame.

In the mid-1920s Timiriazev's hostility to Einstein took on such proportions that he sarcastically denied in a public meeting having suggested that Einstein be shot.[40] He engaged A. A. Bodganov in rancorous debate within the walls of the Communist Academy of Sciences,[41] and later took on physicists I. E. Tamm, S. I. Vavilov, Ia. I. Frenkel', and A. F. Ioffe, an effort which culminated in his *Vvedenie v teoreticheskuiu fiziku* (1933), the most complete defense of his positions, which was published after a two-year delay with a preface from the publisher asserting that Timiriazev had not recognized his mechanistic views as mistaken. This generated further criticism from all camps. In the 1930s, Timiriazev was joined by the specialist in electrical technology, V. F. Mitkevich, in debates with the Leningrad physicists to save his sacred ether. Timiriazev argued that the difference between mechanical and electromagnetic processes was merely the participation of the ether in the motion of charged particles in the latter.

The Mechanists were troubled, finally, by the increasing role of mathematics and statistical laws in the new physics. There were three major issues here: the "mathematicization" of matter; the apparent rejection of causality; and the argument of some physicists that the second law of thermodynamics led inevitably to the "thermal death" of the universe.[42] Only if the second law of thermodynamics were interpreted mechanically would it be possible to avoid the increasing decay and disorder of the universe predicted by some physicists who studied entropy and its implications. Somehow, the Mechanists saw the possibility of overcoming these problems and "bridging the abyss" between theoretical, mathematical, and statistical physics and mechanics through the application of the dialectic. As Timiriazev wrote, "the equations of mechanics are applicable to all phenomena of physics" and could be based on Lagrangean functions of kinetic and potential energy. "Now the time has come to reduce all changes and differences of quality to quantity, to mechanical change of place, to conclude that matter consists of very small identical particles, and that all qualitative changes of chemical elements are called forth by quantitative differences in number and spatial group of these particles."[43]

The Mechanists found fault in particular with the three Leningrad "Fs"—Frenkel', Friedmann, and Frederiks—and later extended their criticism to Gessen, Vavilov, Tamm, and Ioffe. They attacked Frenkel' for "dyed-in-the-wool clericalism [*popovshchina*]" and subservience to a "Leibnizian-Kantian-Pearsonian" view of nature. They accused him of propagandizing the idealist monadology of Leibniz and Boskovich in an introduction to James Rice's *Principle of Relativity*, published in Russia in 1928, where Frenkel' interpreted electrons as unextended, finite force centers, that is "without extension," á la Leibniz.[44] The Mechanists linked Friedmann and Frederiks, the latter of whom disappeared in the gulag, to "aesthetic pessimism" for their comments on *Osnovy teorii otnositel'nosti* on the axiomatization of physics.[45] Worse still, relativity theory had led to the "resurrection of belief [in god]" and the rehabilitation of philosophical positions that had preceded Newton.[46] Then, they castigated Gessen for "mistaken and antidialectical" notions in his rejection of absolute space and time for relativistic four-dimensional space-time, an action which was unthinkable in view of some two hundred years of accumulated scientific effort. Why, the Mechanists asked, stop at four dimensions, why not have five, or even six?[47]

In the environment of tension between Timiriazev and a majority of other Communist philosophers and physicists, the program for the history and philosophy of science at the university became a home for intrigue and party intervention. The department was intended to work with all *aspirants* on Marxist methodology, develop research on the history and philosophy of science, and work with other institutes of the university in developing Marxist approaches. But work did not go well, since the members of the department were few in number and at loggerheads over Timiriazev's rabid mechanism. Furthermore, *aspirants* at the Timiriazev Institute and at Perm and Kazan universities were associated with the department and fell under Timiriazev's direct influence, and leading to tension between them and other students.[48]

According to a Party document sent to the Central Committee at the end of October or beginning of November 1929, the "recent worsening of the political situation" in the country, tied to radical reorganization of higher schools during the cultural revolution, demanded consolidation of forces around loyal party and non-party scientific workers, *aspirants*, and students according to the correctness of their political views. The Party commission noted Timiriazev's contribution to the university between 1922 and 1929, but drew attention to "a series of organizational and political mistakes" he had made. He was devoted, it turned out, not to any "defined political position" but to his hatred of relativity. He often carried out his line against the Party organizations of the university, considering their decisions incorrect, and maintained his "fractional position with respect to philosophy," wrongly accusing such other *fizfak* members as Maksimov and Gessen of being "Machist." The Party committee concluded that Timiriazev's sole motive in siding with the "Black Hundred" Kasterin on many issues was that the latter "rejected relativity theory."[49]

It should be clear that the Mechanists were trapped by their strict allegiance to Newtonian laws of mechanics and a reductionist, simplistic interpretation of Engels and Lenin. They lacked an understanding of both the physical and the philosophical content of the new physics. This led to their criticism of relativity theory and quantum mechanics. They were also disturbed by what they perceived to be discrimination, since the Mechanists' articles were often published with great delay or not at all,[50] while the Deborinites, who demonstrated more complete understanding of the new physics, and such "idealists" as Frenkel' had no trouble in getting their works published.

THE DEBORINITES

The Deborinites (or Dialecticians), on the whole, were well prepared to accept the recent advances in theoretical physics and eager to explore how dialectical materialism and the new physics complemented one another. They believed that the epistemological questions which had arisen in response to the major developments in physics in the first third of the twentieth century—the rise of statistical physics, the wave-particle nature of matter-energy, and indeterminacy at the microlevel—demonstrated their compatibility. The dialecticians were, in simple fact, allied to the physicists, if not in public declaration then at least in a shared belief that quantum mechanics and relativity theory accurately described physical phenomena. At the second congress of Marxist scientific research institutes in 1929, they succeeded in repudiating officially the Mechanists' philosophical position.

This victory was short-lived, however. In the same way that Stalin and his supporters had aligned themselves with the Right Opposition of Bukharin and Rykov against the Left Opposition of Preobrazhenskii and after the victory of the former turned against them, so after the censure of the Mechanists, Stalinist ideologues quickly attacked the Dialecticians and physicists alike with accusations of "menshevizing idealism" and failure to acknowledge the role of class struggle in physics—now with the help of the uncontrite Mechanists.

The Deborinites had broader institutional support, a larger number of followers, and published more than the Mechanists, facts which in and of themselves antagonized Timiriazev's allies. Taken with the brilliant successes of physics in relativity and quantum theory, the ever-growing unanimous scientific voice on the side of the "new physics," and the application of physics in technology and industry, these factors assured the Deborinites' initial victory. It helped that only a small number of opponents were capable of rendering sophisticated arguments on the philosophy of quantum mechanics and relativity theory. Fewer still among these were physicists, while Egorshin, Gessen, and Maksimov all had physics training.

There were three institutional strongholds of the Deborinites: the Communist Academy of Sciences (until 1924 the Socialist Academy); the Sverdlov Communist University—based on an almost proletkultist idea for a network of workers' universities which never materialized—in whose department of philosophy Gessen organized a section on the natural sciences in which Tamm, Timiriazev, and Egorshin participated; and the Institute of Red Professoriat (IKP) where, from 1925, Maksimov, Egorshin, and Gessen offered courses in physics—the structure of matter, energy, relativity theory—to complement others in biology, medicine, and chemistry.[51]

The Communist Academy, after a period of "primitive book accumulation" in 1922, extended its work from education into research on philosophy, Marxist thought, and the natural sciences. But the philosophical efforts of its department of natural and exact sciences, which was organized in 1925, lagged behind pedagogical programs. The natural and exact sciences department had three subsections: The psychoneurological, the biological, and the physico-mathematical, in which the mathematician V. F. Kagan, the physicist N. P. Kasterin, a strong opponent of relativity theory, Gessen, Maksimov, and Egorshin worked. Most Marxist scholars were concentrated in the social sciences (a prerevolutionary tradition); only recently had natural scientists with a Marxist orientation joined its staff. Furthermore, the academy failed to attract sufficient numbers of students in the physical sciences.[52]

A. A. Maksimov had been a longtime Party member in the 1930s when he turned against Deborinite colleagues like Gessen and such physicists as Ioffe, Vavilov, Frenkel', and Tamm. He grew more and more Stalinist in the 1940s and 1950s, serving as one of the leading reactionaries in the struggle against idealism in physics from his new base of power, the Institute of Philosophy of the Academy of Sciences, which had absorbed the departments of the

Communist Academy and IKP with which he had been associated. Even after Stalin's death—and the death knell of ideological vigilance in physics—he tried to use his position as an Academician and scholar at the Institute of Philosophy to carry on the struggle. But such physicists as D. V. Skobel'tsyn, by then director of the Physics Institute of the Academy of Sciences, refused to help Maksimov publish anything in *Voprosy filosofii, Novyi mir,* or any other journal.

In the 1920s, indeed until the condemnation of Deborinite "menshevizing idealism," Maksimov had sought a middle ground between the Deborinites and Timiriazev, drawing the latter's wrath until late in the 1930s. When Maksimov applied to be a student at IKP in August of 1924, he was already a scientific worker and teacher in *fizmatfak* at Moscow University. He desired both facility in a "narrow branch of natural science" and training to assume a leadership position in the educational and political life of *vuzy.* He believed that precisely at a place like Moscow University, in terms of both worldview and philosophy, "questions of ideological struggle and revolution in nature science . . . play a great role." Training in IKP would help him participate in this struggle and revolution.

Maksimov was born in 1891. His father had been a rural priest, his mother a Kazak laborer. He entered Kazan University in 1910 but in connection with demonstrations against Stolypin was thrown out in 1911. In 1913 he organized a legal student circle and then became involved in underground activities. But he did not have much political *savoir-faire* and "fell under the influence" of right-wing socialists and anarchists. He admitted, "I recognize my petit bourgeois origins." Maksimov played no role in the February or October revolutions, but, showing chameleonlike characteristics which would serve him well in the 1930s, had already come "close to the Left SR and Bolshevik positions. As conditions for the intelligentsia worsened . . . , I then went over to work in the cultural-education department of the Kazan Soviet of Workers' and Soldiers' Deputies." Working in public education from that time, he entered the Party in 1918, and spent the next two years fighting in the Civil War forty or fifty miles from the front against Kolchak, in the Caucasus, Kuban, and near Kazan. Eventually he was evacuated to Moscow and in October 1920 demobilized. He worked as a deputy director in a Narkompros *rabfak*, then in Moscow University as a scientist and teacher, and then in Sverdlov Communist University. During this time he began to publish in *Pod znamenem marksizma*.

His career at IKP was quite successful. He worked at the physics institute at the university, joined the Russian Physico-Chemical Society, and as a docent at the university taught a course entitled "Introduction to the History and Philosophy of Natural Sciences." His own work consisted of a book on the "Naturphilosophie of Hegel in light of Marxist theory and contemporary science," written under A. M. Deborin. He was then appointed director of the division of natural sciences of IKP. When sent to Berlin on *komandirovka* in 1928 he suggested that "the most logical candidate" to replace him in the meantime "in the opinion of students and my personal opinion is the Comrade Gessen." While in Germany he worked at the Deutsches Museum on the Naturphilosophie of Kant and agitprop work in the natural sciences. He returned to the USSR in 1929 to lead the division of the natural sciences of IKP and the Communist Academy through the years of cultural revolution, collectivization, and rapid industrialization.[53]

After the cultural revolution, the Party attempted to turn the Communist Academy into its methodological center, restructure its work, and "preserve research which has real leading methodological significance," establishing four departments (planning of science and technology; training of cadres; technological propaganda; and literature), and transferring remaining staff to other institutes.[54] Indeed, the institute was radicalized by the Great Break. Its party cell voted on June 1, 1932, to mobilize to convert the old scientific and technological intelligentsia to Soviet power and dialectical materialism, to call for *"partiinost'* [party-ness]" in science, and to condemn neutrality in philosophy, *"khvostizm* [tailism]," "ultraleft tendencies," conceit, mechanism, vulgar materialism, clericalism, idealism, and a host of other clearly dangerous tendencies.[55]

The Communist Academy had ambitious goals for the Second Five-Year Plan, including scientific expeditions, conferences on topics which ran the gamut from physics to botany to Taylorism, publication of a series of articles and books by Egorshin, Kol'man, Gessen, and Maksimov, and help to publishing houses, journals, and newspapers, on textbooks, articles, texts, even the *Bol'shaia sovetskaia entsiklopedia,* with a section on physics to follow developments closely through the organization of reviews and inspection.[56] But both the academy and the institute were subsumed within the institutes of Philosophy, History of Science and Technology, and other institutes of the Soviet Academy of Sciences by the end of 1934.

Maksimov founded IKP in 1921 with the help of the deputy commissar of Narkompros, the radical historian M. N. Pokrovskii. Two developments in 1922 gave impetus to the formation of a department of natural sciences in the institute: the establishment of the journal *Pod znamenem marksizma* (*PZM*) and the commencement of the systematic study of such scientists as I. P. Pavlov "from a Marxist point of view." The faculty of this department were drawn largely from the Physics Institute of Moscow University. IKP was independent in name only, serving as an adjunct to the Communist Academy, and gaining independence only in July 1931.

IKP planned in the mid-1920s to produce a definitive, multivolume work on the history of the natural sciences during the era of early capitalism with chapters by Maksimov, Iakovskaia, Egorshin, and Gessen (the latter on physics, mechanics, and electrodynamics).[57] But like many other plans, this joint endeavor yielded nothing; the work of the institute was loosely coordinated, and seldom reflected more than the sum of its disparate parts. During the height of cultural revolution, the institute organized a series of lectures to elucidate the history of science and technology, ranging from Greek atomism and the dialectics of Heraclitus to the Naturphilosophie of Kant and Hegel, and of course the philosophy of Marx and Engels.[58] These lectures were intended to complement the agitprop activities of the institute in factories and scientific research institutes.

As director of the natural science division of the institute, Maksimov had great plans for its growth during the cultural revolution. He organized two sectors, the "inorganic" and the biological, with the former including mathematics, physics, chemistry, mineralogy and geology, and technology, and the latter biology, agronomy, and medicine. As for all other organizations caught up in the frenzy of rapid industrialization, he foresaw the number of students

A. F. Ioffe in his study in the late 1930s.

A. F. Ioffe standing by the X-ray chamber of his laboratory in LFTI in the 1920s.

A. F. Ioffe in his laboratory in the 1930s. *From left to right*: Iu. P. Maslano-
vets, A. F. Ioffe, B. T. Kolomiets, and B. V. Kurchatov.

Kurchatov and Kobeko with their co-workers in the park of the Polytechnical Institute across the street from LFTI, in the early 1930s. *From left to
right*: G. Ia. Shchepkin, I. V. Kurchatov, E. V. Kuvshinskii, P. P. Kobeko,
B. I. Bernashevskii, and B. V. Kurchatov.

F. F. Vitman and N. N. Davidenkov, Davidenkov's laboratory, LFTI, in
the mid-late 1930s.

A. A. Chernyshev, Leningrad, at the end of the 1930s.

A. F. Ioffe in the mid-1930s.

Ia. I. Frenkel', Leningrad, 1926.

quadrupling from 127 in 1931 to 514 in 1933.[59] But these plans were not to any great extent fulfilled.

The Communist Academy and IKP both suffered from an inability to attract sufficient cadres in the natural sciences or individuals of peasant or working-class background. This told upon their agit-prop work in factories and higher-educational institutions, and among natural scientists. After cultural revolution and the seeming resolution of the Deborinite-Mechanist debate, the academy became more and more a center for Party intellectuals, never fulfilling the goals of its founders and movers, Maksimov, Deborin, Gessen, Egorshin, and others.

In addition to numerous articles and pamphlets in such journals as *PZM*, the Deborinites authored the three major treatises on relativity theory: V. P. Egorshin's *Estestvoznanie, filosofiia i marksizm* (Moscow, 1930); S. Iu. Semkovskii's *Teoriia otnositel'nosti i materializm* (Kharkov, 1924), and B. M. Gessen's *Osnovnye idei teorii otnositel'nosti* (Moscow, 1928).[60] These works share a common respect for relativity theory, the recognition that classical explanations regarding matter, electricity, and motion were inadequate in certain cases, a belief that relativity theory and dialectical materialism were compatible, and a conviction that physicists, not philosophers, should be the final arbiters in matters of physics.

V. P. Egorshin, born into a peasant family in 1898, entered the Russian Social Democratic Labor Party in 1915. After the revolution he served in various cultural, educational, and agitprop organizations in small villages. By 1921 he had made his way to Moscow University, where he served as a lecturer in the physico-mathematics department in evening Party schools through 1923. He was also active at the Communist University at this time as a member of the educational bureau of the collective of scientific workers, head of the circle of dialectical materialism, and a member of the workers' council. From September 1924 he taught physics at the Communist University.

He joined the Institute of Red Professoriat in the summer of 1924 upon the opening of its natural sciences department at the recommendation of the dean of Sverdlov Communist University. He had entered the Party cell of *fizmatfak* at Moscow State University (MGU) in February of the previous year and was given very good credentials as an active member. At IKP he studied analytical geometry, differential calculus, higher algebra, physics 1 and 2, differential geometry, theory of probability, calculus, and chemistry. He seems to have worked closely with Timiriazev at this time.

In March 1928, like many physicists, Egorshin and Gessen requested permission for *komandirovka* to Germany. "Our work in the area of theoretical physics calls forth the necessity of direct familiarization with scientific work in Germany," Egorshin wrote. He hoped to work in particular with von Mises on statistics and probability, and Born on quantum mechanics. It is not known if Egorshin and Gessen succeeded in getting to Germany. Egorshin graduated from IKP in 1929 and engaged in propaganda activities in the countryside while working on relativity theory, which he initially embraced; he wrote an article "on the reactionary tendencies of the Mechanists and Timiriazevists" and completed a handbook for the scientific worker on Marxist theory in the natural sciences to be published by the state publishing house.[61]

Egorshin had already begun to abandon his earlier Deborinite allies when *Estestvoznanie, filosofiia, i marksizm* (*Science, Philosophy and Marxism*) was published in 1930. He had written the book for Deborin's seminar at IKP in 1926, while still a Dialectician. In this volume, Egorshin did not deal directly with the scientific issues surrounding the reception of relativity theory. Rather, he sought to publicize the crisis in the natural sciences and the impetus it gave to mysticism and belief in the supernatural, especially in capitalist society; to stress the advantages of a proletarian, class-based view of science, and *partiinost'* in philosophy; and to attack the insufficiencies of mechanism and empiricism.

While mechanistic materialism had been progressive historically, and had emancipated matter from spirit, Egorshin argued, it now demanded revision. Developments in chemistry, electricity, and magnetism required a shift from mechanical explanations to a dialectical synthesis of old and new concepts: instead of force, acceleration; instead of heat, motion of molecules; instead of color, length of waves. This would avoid reductionism (*svodizm*) of psychiatry to reflexology, of biology and chemistry to physics, and of physics to matter in motion. Egorshin rightly accused a number of physicists of reductionism, including P. P. Lazarev and Timiriazev, but held Khvol'son and Frenkel' guilty as well. He argued that in *Stroenie materii* (1923) Frenkel' had reduced the world to motion and interaction of unchanging positive and negative particles. Timiriazev, needless to say, was the most guilty of *svodizm*, especially in his *Kineticheskaia teoriia materii* (1925), where he described all thermal processes phenomenologically, and thermodynamics as mere molecular mechanics.[62]

Semen Iulevich Semkovskii offered a more sophisticated treatment of the "crisis in the natural sciences" in *Teoriia otnositel'nosti i materializm* (*The Theory of Relativity and Materialism*). The leading Marxist philosopher in the Ukraine, Semkovskii tried to enlist natural scientists in the cause of the Revolution. An opponent of mechanism, he nonetheless refused to join the Deborinite faction until 1929 because of Deborin's alleged positivism. But his views on relativity theory turned out to be rather close to those of the Deborinites, who came to adopt them as their own. He believed that relativity theory in physics, Darwinism in biology, and Marxism in the study of man and society were "concrete forms" of dialectical materialism.[63] In February 1925, he read a paper in Kharkov before an audience of scientists in which he argued that relativity theory was the brilliant confirmation of dialectical materialism. This paper was based on his book, which had been published in Kharkov the previous year.

In *The Theory of Relativity and Materialism*, Semkovskii set out to counter attempts to use relativity theory in the spirit of idealism or relativism, or without understanding, as the Mechanists had done. After outlining the principles of the electrodynamics of moving bodies and relativity, its experimental verification (the calculation of the advance of the perihelion of mercury, and the bending of light), and the historical background to its development, Semkovskii turned to an examination of the philosophical and physical implications of such developments as the rejection of absolute space and time as the universal frame of reference. He contended that the geometric characteristics of matter in the four-dimensional space-time continuum of the general theory of relativity were not independent of, but conditioned by the matter in it, and hence not a Kantian synthetic a priori, but "materialistic." Since there was no matter without motion, how was it possible to have absolute motion? he asked. This had "no physical meaning," contrary to Timiriazev's protestations of "uneducated muddle."[64]

In a revised and expanded version of this work, *Dialekticheskii materializm i printsip otnositel'nosti* (*Dialectical Materialism and the Principle of Relativity*) (1926), Semkovskii rejected Timiriazev's contention that relativity theory was "Machist." In spite of the personal philosophical sympathies of Einstein and the historical influence of Mach's *Die Mechanik in ihrer Entwicklung historisch-kritisch dargestellt* on him, Semkovskii argued, "Relativity theory breaks decisively with 'Machism' for it builds not on 'elementary sensations' but on matter."[65]

Nonetheless, Semkovskii had difficulty in reconciling the establishment of an electromagnetic world picture with the need to abandon the concept of the ether. He hesitated to side with Frenkel', who had ridiculed persistent efforts to resurrect the mystical and undetectable medium, although Semkovskii castigated the "vulgar materialists [Mechanists]," in particular Tseitlin, for their reliance on a Cartesian mechanical ether, and Timiriazev for "binding his fate" to it. He did argue, however, that dialectical materialism could not resolve the issue: it did not require recognition or the rejection of the ether. This was a question for physicists to resolve, not philosophers.[66] In sum, in view of the convergence of relativity theory and dialectical materialism on many issues, the former's experimental verification, and the thoroughly materialist kernel of the theory, Semkovskii concluded that relativity theory was "brilliant confirmation" of dialectical materialism. Physicists would soon resolve any unanswered questions.[67]

Boris Gessen, active in the Communist Academy, IKP, and Moscow State University, and editor of *UFN*, is a major figure of the history of Soviet physics. He was influential in spreading the gospel of Marxist philosophy among physicists. Gessen is best known for a paper he delivered to the second International Congress of the History of Science in London in 1931, entitled "On the Social and Economic Roots of Newton's *Principia.*" This work stimulated the development of "externalism" in the history of science. Gessen's works on quantum theory and relativity theory, which represent the first major dialectical materialist defenses of the new physics, are far more important to this analysis.[68]

At the end of 1928 Gessen began to face attack from his former allies, Egorshin and Maksimov, at a general meeting of the Society of Militant Materialist Dialecticians for his supposed "menshevizing idealism" and "muddle."[69] Gessen, they asserted, had not addressed the concrete political questions of the time in an article on the ergodic theorem in *UFN*;[70] had not demonstrated the tie between theory and practice; was inclined to follow uncritically the lead of bourgeois scholars, especially regarding a series of articles by the statistician von Mises;[71] and in his work on relativity substituted a synthesis of space and time for matter.[72] This attack marked the beginning of repeated criticism in the press and Gessen's rapid loss of favor.

According to his autobiographical statement in the archives of IKP, Gessen spent the 1913–14 academic year at Edinburgh University in the mathematics division of the department of pure science, where he worked with professors Whitaker, Carse, and Bazkla. He then entered the economics department of Petrograd Polytechnical Institute, where he studied statistics under A. A. Chuirov. He also audited courses in calculus at the university; because he was a Jew, he was not admitted as a regular student. Until October 1917, Gessen was secretary of an organization of internationalists in Elisavetgrad, a position that no doubt opened him to attack as a Trotskiite in the 1930s; after the Bolshevik coup, he became secretary of the Soviet of Workers' and Peasants' Deputies.

Gessen continued his involvement in education in the first years of the Revolution. In 1919–1921 he was an instructor in politics and from 1921 through 1924 an instructor in political economy at the Communist University. At this time Gessen decided to enter IKP for further training in mathematics and physics, after a seven-year hiatus from mathematics and having "no systematic knowledge in physics." Specifically, he asked to study the structure of matter, the physics of heat and electricity, and theoreticial physics. It is likely that Gessen studied closely with L. I. Mandel'shtam. He "participated in a seminar of Akademician Mandel'shtam where he gave a paper on the law of big numbers in connection with the works of [von] Mises" which was later published in *UFN*.

Like Maksimov and Egorshin, Gessen asked for permission to travel abroad in 1928 to expand the work on statistical mechanics he had completed under Mandel'shtam, study with Richard von Mises at Berlin University, attend a special summer course for foreign physicists to be offered by Eddington and Plank in Berlin (attended by Ehrenfest and perhaps even Einstein), and return to his own work after serving as director of the natural science division during Maksimov's absence. Apparently the government approved his trip for four months, but there is no record of his Berlin activities.

Gessen published regularly in such journals as *Estestvoznanie i marksizma, UFN, PZM,* and even *Molodaia gvardiia,* while completing his monograph on relativity theory and an essay on quantum mechanics. He later headed up the physics section of the Communist Academy, was active in agit- and techprop work at Dinamo, Elektrozavod, and Moselektrik factories, where he organized brigades on modern physics, and lectured frequently at the Institute of Philosophy, the Institute of History of Science and Technology, and Moscow University, where he was dean of *fizfak* until his arrest in the fall of 1936.[73] He was removed from editorship of *UFN* in 1936, and disappeared in the purges in 1937. He also participated in the VAF conference in Odessa in 1931 where he read a speech entitled "The Materialist Dialectic and Contemporary Physics," which summarized his views on the compatibility of the new physics and dialectical materialism.[74]

Gessen's most complete treatment of relativity, *Osnovnye idei teorii otnositel'nosti,* was published in 1928 by the Moscow Worker publishing house for the reader with rudimentary knowledge of physics. It appears to be based largely on *Space, Time and Gravitation* by Arthur Eddington (first published in 1921) in terms of structure and presentation, especially concerning chapters on the Fitzgerald-Lorentz contraction and fields of force. This reliance on a British scholar perceived to be a representative of idealism in physics, the relatively small space devoted to dialectical materialism, and the failure to deal conclusively with such opponents as Timiriazev would leave Gessen open to criticism. However, his discussion of Newtonian absolute and relativistic space and time, simultaneity, the ether, and physical versus philosophical relativism placed the work firmly within the dialectical materialist tradition.

Gessen's main argument centered on a demonstration that the epistemological propositions implicit in relativity theory—both general and special—were compatible with dialectical materialism. These propositions were based on the recognition of four objective properties of the external world: the synthesis and unity of space and time; matter in motion; rejection of absolute, empty, motion-less space as the universal frame of reference; and relative motion of matter with respect to matter.[75]

Gessen's treatise on relativity theory begins with a discussion of Galilean relativity and then turns to the Fitzgerald-Lorentz contrac-tion as a possible way out of the negative result of the Michelson-Morley experiment. But it is a discussion of Newtonian absolute space and time which serves as the real focus of the first third of *Osnovnye idei teorii otnositel'nosti.* Gessen criticized Newtonian absolute space and time for having taken on independent objective reality. This was a metaphysical position, an abstraction, since space and time were inseparable from matter.[76] Relative space and time, on the other hand, were confirmed by dialectical materialism. All motion was matter in motion with respect to matter, not abso-lute space. Similarly, there was no absolute time but processes by which we can compare the course of one process with another. In sum, dialectical materialism recognized the synthesis or unity of space and time in matter in motion made possible under the theory of relativity, and rejected empty, unmoving Newtonian absolute space.[77]

While rejecting Newtonian absolute space and time as a meta-physical objectivization of an abstract reality, Gessen nonethe-less introduced an "ether" with rather striking properties to serve as an absolute frame of reference. Since electromagnetic phenom-ena are dispersed over time, Gessen concluded that there must be a medium to carry them. "'Empty space'" he wrote, "means space that is filled with ether. We do not know any matter more elementary than the ether."[78] To reject the ether would have been pure phenomenalism or idealism, for "to speak about waves and oscillations which occur without a material carrier [in a vacuum] means to speak about motion without matter. Motion without that which moves is an empty abstraction of motion, like empty space. Real motion is always connected with matter."[79]

How did Gessen's ether differ from that of the mechanistic physicists? The ether of relativity did not consist of particles or have molecular structure, and therefore the mechanistic concep-tion of motion meaning "mechanical change of place" had no meaning. Furthermore, it was impossible to observe the motion of a body with respect to the relativistic ether.[80] It is unclear how Gessen came to these conclusions about an ether which apparently it was impossible to sense.

While Gessen's work demonstrates a fairly complete mastery of some of the implications of relativity, he failed to see that its acceptance should have led to the rejection of the ether. Acknowledging that the question of the properties of the ether remained open, however, Gessen argued that the resolution of this issue was a matter for physicists and not philosophers to decide: "It would be improper and undialectical to demand mechanical explanations beforehand."[81] V. P. Egorshin, on the other hand, had welcomed the ether as "matter of a special kind, unlike any other," which permitted the transformation of electrons, and by implication protons, into ether and back in the process of the development of matter![82]

Another major focus of *Osnovnye idei* concerned the epistemological issues surrounding the relationship between subject and object. Such philosophers and physicists as Maksimov and Timiriazev had interpreted relativity to mean that all knowledge was relative and dependent upon the condition of the subject;[83] this would violate one of the primary postulates of materialism. Einstein's theory did assert relativity of measurement concerning both the system in question and the condition of the observer. Was it possible to arrive at absolute cognition of nature? Or was all knowledge relative? Gessen's analysis was intended to provide a way out of this difficulty. He agreed that while perception depends on the condition of subject and object, the form of existence of the object in no way depended upon the subject and its "facilities," that is, for cognition a subject is necessary; no kind of subject is necessary for the existence of the object.[84] Dialectical materialism offered a way out of subjective idealism through the mutual interaction of the subject and object. While there was a moment of relativity and subjectivism in cognition,

> the subject and the object acquire true reality, vital reality only in the process of mutual interaction. The object does not stand in opposition to the subject as some kind of indifferent *thing-in-itself*. No, it is exposed to the subject through mutual interaction. It is cognizable, and this cognition is an eternal process of mutual interaction.[85]

As Gessen wrote, "through this interaction we come to the objective through the subjective, through relative knowledge we approach absolute knowledge."[86]

Like other Marxist philosophers, Gessen accepted the view that the history of cognition is not in the least relativistic but an asymptotic approach to the absolute truth. For him, relativity theory, as the substitution of relativistic four-dimensional space-time for the Newtonian metaphysical and abstract conception of absolute space and time, was a stage in the cognition of the absolute. Gessen rejected the view that relativity theory implied the subjectivity of knowledge or led to philosophical relativism. Granted, cognition was dependent upon the individual observer.

> All knowledge of necessity includes a subjective element. The process of cognition is the process of the interaction between the subject and the object. In this interaction the object is revealed to the subject more and more fully. At each given degree knowledge is relative, but relative knowledge is a stage on the path to absolute knowledge. The process of cognition is a historical process.[87]

Gessen thus argued that the external world exists independently from the cognition of the subject, adopting a Plekhanovite, if not Bogdanovite, view of cognition as a historical process.

For Gessen, relativity theory provided a way out of a series of epistemological problems generated by nineteenth- and twentieth-century physics. Relativity theory was neither metaphysical nor mystical, nor a philosophical system, nor a complete worldview as the Newtonian system had become, but a definite conception of space and time based on epistemological propositions recognizing the materiality of four dimensions. There was a great difference between philosophical and physical relativism, with relativity theory contending only the latter. It was not only a new methodological tool but also a rigorous physical theory which explained more than Newtonian physics. Newtonian mechanics were still valid in many areas; the theory of relativity included classical mechanics as a special limiting case. Finally, relativity was not a Machist rejection of materialist space. Yes, Mach had influenced Einstein to abandon absolute space and time, but relativistic notions in fact represented objective reality.

While many mechanistic Marxist philosophers and several scientists had rejected several of the fundamental theories of physical science in the twentieth century, the Deborinites and their allies embraced relativity and quantum mechanics. They recognized that mechanical explanations had had their role in history but could no longer be applied to all physical processes. The reductionism of the Mechanists, while seeming to provide Marxian interpretations of the new physics, simply could not accurately describe physical phenomena of the microworld or at velocities approaching the speed of light. The Deborinites futhermore recognized the dialectical, if not revisionist, qualities of Marxism; Marxism was not a stagnant, unchanging "stencil [*shablon*]" to be applied to interpretations of the sciences but should reflect advances in science.

In the main it was for this viewpoint that the Dialecticians encountered the opposition of Stalinist ideologues. Gessen, Semkovskii, and scientists themselves believed that physicists should determine the correct interpretation of advances in the sciences; ideologues thought that they themselves should determine the truth for scientists to follow: which theories were "idealist" and which "materialist"; which Western scholars' works were acceptable; in a word, what subjects should be studied and how.

In the early 1930s, after the defeat of the Mechanists at the second conference of the Marxist-Leninist scientific research institutes, Stalinist ideologues and philosophers turned against the Deborinites. They saw idealism in the increasing role of probability and statistics in science, the replacement of matter by mathematics in quantum descriptions of reality, and the growing reliance of Soviet physicists upon such Western scholars as Bohr, Schrödinger, and Eddington.

Furthermore, they dismissed out of hand the view that Marxism was or should be a revisionist doctrine. In the environment of cultural revolution and Stalinist politics of science, the debates between ideologues and physicists—and between physicist and physicist—became politicized. Although physicists had adroitly managed the discussions which took place in study circles in their institutes, they could no longer do so when philosophical disputes rocked the discipline from top to bottom, and from Moscow to Leningrad.

8

Theoretical Physics: Dialectical Materialism and Philosophical Disputes

The siege mentality engendered by the Stalinist revolution left no region of Soviet society untouched. Even the philosophy of science and epistemology, areas that at first glance seem remote from the Soviet history and politics of the 1930s, fell into the maelstrom of party politics, intrigues, and backstabbing so pronounced in industry, agriculture, and education. In physics, the polemics between scientists and philosophers resulted in the destruction of careers, in arrests, and executions.

Marxist philosophers devoted excessive scrutiny to the work of Soviet physicists, never hesitating to criticize those works which they deemed to be "idealist," ignorant of Leninist doctrine, or in some other way anti-Soviet. In an environment of class war against the old intelligentsia; advancement of workers into positions of responsibility; war on the countryside and forced collectivization; and dizzyingly rapid industrialization, questions of the philosophy of science assumed a political overtone. Most ideologues believed that the failure to fight idealism in science would lead inevitably to the victory of forces hostile to Soviet socialism; idealist physicists within the USSR were seen as being more dangerous than "bourgeois" scholars abroad.

The heightened philosophical disputes over the proper interpretation of the new physics turned Marxist against Marxist, physicist against physicist, and comrades against each other. After the condemnation of mechanism and the castigation of Deborinites for "menshevizing idealism," Soviet Marxism became ossified in vulgar materialist pronouncements about the properties of matter in motion, the importance of "party-ness" in philosophy, and the role of class struggle in science.

The Communist Academy stood at the forefront of the effort to bring about the penetration of "correct" dialectical materialist philosophy into physics, chemistry, and biology. At first, those who insisted upon this new approach to science took the side of the Dialecticians concerning the "crisis" in physics, and suggested that physicists themselves could best resolve the issues.[1] A period of brief calm on the philosophical front followed the defeat of the Mechanists by the Deborinites in 1929. But the publication of an article in *Pravda* on January 26, 1930, signaled the end to the Dialecticians' brief hegemony in the philosophical sphere and an assault on scientists who were seen as representatives of idealist philosophy.

A year later, on January 6, 1931, the Communist Academy passed a series of resolutions condemning "menshevizing idealism."[2] The criticism centered on charges of using materialist phraseology but embracing counterrevolutionary ideas, of avoiding self-criticism, adopting eclectic, agnostic, and philistine views, and voicing the opinion that philosophy could be "apolitical" or neutral.[3] It was alleged that physicists and Deborinites alike had insufficiently recognized Lenin's role in the development of Marxist philosophy; avoided criticism of Hegel, Feuerbach, and Plekhanov; failed to take into consideration the presence of the "class war" being waged by the enemies of the USSR; and took an uncritical attitude toward bourgeois philosophy and science.[4]

With regard to physics, the resolution of the question of what was idealist and what materialist in contemporary science centered on three issues: the conservation of energy, action at a distance, and quantum mechanics. But the crux of the matter was the question of who would ultimately resolve these issues, physicists or philosophers.

In Stalin's Russia the process of resolving philosophical questions became highly politicized. In an effort to regain their self-respect, and no doubt to settle a few old scores, the mechanists joined in the chorus of "idealism" to draw Stalinist philosophers into discussions on their side. By the end of the decade they had attempted to turn the Academy of Sciences against their idealist enemies. The Academy's division of mathematical and natural sciences, which had briefly become a refuge from the worst sorts of vitriol, held a series of special sessions in 1936, 1937, and 1938 which went beyond the bounds of physics to put such physicists as Ioffe, Frenkel, and Rozhdestvenskii under the Stalinist spotlight. What was the philosophical basis of these attacks on Soviet physicists? And did physicists manage to parry them?

SOVIET DIALECTICAL MATERIALISM

One of the most important outcomes of the Stalin years was the development of an officially sanctioned philosophy, dialectical materialism, which had its roots in the nineteenth century. G. V. Plekhanov, often called the father of Russian Marxism, is credited with first using the term. Dialectical materialism, according to Graham, is based on three principles: "all that exists is real; this real world consists of matter-energy; and this matter-energy develops in accordance with universal regularities or laws."[5]

As mentioned earlier, Marx himself wrote very little on science; Engels's *Anti-Dühring* and *Dialectics of Nature* are more central to the development of dialectical materialism, and with Lenin's *Materialism and Empiriocriticism* served as a basis for discussions between the Mechanists and Deborinites. Lenin's entry into the field of philosophy was provoked by his disagreement with such Left Bolsheviks as A. A. Bogdanov over the latter's so-called Machism.

The sensationalist philosophy of the physicist Ernst Mach had attracted a diverse group—from Einstein to logical positivists, to the Left Bolsheviks. Mach believed that "the world consists of our sensations."[6] This contradicted the views of Lenin, who insisted that sensations are reflections of objects which exist independently of the observer. Mach's views, taken together with the discoveries in physics of X rays and the electron, seemed to refute the notion that matter is real or immutable. Lenin grew convinced of the need to take a firm stand on the side of materialism against the alleged idealism of the Bogdanov group. On the basis of a year of research conducted at the British Museum in 1908, Lenin set out to advance a rigorously materialist doctrine. The result was *Materialism and Empiriocriticism*, where Lenin argued that the world does not consist of sensations but that objects in themselves exist; our ideas and sensations are copies of them. *Materialism and Empiriocriticism* is characterized by vulgar materialist generalizations, lengthy diatribes, and excessively long quotations of opponents followed by brief, haughty ridicule. Had the Bolsheviks not succeeded in 1917, it is doubtful that Lenin's book would be read today.

Soviet Marxists added several aspects of Engels's writings to the vulgar materialist Leninism to develop dialectical materialism. The first was the suggestion that "reality" consists of matter-energy in motion. The second was Engels's contention that there are three dialectical laws which operate in nature: (1) the interpenetration or unity of opposites; (2) the negation of the negation; and (3) the transformation of quantitative changes into qualitative ones, and vice versa.[7]

An example of the first law might be a magnet with a north and south pole or the wave-particle duality of light. An example of the second would include any birth-life-death-rebirth process (an annual flower which blooms, withers, dies, and blooms again). And in the third category, examples from chemistry abound: the methane series C_nH_{2n+2}, where the same quantitative formula leads to different properties of compounds. Not all of the philosophical disputes of the 1930s so neatly reduced to a consideration of the applicability of these ideas to modern science. But the important thing is that many of the participants in the debates believed that they did.

THE BATTLE JOINED:
MECHANISM, MODERN PHYSICS,
AND ACTION AT A DISTANCE

In the 1930s, philosophical discussions between physicists and Marxist ideologues centered on three major issues. First, the philosophers E. Kol'man, Egorshin, and Maksimov contested the "idealism" inherent in quantum mechanics and the "mathematical formalism" which they believed pervaded the exact sciences. V. A. Fock was a leading defender of quantum mechanics and physicists' right to decide epistemological questions without needless assistance from ideologues. Inasmuch as this philosophical controversy is discussed at length elsewhere, it will not be treated at any length here.

Second, the philosopher F. Galperin and the quantum mechanician D. I. Blokhintsev attacked what they saw to be the rejection of the law of conservation of energy in new theories of nuclear transformations (beta and alpha decay, the Pauli hypothesis on neutrinos, and so on) and stellar phenomena. A promising young researcher at LFTI, Matvei Petrovich Bronshtein, was caught in the sights of this controversy.

Blokhintsev and Galperin cited experimental evidence and new theories of the neutrino (the work of Pauli and Fermi) and beta decay to show that the conversation of energy applied to nuclear physics, the seemingly arbitrary jump of the atom from one energy level to another notwithstanding.[8] Galperin in particular was upset with the fact that "a series of Soviet physicists introduced idealist conclusions into the country of the dictatorship of the proletariat" when they rejected the conservation of energy for nuclear processes. The offenders included L. D. Landau, George Gamov, and M. P. Bronshtein.

Bronshtein was a brilliant theoretician who was only thirty-two when his life was cut short by the purge. Born in 1906 to a Jewish family, he already had publications in *Zeitschrift für Physik* and *ZhRFKhO* when he finished Leningrad University in 1930 and was taken on as a graduate student by Frenkel', who saw in him an "extremely talented" and promising scholar with "broad interests." He knew French, German, English, Ukrainian, and Russian, and was at the cutting edge of developments in relativistic cosmology. His doctoral dissertation drew rave reviews from Tamm and Frenkel'. He taught at the polytechnic institute and worked in the theoretical department of the Ioffe Institute. Bronshtein was an active popularizer of science and inclined toward Trotskiism, both of which made him visible and vulnerable. He was arrested shortly after becoming a senior physicist at LFTI in 1937, and was shot on February 18, 1938. Only in 1957 was he posthumously rehabilitated.[9]

The most guilty party in Galperin's eyes, Bronshtein conducted a series of seminars on such contemporary research topics as the work of the Fermi group on radioactive decay and the applicability of nuclear processes for understanding stellar phenomena. Already in 1931 the editorial board of *UFN* had concluded that some of Bronshtein's conclusions contradicted materialism.[10] In Galperin's view, Bronshtein had adopted a Copenhagen interpretation of quantum and relativity theory when he argued that the energy of some stars increased whereas others decreased.[11] This was an indication of the battle that bourgeois physicists and their Soviet sympathizers were carrying out against conservation of energy, under the flag of a struggle with "dogmatism and canonization."[12] This controversy was not as long-lived as the ones over quantum mechanics and action at a distance, since philosophers of science found it difficult to keep up with the rapid new developments in nuclear physics.

A third major disagreement developed between leading theoretical physicists and those who, trapped by allegiance to a nineteenth-century worldview, supported a mechanical medium to explain electromagnetic radiation and rejected the twentieth-century concepts of electromagnetic "field" and action at a distance. This dispute pitted the physicists V. F. Mitkevich, N. P. Kasterin, and K. V. Nikol'skii (all with the support of Timiriazev) against scholars at LFTI (primary among them Frenkel') and such Moscow researchers as Tamm. If physicists assumed in 1930 that they had seen the last of mechanism with the victory of the Deborinites, then they were sorely disappointed, for the Mechanists renewed their attack on the physical "idealism" inherent in the concept of action at a distance and on its adherents, in particular scholars at the LFTI.

The mechanistic physicists developed a curious and anachronistic view of electromagnetic phenomena and posited the existence of an ether with mechanical properties to explain them. V. F. Mitkevich, the Leningrad specialist in electrical technology who first headed the Academy's newly created division of technological sciences (1931), published a series of articles in which he posited an ether that was an anomalous magnetic flux whose transformations were based on Faraday's electrical force lines.[13] Mitkevich (1872–1951) was one of the corresponding members of the Academy to be swept into full membership by the elections of 1929. He was strictly an electrical technologist, working for GOELRO, the state program for electrification, and from 1921 to 1937 headed a special technical bureau on military inventions for the Commissariat of Defense. Mitkevich was so invested in classical electrodynamics in his work that he viewed all things from the point of view of Faraday, Maxwell, force lines, and the ether. According to historian of physics G. E. Gorelik, he believed every change in the nature of physics was to be met with great trepidation.[14]

Mitkevich's research on electromagnetic force lines produced wonderment among the Leningrad physicists. In December 1929 and January and March of 1930 they held a series of discussions at the polytechnical institute at the request of students to debate Mitkevich's theories. Frenkel', Ioffe, Bursian, V. K. Lebedinskii, and Paul Ehrenfest hoped to set the students straight. Two philosophers from the LFTI circle on dialectical materialism, M. L. Shirvindt and T. A. Gornshtein, took part.

Ioffe introduced the discussants at the first session, and set the tone for future discussions by noting that Mitkevich's "new and very curious views" were based on Maxwell's theory of electromagnetism and recognized magnetic force lines as real physical entities.[15] Mitkevich opened each of the three sessions with a statement based on lengthy investigations he had published in *Fizicheskie osnovy elektrotekhniki*, where he concluded that an ether was fundamental to understanding all properties of electrical current; real current could not pass through an absolute vacuum.[16] Mitkevich accused Frenkel' of mysticism, mathematical formalism, and idealism for believing that electromagnetism and gravity may act at a distance and for using mathematics to describe this "reality."

Frenkel' and his supporters refused to recognize the existence of a weightless, imponderable and unknowable medium to explain electromagnetic phenomena. Mitkevich, according to Frenkel', had confused terminology, returned anachronistically to the ether of the English school of Maxwell, Poynting, and Thomson, and simply failed to offer a physical theory of any kind. Mitkevich's query about whether electromagnetic current could pass through an impermeable membrane was a "*Scheinproblem* [mock problem]," like asking if the devil had a tail. "We are often the victim of such an incorrectly posed question which leads us to fictional problems," Frenkel' declared.[17] Relativity theory provided a way out. It reconciled electromagnetic phenomena with classical mechanics, avoided the materialization of force lines, and permitted action at a distance through the existence of electromagnetic fields. While Frenkel' clearly found all of the interest in dialectical materialism boring, he tried to give his presentation the appropriate Marxist flavor by noting, "I think, however, that we must consider not the field but matter, that is, the motion and interaction of material particles, as the fundamental reality, and to look at the electro-magnetic field as an auxiliary construction which serves as a more convenient description of this interaction."[18]

While Ehrenfest, Ioffe, and several others attempted to smooth the waters, contending that the disagreements between Mitkevich and Frenkel' were semantic or terminological ones, Mitkevich himself and several Marxist philosophers saw their dispute as being a fundamental one between idealism and materialism. The philosophers were unwilling, however, to find for the plaintiff, Mitkevich. Unlike most Stalinist ideologues they did not believe that action at a distance required rejection of the universal password "everything is matter in motion." The Hegelian T. A. Gornshtein, who often ran the circle on dialectical materialism at LFTI, was a last-minute substitute for B. M. Gessen, who for some reason could not attend. She observed that "today's debate" was "in essence . . . a philosophical debate" which involved general philosophical categories. It was a debate over the nature of "objective reality" with reference to the basic concepts of dialectical materialism: matter, reality, and truth.[19]

M. L. Shirvindt agreed that the discussions were significant for dialectical materialism, but he believed that the source of the disagreement between Frenkel' and Mitkevich was the former's reliance on mathematics and the latter's on models. The difference between those who used models in pursuit of clarity and experimental verification and those who used mathematics in pursuit of rigor and theoretical basing was merely one of approach, since both shared the goal of cognition and objectivization of experiment. Nevertheless, Shirvindt sided with Frenkel'; Academician Mitkevich's force lines were so subjective that they "hardly corresponded with reality."[20]

Taking Gornshtein's comments to heart, Mitkevich began his presentation at the second session on January 3, 1930, with a statement that the goal of physics is to describe objective reality. A historical illustration would show that action at a distance had no physical meaning:

> Gravity should depend upon some agent which acts continuously in agreement with a known "law." Thus proclaimed the physicist Newton. Unfortunately, until recently there has been no sufficiently clear understanding of these words of the physicist Newton among indi-

viduals who study physics; I therefore permit myself to show with a simple example what we should understand by "absurd," and what it consists of. Let us imagine a mass m_1 and mass m_2 (figure 1). Let us suppose that these masses

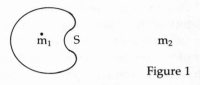

$\dot{m}_1$ S m_2

Figure 1

actively may act upon each other at a distance. It would signify the following. Let there be some kind of sphere, S, which is completely closed, which surrounds all sides of mass m_1. If someone assumes that mass m_1 *acts at a distance* on mass m_2, then consequently he considers possible the existence of this action without some sort of physical agent penetrating through any part of the sphere S. Of course, we may imagine this if we call for the help of some spiritual- . . . phenomenon, but physicists until recently usually did not use such things and managed without them. Therefore, for everyone it is clear that the physicist cannot speak about this scheme of a physical concept other than as being extremely absurd. Thus the mathematician Ia. I. Frenkel', like the mathematician Newton, has the complete right and, sometimes, perhaps, should utilize the ideas of action at a distance in mathematical exploration of questions from the region of electromagnetism. But if Ia. I. Frenkel' wishes to speak as a physicist, he should remember the golden words of the physicist Newton which I have read, and should consider the complete physical absurdity of the idea of action at a distance and put *actio in distans* in a proper place.[21]

For Mitkevich there were only two points of view, one correct and one incorrect: belief in the reality of force lines and belief in action at a distance, respectively.

Frenkel' responded by demonstrating that Mitkevich's scheme required an ether which acted like the particulate plenum in Descartes's vortex theory. Why, Frenkel' asked, had some scholars turned to short-range action (*blizkodeistvie*—the mechanism of Mitkevich's ether)? First, psychologically and anthropomorphically it was easier to visualize. Second, classical physics was the physics of the macroworld, where we are accustomed to considering the interaction of two bodies as a result of their direct contact. In the microworld, however, new laws came into play, and Mitkevich's system was nothing more than "a physical absurdity and a working hypothesis." Frenkel', setting a standard for physicists' behavior throughout the decade, refused to answer whether some force could act upon m_1 and m_2 through sphere S. Mitkevich threatened to leave the hall, demanding a response of "yes" or "no."[22] The final session, on March 14, 1930, considered these same issues and debated them in the same terms, with Frenkel' adding only a history of the theory of electromagnetic phenomena from Maxwell through Einstein and the advent of relativity theory, and a discussion of its significance with respect to action at a distance. Mitkevich merely repeated past arguments, and insisted upon "yes" or "no" answers to his questions.[23]

The centennial of Faraday's work on electromagnetic induction (1931) triggered further discussion, with I. E Tamm entering the deliberations as a spokesman for the reality of action at a distance. Tamm (1895–1971), a leading Moscow theoretician and Nobel Prize winner for his theoretical explanation (with I. M. Frank) of the Cherenkov effect—the interaction with radiation of fast particles moving in a medium (crystals, plasma, and so on), was involved with socialist causes from early in his life. He grew up in the town of Elizavetgrad, from which he was elected a delegate to the first congress of Soviets. He also studied at Edinburgh University just before World War I and therefore must have known Boris Gessen personally, from either Edinburgh or Elizavetgrad, even before they both joined up again at Moscow University. Tamm taught physics at the University of the Crimea in Simferopol (where Frenkel' worked) and then at Odessa Polytechnical Institute (where Mandel'shtam organized a new department). From 1922 on, he lived in Moscow, where for some time he headed the Moscow State University theoretical physics program and in 1934, at the invitation of S. I. Vavilov, organized the theoretical physics department at the newly created Physics Institute of the Academy of Sciences. D. I. Blokhintsev, M. A. Leontovich, Iu. B. Rumer, and M. A. Markov joined him, and V. A. Fock played an active role in the development of theoretical physics in the department.[24] Tamm's specialties included relativistic quantum mechanics, the modern theory of metals, and the nature of nuclear forces.

Tamm was a popularizer of action at a distance; he often gave lectures to Moscow factory workers, *vydvizhentsy*, and students.[25] In a talk before the Palace of Unions on September 23, 1931, Tamm acknowledged the "impossibility of motion without matter." For

Faraday the force lines had a hypothetical nature; in contemporary physics they served no function in explaining polarity of electricity and magnetism, gravity, or the ether. Action at a distance provided all these functions.[26] Tamm published a prominent article about Faraday's works in *UFN*, and used the occasion to admonish Mitkevich and others (although not by name), and later faced Timiriazev down at a university forum in June 1935.

After the initial thrusts and parries at the polytechnical institute in 1929–1930 and over the Faraday Centennial, it fell to D. B. Gogoberidze, a researcher at LFTI who defended his candidate's dissertation on "Mechanical Twinning" in January 1937, to counter Mitkevich's incessant cries of "Answer me, 'yes' or 'no'! Can there be action at a distance?" Gogoberidze's refutation was quick and to the point. He noted that there were two extreme points of view of electromagnetic phenomena, the mechanical view of Mitkevich and an "idealistic" one. The majority of physicists fell between the two positions. But virtually all rejected Mitkevich's approach out of hand, since it ignored difficulties involved in describing the behavior of radiant energy and the electrodynamics of bodies moving through an ether. First, it was based on antiquated ideas of force lines. Second, the ether itself was another kind of matter, with very small mass, great elasticity, and absolute incompressibility, which avoided quantization and relativistic concepts of space and time. Gogoberidze condemned Mitkevich's wreckless scheme:

> The real existence of some kind of objective reality, of some kind of agent which is the carrier of electromagnetic forces, is undoubted for every materialistically thinking physicist, and the recognition of the objective existence of this reality for us is obligatory. It is necessary to note, however, that properties of this reality, an "ether" if you please, have nothing in common with the properties of that ether about which Academician Mitkevich speaks and which he utilizes.[27]

What *was* Mitkevich's medium? What processes occurred in it? The answer "Faraday-Maxwell force lines" was not enough. Wasn't Faraday himself more cautious than Mitkevich in coming to the idea of the universal ether, and hadn't he decided that there was no basis for a physical theory of an ether? And, if an ether were posited, then the entire system of ideas upon which relativity theory was based would have to be rejected. Gogoberidze concluded that Mitkevich charged his opponents with idealism not because of their philosophical failings but because they had attacked his mechanism.[28] Mitkevich merely responded with the usual arguments and the contention that Gogoberidze, like Frenkel' and Ia. N. Shpil'rein before him, was a "sly idealist."[29] The physical concepts of action at a distance were "only a reflection of mathematical symbolization."[30]

Mitkevich had not rested during this time. He read a lengthy paper at the Academy in 1933 "On 'Physical' Action at a Distance," wrote articles in *Sorena* and *PZM* to defend his work, and sought the support of Maksimov and other philosophers. He seems to have been particularly stung by Fock's bitter condemnation of his book, *Osnovnye fizicheskie vozzreniia*.[31] In meetings in 1936 at the physics department, at the university, and at the Academy, Mitkevich again took his physicist opponents to task. In the former, Tamm, G. S. Landsberg, and Gessen spoke in favor of the electromagnetic propagation of energy without mechanical transference of some sort, while the "opponents pelted them with a whole variety of accusations, demagogical claims, and even indecent comments."[32]

At a March 1936 session of the Academy devoted to criticism of the performance of LFTI (see chap. 9), Mitkevich used the opportunity to attack the electromagnetic theory again. Here, once again, Ioffe, Frenkel', and Tamm refused to answer Mitkevich's baiting questions. An eyewitness, the astrophysicist V. L. Ginzburg, who attended the session recalled, "Tamm said, 'There are questions for which no intelligent answer can be given; for instance, what colour is the meridian which passes through Pulkovo? Red or Green?' Akademician Mitkevich burst in with: 'Professor Tamm does not know the colour of the meridian on which he is standing, but I know I am on a red meridian.'"[33]

Other Mechanists joined Mitkevich in hounding the Leningrad physics community for their "idealism." Building upon Maxwell's electrodynamics, N. P. Kasterin advanced mechanical explanations to counter relativity theory and quantum mechanics. From the point of view of Mitkevich and Timiriazev, Kasterin was a materialist. But Kasterin did not intend to reject relativity as part of the Mechanists' program. He had been an antirelativist since before the Revolution, and kept to the same program until the late 1930s. He never turned against his fellow physicists, although they attacked him violently and denied him many of the perquisites of Academy membership.[34]

Kasterin's major effort involved a generalization of the basic equations of aerodynamics and electrodynamics into one system in an effort to undermine Einstein's theories. He presented this theory at a special session of the Academy of Sciences in December 1936, where he argued that all physics, aerodynamics included, is based upon experimental measurements whose exactness was always limited; virtually all laws of contemporary physics were "first approximations," including the equations of the electromagnetic field of Maxwell. Since the time of Faraday, Maxwell, and Hertz, as the preciseness of measurements in electrodynamics had grown, so had the insufficiency of the equations.[35] The need therefore arose to find a second approximation to determine "whether these more general equations can embrace the entire aggregate of facts" in electromagnetism and aero- and electrodynamics.[36] To bring about this second-order approximation, Kasterin proposed equations of an electromagnetic field which differed from those of Maxwell not only by virtue of the fact that they were nonlinear, but also because "the speed of light within this field is variable." Kasterin also relied heavily on Faraday force lines to describe electromagnetic activity.[37]

Leading Soviet theoreticians wasted little time in answering Kasterin's work. Having found philosophers' activities tedious at best, they were not about to tolerate another Mechanist in their midst. At the suggestion of the physical group of the Academy of Sciences and the academic council of S. I. Vavilov's Physics Institute, a committee composed of Frenkel', Fock, Tamm, Blokhintsev, and others met to consider a response to Kasterin's work. The committee members revealed nothing but unrestrained hostility, a fact which troubled many of the older members of the Academy, who expected civility in scientific disputes.

The committee had expected to encounter a clear line of reasoning, fully developed ideas, a methodology, and "concrete proof," even though Kasterin had stated his work was "preliminary," but there was "nothing of the kind." Kasterin provided enough material, however, for the committee to judge his theory totally unfounded: "It is completely sufficient to show the internal contradictions of its basic equations and [for us to] be convinced of the contradictions of the given theory with experiment." Mathematical mistakes had occurred so frequently that "they comprise the major part of the contents of the article."[38] Kasterin had ignored the viscosity of gases; his aerodynamical equations contained no magnitude which characterized the atomic structure of gases; and he spoke about the electromagnetic field as a continuous whole, although there was no evidence for this view. Finally, the author had set forth his generalizations of the equations of aero- and electrodynamics as if he were the first (like Maxwell or Euler), whereas in fact Heisenberg and Born had developed nonlinear electrodynamics, and Boltzman, Lorentz, and Hilbert had done the same for aerodynamics. In sum, Kasterin's purely mechanical hypothesis of the aerodynamical nature of an electromagnetic ether was a "fiction" and a "shoddy piece of work."[39]

The Academy thereupon seems to have held up payments from the Council of Ministers (Sovnarkom) to Kasterin through their coffers, although the Sovnarkom had approved a special second year of funding for Kasterin's work. It denied him access to KSU (Komissiia sodeistviia uchenykh, the successor of TsEKUBU) sanitaria, and made it difficult for him to gain access to a typewriter (printing presses and typewriters had had to be registered with the government for some time now), foreign book, and journals.[40]

Kasterin, Timiriazev, and Mitkevich were shocked by the intensity of the physicists' response. Kasterin protested the speed of the Academy's decision, but modestly stood aside to let the Mechanists fight onward.[41] Timiriazev wrote several letters to the Academy commission on Kasterin with the usual slander of Frenkel' and, finding no response, turned to the academic secretary of the division of mathematics and natural sciences. He characterized Kasterin's work as being on the level of achievement of Maxwell, Thomson, and Zhukovskii in aerodynamics. Between 1934 and 1936, with a small grant from the Sovnarkom, Timiriazev asserted, Kasterin had produced equations which required, once and for all, for idealist enemics to keep their "hands off!" Timiriazev further suggested that the attack on Kasterin was like those on Galileo and Bruno, who had tried to defend Copernicus.[42] Did Timiriazev see himself as Bruno or Copernicus?

Mitkevich turned directly to V. M. Molotov, then chair of the Sovnarkom, member of the Central Committee, and one of Stalin's closest associates, with a request for a laboratory to develop Kasterin's work further. This new region of aerodynamics and electrodynamics required a group of ten to fifteen individuals but not a large facility, "in order to be delivered from the dangerous clashes with physicist-idealists who now control the group of physics of the Academy of Sciences." Mitkevich hoped that Kasterin, Chaplygin, Timiriazev, Mitkevich himself, and others from the university would staff the new organization. "Without resolution of this question," Mitkevich informed Molotov, "the matter of assimilation and application of this great theory . . . may be slowed, and the initiative *will easily leave the USSR for abroad.*"[43] The Sovnarkom apparently provided additional funding for Kasterin for one year, but the matter got no further.

In 1937 Mitkevich would try one more time to bring about victory in the struggle with his hated adversaries. In January he wrote N. P. Gorbunov, head of the Scientific-Technical Department of Vesenkha in its first days, and now permanent secretary of the Academy, requesting an Academy session devoted to proving once and for all the idealism inherent in the concept of action at a distance. (Less than a week later, Mitkevich wrote Gorbunov again, nominating Kasterin and Timiriazev for membership in the Academy.) Mitkevich also hoped to obtain Gorbunov's help in forcing Vavilov, Ioffe, Frenkel', and Tamm to answer his questions. At Gorbunov's order, *Pravda* published a note about the forthcoming conference.

The council of the division of mathematical and natural sciences of the Academy appointed a committee of physicists and philosophers under the direction of Maksimov and Blokhintsev to organize the special session. Mitkevich at this point refused to give the keynote speech unless Ioffe and Vavilov answered him in print. Vavilov published an article in *PZM* in 1937,[44] while Ioffe rejected the idea out of hand (but responded later that year in *PZM*).[45]

In January 1938, the special committee, which had met by this time on several occasions, had agreed upon the format for the session. There would be a speech on idealism in the West and in the Soviet Union by Mitkevich, and one on causality and indeterminacy by Maksimov. But within a month, the meeting was canceled. In the interim Fock had written a scathing attack on Mitkevich and his so-called physics, which he directed to the presidium of the Academy. He declared that the "scandal" of Kasterin's presentation to the Academy in December 1936 had been more than enough. The fact that the session was not held is now known as the "wonder of Saint Vladimir [Fock]."[46]

It is unlikely that Fock alone could have provoked the cancellation of Mitkevich's hoped-for witch-hunt. Fock himself had recently returned from arrest. More likely, the absence of complete support from philosophers, combined with nearly universal condemnation from the side of physicists, did the trick. Indeed, Maksimov himself, while attacking Frenkel', Tamm, and Shpil'rein over their alleged idealism, could not help but criticize Mitkevich for his mechanism in the same issue of *PZM* in which Vavilov had raked Mitkevich over the coals.[47]

Maksimov tried to explain his apparent ambivalence on the matter. In a letter to Mitkevich written just after the decision to cancel the special session, Maksimov thanked Mitkevich for a complimentary copy of *Osnovnye fizicheskie vozzreniia*, but reminded Mitkevich that he had unthinkingly defended Maxwell and Faraday's views, rather than just their mechanism. Maksimov did not accuse Mitkevich of mechanism, which "Fock, Tamm, and other reactionaries in science do. They are generally hostile to materialism." But, Maksimov wrote, "I personally considered publication of your work incorrect and not in the interests of the general state of affairs."[48] Maksimov also explained to Mitkevich that the world was not against him. *PZM*, Maksimov, and two philosophers from the Institute of Philosophy, Mitin and Iudin, repeatedly spoke out against Tamm, Frenkel', and the others. But Maksimov urged Mitkevich to be wary of idealism in view of the 1931 Central Committee ruling against mechanism.[49]

The debates between Leningrad physicists and the Mechanists and their allies over action at a distance, the reality of a universal ether, and conservation of energy lasted a decade in spite of the fact that a vast majority of physicists sided with their Leningrad colleagues. The Mechanists were able to draw on Stalinist ideologues to keep discussions alive, since the latter were more concerned about supposed idealism than about mechanism. Similarly, the philosophers found it simpler to reconcile a reductionistic view of all physical law—"everything is matter in motion"—with vulgar materialist notions than to recognize that experimental and theoretical advances required fundamentally new understandings of the relationship between matter and energy which it was best to allow physicists to work out.

PHILOSOPHICAL DEBATES:
PHILOSOPHERS AND PHYSICISTS

After the Communist Academy served notice in 1931 that all forms of mechanism, menshevizing idealism, and Trotskiism would be met with the tools of class struggle, the leaders of the Leningrad physics community began to face intense scrutiny. The Communist Academy condemned "the manifestation of bourgeois tendencies which are hostile to us." It singled out Frenkel' for publicly rejecting Leninism in the natural sciences in a talk at an all-union conference on chemistry in 1931, and the Nobel laureate and physiologist I. P. Pavlov for his "demonstrative refusal in the Soviet scientific press to cooperate" with Soviet power. Darwinism; the uncritical adoption of the achievements of contemporary bourgeois science, in particular by the physicist Sphil'rein; and "the penetration of Trotskiites within our midst (Gogoberidze, Kagan)" received honorable mention.[50]

The Academy tried to make an example of Frenkel'. In his speech, Frenkel' showed too much contempt for his surroundings. He described the electron as a function of probability, rejected the ether, saw the electromagnetic field as objective reality, and did not evince any compulsion to acknowledge the power of dialectical materialism in his work. He had argued, in fact, that the "dogmatic cultivation of dialectical materialism among the studying youth and scientists is reactionary and arrests [the development] of science." While acknowledging some value for the social sciences, having read both Lenin and Marx and found them interesting, if not compelling, he fought the penetration of dialectical materialist thought into the natural sciences, believing it incapable of giving anything to physics.[51]

In 1949 at an in-house LFTI session devoted to criticism of idealism in physics, Frenkel' admitted mistakes in his 1931 views of dialectical materialism. At the chemical conference,

> in connection with discussion on my paper on the nature of chemical bonding, I presented criticism of dialectical materialism and expressed doubts about its applicability to the principal problems of physicists. In the last twenty years my position in this question has essentially changed. In the first place, in my scientific work I began all the more consciously to utilize the methods of the materialist dialectic. . . . Second, I began distinctly to understand the connection between the Marxist worldview and communism, from the one side, and all different kinds of anti-Marxist views and political reaction from the other.[52]

Taking the Leninist point of view about scientists being "natural" materialists, he argued that "as a physicist for whom the occupation of physics is not a distraction but is a matter of honor and duty, I never was and cannot be an idealist. The external world which is studied by the physicist always presented itself to me as objective reality, and not a figment of my imagination."[53]

The three philosophers who carried out the first stages of the struggle against alleged representatives of idealism in the USSR such as Frenkel' were Egorshin, Maksimov, and Kol'man. It fell to Egorshin, a former Deborinite ally, to begin the attack on "wrecking," "Machism," and idealism in physics. As Egorshin saw it, idealism had four manifestations: (1) the mathematicization of physics—the overuse of the tools of mathematics outside of available empirical evidence. Egorshin claimed that this resulted in the substitution of formulas for "concrete reality"; (2) the destruction of Newtonian concepts of space and time and their replacement by philosophically relativistic ones; (3) the elimination of causal notions and their replacement by indeterminacy and uncertainty in quantum mechanics; and (4) such external factors as the crisis of capitalism abroad and the presence of a class enemy at home.[54]

Egorshin admitted that there had been achievements in the struggle for the proper Marxian worldview, the training of new cadres through *vydvizhenie*, and the publication of progressive journals. But the failures were far more serious, especially with regard to physics. First of all, physics activity remained unplanned. Second, some institutions resembled "feudal principalities"; this was a not-so-veiled reference to Ioffe and his Leningrad Physico-Technical Institute. Finally, and most importantly, many philosophical controversies persisted: Soviet scholars had fallen under the influence of "bourgeois eclecticism and philistinism." "Frenkel', Khvol'son, and Professor Andreev," Egorshin asserted, "advance under an idealist flag, in spite of the fact that they live under a Soviet proletarian government and print their work with Soviet money."[55] No longer could scientists be "spontaneous" or "natural [*stikhiinyi*]" materialists, or remain neutral in philosophical matters; dialectical materialism had to be applied methodically and consciously in all scientific institutions, and especially in Soviet textbooks and journals. Egorshin singled out *ZPF, ZhRFKhO*, and *UFN* for failing "to have a Soviet face," publishing no articles on "socialist construction" or class struggle, and differing little from bourgeois journals.[56] To be sure, the articles which were published dealt exclusively with physics, but Egorshin seems to have missed the point. Soviet physics texts were "no better than physics journals." Egorshin singled out Gessen and Frenkel' for "muddle [*putanitsa*]," subjective idealism, and relativism.[57] Such blatant disdain for dialectical materialism as that expressed by Frenkel' drew the philosopher Maksimov and the mathematician and epistemologist Ernst Kol'man into the fray.

Like Egorshin, Maksimov quickly turned against his former allies and raised a Stalinist flag. He showed little understanding of the issues, confused the use of mathematics with idealism, and saw Machism, "mathematical formalism" and "working hypothesis" as one and the same. As Lenin had before him, he divided the world into two camps, the idealist and the materialist. The struggle between these two camps reflected class struggle; only vigilance against representatives of idealism in class war would guarantee the victory of the Soviet social system. Maksimov did not limit his criticism to "idealists," but also included non-repentant Mechanists.[58]

Ernst Kol'man, who was perhaps more prolific than Maksimov, displayed a more sophisticated and complete understanding of quantum mechanics. A mathematician and philosopher of science, he played a vital yet conservative role in the development of dialectical materialist analysis of quantum mechanics and cybernetics, until mellowing in the 1960s.[59] He objected strenuously to the use of mathematical formalisms which seemed to serve as a substitute for, if not deny the existence of, an objective reality independent of the observer; he criticized the indeterminacy of Heisenberg; he found fault with many of the philosophical views of Einstein; and he rejected the "idealist" understandings of probability in the wave mechanics of de Broglie and Schrödinger. Yet he unceasingly criticized Timiriazev's mechanism, and argued for the rejection not of quantum mechanics but of its idealistic epistemological conclusions. If one embraced and interpreted dialectical materialism properly, he would see that quantum mechanics fit quite well within the framework of the Soviet philosophy of science.

Schrödinger, for example, was not far from materialism in his struggle against the agnosticism and phenomenalism of the Copenhagen school.[60] Similarly, "the rational nucleus of relativity theory was sufficiently grounded and verified by practice," although some of its adherents had jumped to idealistic conclusions when they attempted to create multidimensional space with mathematics.[61] But resolution of the philosophical questions advanced by quantum and relativity theory was possible only on the basis of dialectical materialism.[62] Fock and Blokhintsev joined with Kol'man in stressing the importance of dialectical materialism for understanding the materialist bases of quantum mechanics.[63] For Kol'man, all of this meant that the law of the transformation and conservation of energy holds true in the microworld; that atoms must be examined in relation to each other in an "unbreakable organic tie"; and finally that mathematics had to have an auxiliary, not central, role in physics. The "Leningrad school" of Ioffe, Frenkel', Bronshtein and the Moscovites Tamm and Vavilov followed a path of radical mathematization and formalism, with the result that energy was materialized and matter annihilated, there was action at a distance, and causality had disappeared from modern physics.[64]

Several Deborinites and physicists set out to demonstrate that there was a place for quantum mechanics and relativity theory in dialectical materialism. Deborin himself wrote little on quantum mechanics, but he nevertheless recognized its compatibility with dialectical materialism.[65] As he had for relativity theory, Gessen wrote the first major dialectical materialist defense of quantum mechanics, published as the introduction to the translation of Artur Haas's *Materiewellen i Quantummekanik*. Vavilov had reviewed Haas's study, first published in Leipzig in 1928, and called for its translation.[66] The translation had appeared two years later. In his essay Gessen addressed many of the philosophical and physical issues raised by quantum mechanics, the development of statistical laws, the wave-particle duality of matter-energy, the uncertainty principle, and the inherent difficulty of accounting for the interaction of the subject and the object in subatomic processes, including measurement. In all cases, Gessen concluded that quantum mechanics and dialectical materialism were consonant.

Gessen suggested that the crisis in physics, which dated to the turn of the century and required the transformation of the concepts of mass and energy, and of the classical notion of the indestructible and unchanging atom, had its roots in certain metaphysical positions first suggested by Newton. The relationship between statistical and dynamic laws, and epistemology in general, had to be reevaluated. Only dialectical materialism introduced the concept of change and development into the microcosmos, pointed to the proper understanding of the relationship between subject and object, and provided a way out of this crisis.

Quantum mechanics, Gessen argued, was a verification of the central law, the unity of opposites, in three major cases: the interrelation of continuous and discrete or discontinous phenomena, the existence of matter-energy, and the synthesis of statistical and dynamic laws. In each case, Gessen stressed the need for synthesis, not the primacy of one over the other, through the principle of "mutual interaction." Having an overriding interest in the mathematical aspects of quantum mechanics, Gessen devoted most of his attention in the essay to the role of statistical laws and probability in modern physics.

Gessen first explained how the classical notion of deterministic causality was based on metaphysical assumptions advanced by Newton and others. Dynamics was based on simple mechanical determinism where subsequent conditions were determined by prior ones; this was a real possibility, Gessen asserted, only when the system being examined was completely isolated. Dynamics was "an abstraction, an idealization of real relationships."[67] The transition to the study of phenomena of the microworld required a new approach: statistical laws with stochastic properties. The end of the classical notion of determinism did not require abandonment of causality, however, as Gessen showed through an analysis of the relationship between dynamic and statistical laws.[68]

Had causality been rejected? Did indeterminism reign in the microcosmos? Gessen answered "no" to each question:

> The direction of motion of an individual quantum is random not in the sense that it is indeterminate but in the sense that the behavior of one quantum is not essential for the entire aggregate of quanta, only as a whole which gives the statistical law. Of course, knowledge of this statistical law does not give the possibility to deduce from it the behavior of an individual quantum or the behavior of a coin in an individual toss. But this in no way is proof of acausality of a single phenomenon. The fact that the statistical law does not predict the behavior of individual phenomena which are a part of the aggregate cannot be attributed to the inadequacy of the law and still less can serve as proof of acausality of an individual phenomenon.[69]

And further:

> A statistical law . . . is closely tied with the concept of randomness, that is, with the differentiation of essential from inessential laws. Statistical laws of a collective of molecules have at their foundation the interaction of molecules. The behavior of an individual molecule is random with respect to the law of the collective as a whole, since the motion and position of the individual molecule are not essential for the behavior of the collective. . . . A statistical law is richer in content than a dynamic law, inasmuch as it enhances the concept of abstract determinism and proposes a concept of randomness and mutual interaction. Thus, if we reject the fatalistic conception of determinism on the one hand and recognize randomness not only as a consequence of our ignorance but as an objective category, then the contradiction between dynamic and statistical laws is destroyed.[70]

Gessen concluded his analysis of the compatibility of quantum mechanics and Marxist philosophy of science by pointing out that just as some physicists had incorrectly opposed causality to statistical laws, so a number had taken Heisenberg's principle of indeterminacy—the essence of which consisted in the fact that the position and impulse of elementary particles could not be known simultaneously with any exactitude—to require the abandonment of any causal relationships in modern physics. Indeterminacy was connected with the limitations placed on contemporary experimental physics in terms of measurement.[71] In considering all of these issues, Gessen warned that physicists and not philosophers should decide whether a given physical magnitude is observable or not. It "should be solved in correspondence with physical theory, not a priori."[72]

In his analysis of relativity theory and quantum mechanics Gessen had shown that philosophical issues raised by the new physics in no way contradicted Marxist epistemology. Both required a new understanding of the relationship between the subject and the object; both demanded the rejection of concepts which had been objectivized as absolute realities; and both raised the specter of relativism and idealism. But both were modern theories which provided more complete understanding of physical phenomena than their predecessors on the road toward cognition of absolute truth. Relativity theory and quantum mechanics, Gessen concluded, verified dialectical materialism, and in turn were verified by it.

THE CONFLICT JOINED:
PHYSICISTS DEFEND THEIR DISCIPLINE

When *vydvizhenie* was abandoned in 1934 the Party turned its attention toward rooting out idealism, which it believed permeated the natural and exact sciences. The nearly simultaneous occurrence of the fiftieth anniversary of Marx's death, the tenth anniversary of Lenin's death, and the twenty-fifth Anniversary of the publication of *Materialism and Empiriocriticism* served to focus attention on dialectical materialism and the philosophy of science and contributed to the increasingly polemical and personal nature of philosophical discussions. At several points over the next half-decade, Ioffe assumed a prominent role in the defense of the physics discipline. His defense was courageous, since a seige mentality permeated the society as it faced the increasing momentum of the purges. What is more, many ideologues and physicists alike had interpreted the criticism of Ioffe's institute at a special session of the Academy of Sciences in March 1936 as an indication that he, too, was fair game.

In June 1934, a special session at the Communist Academy held to commemorate the twenty-fifth anniversary of the publication of *Materialism and Empiriocriticism* stressed the dangers of idealism.[73] During the months preceding the conference, such journals as *FNiT* (*Front nauki i tekhniki*), *Sorena,* and *PZM* ran a series of articles which exposed Heisenberg, Schrödinger, Bohr, and Born as its Western representatives and subjected its alleged

Soviet leaders—Gessen, Ioffe, Frenkel', and Tamm—to abuse. Physicists sought to head off criticism, which they believed to be misdirected at best, and out-and-out madness at worst. They, too, published widely, in defense of the discipline. More important, Ioffe and Vavilov participated in the commemoration. They argued that contemporary physics, far from being idealist, confirmed the presence of dialectical laws in nature.

In his talk on twentieth-century atomism, Ioffe demonstrated that indeterminacy, the major stumbling block to Marxist philosophers' acceptance of the new physics, actually coincided with dialectical materialism. Indeterminacy was "just a new form of description of new facts." The problem was that ideal or exact measurements were impossible; indeterminacy provided the most complete picture of reality. Ioffe saw the verification of dialectical materialism in six major areas of physics, including quantum mechanics, a new conception of space and time, and transitions from macro- to microworld magnitudes, and from atomic to nuclear physics. Taking a chapter from Gessen's work, Ioffe argued that in each case seeming contradictions achieved a dialectical synthesis in the unity of opposites.

Ioffe admonished Marxist philosophers who found idealism under every stone, although he stopped short of calling them incompetent to judge the true state of affairs in physics. He sought to ward off the incursion of Stalinist ideologues. He urged physicists to stick to physics, and philosophers to philosophy. In establishing provinces of expertise—and, he hoped, autonomy—for philosophers and physicists alike, however, he came to the unfortunate conclusion that physicists were materialist until they began to philosophize. This opened the way for philosophers to assail physicists for "idealism" whenever the latter ventured into the philosophy of physics.[74]

Gessen appears to have been the only Marxist philosopher in attendance at the Communist Academy sessions who defended physics from charges of idealism. All others tarred contemporary theoretical physics with the brush of Stalinist epithets: Galperin attacked "bourgeois physicists" and their Soviet sympathizers for their rejection of the conservation of energy; Egorshin rejected the relativism, mathematical formalism, and subjectivism of Frank, von Mises, Bohr, N. N. Andreev, V. K. Frederiks, M. P. Bronshtein, and I. E. Tamm; and Kol'man granted Ioffe and Vavilov very little for their efforts.[75] He welcomed Ioffe's criticism of the Mechanists, "who stand in the way of progress and see idealism in relativity and quantum theory and everything not in agreement with Maxwell and Faraday," and he admitted that "the very presence of Ioffe and Vavilov at the session is a result of the fact that both our great scholar–natural scientists and our philosopher-dialecticians are now seriously attempting to come together, work, and help each other." But Kol'man singled out physicists at LFTI for special criticism.[76]

As they had in response to the Mechanists Timiriazev, Mitkevich, and Kasterin, leading physicists united in an effort to lessen the impact of ideological intrusions from philosophers. Ioffe assumed a leading role in this process in a remarkable article, "The Situation on the Philosophical Front of Soviet Physics," in which he took issue with both individuals and tactics, abusing the Mechanists with a vehemence that resembled theirs, while saving enough vitriol for the philosophers.

Ioffe took issue first of all with the traditional rule of Bolshevik disputation, "He who is not with us is against us." He asked, not so rhetorically, "Is everyone who does not recognize the reality of magnetic force lines and who hopes to go beyond the limits of physics of the middle of the last century an idealist?"[77] It was strange to be told constantly that those who had not accepted Mitkevich's mistaken views were guilty of idealism. Next he demonstrated the physical errors in the ethero-Mechanists' system, their fear of mathematics, and their anachronistic position with respect to Maxwell and Faraday. Finally, Ioffe accused the antirelativists in Russia (Kasterin, Timiriazev, and Mitkevich), their foreign allies (Lenard and Stark), and their fellow travelers in accusations of idealism (such Stalinist ideologues as Maksimov) of doing much more damage to Soviet physics than good. The danger lay, in fact, with these supposed defenders of materialism.

Ioffe left no doubt about the essence of the last charge. The antirelativists were anti-Semitic and reactionary. Why did Timiriazev and Kasterin side with Lenard and Stark, Nazis who saw relativity theory as non-Aryan physics? What did they have to say about the label applied to Heisenberg for his views on indeterminacy, "Weisser Jude"? Why did they rail against the relativity of Einstein "who, in spite of his pacifist and Zionist leanings is undoubtedly antifascist and democratic"?[78] Second, Ioffe strongly objected to charges of idealism. Fock, Frenkel', Tamm, Mandel'shtam, and Landau had produced theories of the solid state and metals, of the photoeffect, and of diamagnetism, and had contributed to the development of quantum mechanics; the Mechanists, on the other hand, had created only the fetishes of the ether, force tubes, and "electrical bagels [*bubliki*]."[79] They, and not mainstream physicists, were guilty of philosophical errors:

> Timiriazev, Maksimov, Mitkevich, considering themselves materialist, are in reality scientific reactionaries, for they either do not know or do not desire to know the real world as it is, and they strive to change it to be as they wish to see it. On the other hand, Tamm, Frenkel', Fock, starting from existence of phenomena of the real world beyond our sensations, are real materialists, although several of them are still far from a dialectical understanding of physics and incorrectly evaluate the significance of dialectical methodology for the philosophical interpretation of physics.[80]

Having finished with the physicists, Ioffe took the philosophers to task.

Ioffe held Maksimov responsible for the sad state of affairs on the philosophical front of Soviet physics. At the June 1934 Communist Academy session, Ioffe had promised that he and other physicists would study philosophy. They would investigate the relationship between dialectical materialism and physics; Ioffe himself had demonstrated that they went hand in hand. In exchange, Maksimov would study physics. In fact, he had done nothing of the kind. Maksimov, the "mouthpiece" of the Mechanists, merely repeated past accusations, charging that there was some sort of philosophical "group" intent on doing damage to Soviet physics. Yes, Ioffe responded, he, Tamm, Frenkel', and the others were a "group," since they all studied physics and participated in its development, and they agreed about the philosophical and physical content of quantum mechanics and atomic and nuclear physics. But were they idealists? Ioffe believed that Maksimov attacked physicists for idealism simply because he didn't like them, for he clearly did not understand physics. In fact, Maksimov was an idealist: "The a priori certainty of Comrade Maksimov in relation to physical theory and the rejection of its experimental bases—now *this* is idealism, if not naive ignorance of the contents of the question."[81] A dismayed Ioffe concluded:

> I see clearly the objective harm of their activity. They strive to create at [Moscow University] a center for reactionary physics. They carry on intrigues in the best journal on which Soviet physicists are brought up—*UFN*, striving to hide from Soviet youth the progressive ideas of the leading scholars, thus preserving their authority.[82]

While Timiriazev and others argued that Gessen, the enemy of the people, had created a branch center for bourgeois physics at the university, Ioffe tried to turn the tables to show, on the contrary, that the witch-hunt for idealism was by far more dangerous for Soviet physics.

It is true that physicists were not always successful in interceding on behalf of their colleagues, a number of whom would perish during the Great Terror. This was because fundamental questions had been resolved by the mid-1930s, beyond the epistemological questions which physicists and ideologues debated. Primary among these was the decision that Party philosophers, not physicists, would determine the physical content of theories. Such leading representatives of physics as Ioffe and Vavilov, and the Marxist physicist Gessen had tried to limit the damage, the former by fixing boundaries of expertise, and the latter by showing how dialectical materialism and the exact sciences were indeed commensurate. By 1934, however, the Party had decided that it would control the research agenda through planning and central administration; it would determine the extent of Soviet physicists' contact with Western scholars and ideas; and it would inform physicists where idealism lurked in the discipline, both at home and abroad.

CONCLUSIONS

What distinguishes the epistemological debates over "the new physics" in the Soviet Union from those in the West? The singling out of Jewish scholars and the representatives of the Leningrad physics community has some parallels with the rise of Aryan physics under Lenard and Stark in Nazi Germany. Many physicists throughout the Western world recognized the idealism of the Copenhagen interpretation and sought ways to construct a realist epistemology. And there were examples in the West, of debates that went far beyond institutional settings to reflect cultural and social processes at work.[83]

What sets the Soviet debates apart is the combination of these factors with the vindictiveness of the participants, the anachronism of many of the claims, and the kind of assault on physicists' autonomy that these disputes entailed. Ideologues and administrators rather than physicists gained the upper hand in determining the philosophical content of physical theories. Only after the death of Stalin, in a series of articles and conferences to commemorate the fiftieth anniversary of relativity theory in 1955, and at an all-union conference held at the Academy of Sciences in Moscow in October of 1958 would scholars reassert their primacy in resolving questions of the philosophy of science.[84]

The philosophical disputes recognized no sanctity of the research environment. Physicists were called out, shouted down, run out of town, arrested, even executed. They were expected to entertain proletarian scientists, Mechanists, and Stalinist ideologues in the open public forums, to tolerate personal attacks, and to consider changing research focus. The level of discourse reached a low in such Marxist journals as *PZM* and *Sovetskaia nauka*. The result was a field divided against itself.

Soviet physicists had participated in the genesis of quantum mechanics and maintained a leading position in the world scientific community, guaranteeing that they remained current in all developments. The financial support given to theoretical departments in a series of new physics institutes was tacit acknowledgment by the authorities of the validity of theoretical pursuits, so long as they were accompanied by applied endeavors. Because of their involvement with Marxist study circles, physicists were thus prepared on many levels to refute charges of "idealism" and primed to enter the debates in Soviet journals and public forums, although they could not anticipate their tone or duration.

In his courageous defense of the autonomy of physicists, Ioffe defined the rightful province of physicists' activity and independence. He suggested that Soviet physicists would tolerate little more interference without battle. With the assistance of such scholars as Vavilov, Tamm, and Frenkel', he proposed boundaries of expertise beyond which philosophers should not tread without having first familiarized themselves with ongoing physics research.

But in spite of the best efforts of leading scholars, the purges came to physics. The strength of theoretical physics departments in such research centers as LFTI, the Physics Institute of the Academy of Sciences, and the Institute of Physical Problems provided both the intellectual leadership and the firm institutional grounding that enabled Ioffe and others to stand up to ideologues, and protected physicists from the kind of ideological and personal onslaught which devastated biology under Lysenko. But they could not save the Leningrad physics community and its leading institute, the Leningrad Physico-Technical Institute, from decline as part of the heavy toll of high Stalinism.

9

The Great Terror and the Assault on
the Leningrad Physics Community

In the mid-1930s the Great Terror spread throughout Soviet society. Stalin set the Terror in motion, personally signing lists which condemned hundreds of thousands to death. Initially used to subjugate all of society to the will of the highest Party organs and Stalin himself, and directed against such visible enemies as the Old Bolsheviks who were seen as his rivals, the Terror gained momentum in 1937–1938, and the center lost control. Most likely, tens of millions died; ten to fifteen million were interned at one point or another in Stalin's gulag. Perhaps an equal number died in the famine caused by forced collectivization; the Soviet people suffered unspeakable losses.

Physicists were not immune to the Terror. Philosophical disputes continued in Soviet physics throughout the early 1950s. But in 1936 the assault on physics left the editorial boardroom and entered the physics community at large. Tens of leading scholars were arrested; it is not clear how many were shot or perished in the camps, or how many "merely" lost careers in midlife, to be rehabilitated and commence physics research again during Khrushchev's de-Stalinization thaw. The Terror struck the entire discipline—in Moscow, Kharkov, Dnepropetrovsk—but was directed above all against Leningrad and leading theoreticians.

The purge of the Leningrad physics community was part of a general attack on the city and culture of Leningrad and its Party apparatus, the transfer of the Academy of Sciences to Moscow, and a shift in the center of gravity of Soviet physics from Leningrad to Moscow.

In moving the Academy and establishing new physics research centers in Moscow in the 1930s, the Party was following through on the logic of Stalinist science policy: centralization of R and D in spite of a desire to bring physics to the provinces; the creation of near total autarky in science; and a top-heavy, inefficient system of supply and training of cadres, all of which forced the new institutes to stumble nearly every step of the way in building up a research base and continue to belabor the performance of R and D to this day. But the physics discipline, the Leningrad physics community, and LFTI showed remarkable resilience, indicating once again the importance of capable leadership and a stable institutional basis in protecting science from misguided, even irrational national science policy.

On the eve of World War II there were roughly one thousand physicists in the USSR, a tenfold increase since 1917. There were six physics institutes within Narkomtiazhprom, three within the Academy of Sciences, and several score physics departments in higher-educational institutions and in other commissariats, replacing the handful of prerevolutionary research centers. Physicists played a major role in the coordination of research and training activities through a series of such scientific, governmental, and Party organs as the Physics Association of Narkomtiazhprom. They issued a series of specialized physics journals which appeared in monthly one-thousand-page volumes, making up for the few sporadically published at the time of the Revolution. And they had become willing participants in the international arena of physics, through both publication in Western journals and attendance at conferences.

These achievements notwithstanding, by 1935 Stalinist politics of science had intruded upon the physics enterprise, shattering physicists' sense of autonomy and accomplishment. The institution of five-year plans subjected scientists to the pressure to undertake applied research at the expense of fundamental research. While cultural revolution had abated, the Party continued to press, through cooptation or coercion, for the positioning of increasing numbers of communists in scientific research institutes to ensure its control over science and technology both by edict from above and from within. Debates in the philosophy of science took on an increasingly bitter, personal, and political tone. And the Great Terror, which would claim a number of physicists, loomed on the horizon.

In this environment, LFTI and its director fell under attack for failure to meet the dictates of Stalinist politics of science, and its influence and preeminence began to wane. In addition to factors explored in earlier chapters, such as philosophical disputes between Leningrad physicists and other scholars and ideologues, four interrelated events played a major role in the decline of the institute: (1) the decline of Leningrad as the scientific and political rival of Moscow; (2) national political intrigues orchestrated by Stalin against the Leningrad Party organization; (3) a special Academy of Sciences session in March 1936 devoted to criticism of LFTI; and (4) the purges in physics.

THE DECLINE OF LENINGRAD

World War II completed the destruction, begun decades earlier, of the entire social and political fabric of Leningrad. The nine-hundred-day seige and blockade led to the deaths of ten times more than died in Hiroshima. The population dropped from around 3.2 million to approximately 640,000, primarily through starvation and cold. Individuals dropped on the street. Some were put on sleds and carted off to burial in immense common graves at the Piskarevskoe Cemetery. Others fell into huge snow drifts and had to be cut out when they reappeared in the spring thaws to protect against the spread of disease. Food stocks disappeared, and residents were relegated to eating shoe leather, wallpaper glue, or spoiled animal fat. The physical plant of the city was destroyed, and many of the major scientific research institutes, their equipment, and personnel were moved wholesale to Kazan and other cities in the east on trains.

The decline of Leningrad as a "charismatic active center" of social and cultural institutions had begun at the turn of the century for political and economic reasons. Then, between 1918 and 1940, the capital of Russia was transferred from Leningrad to Moscow; Leningrad ceased to be as important as a port of entry; "the city's peripheral location and lack of a large, wealthy, and populous hinterland further eroded" its importance as an industrial center, at least in relation to Moscow. While Leningrad had received 4.5 percent of investment in the national economy in the First Five-Year Plan and 4.6 percent in the Second, it has declined ever since.[1]

Moreover, Stalin saw the Leningrad Party apparatus as a last major rival to his uncontested power, and set out to destroy it sometime in 1933 or 1934. Beginning with S. M. Kirov's murder on December 1, 1934, and the immediate purge of the Leningrad Party leadership and deportation of 30,000 to 40,000 Leningraders, and concluding with the 1937 to 1938 all-out assault on the highest echelons of the Party, Leningrad was emasculated as a power rival to Stalin and the Moscow apparatus.[2] Conquest calculates that of 65 members of the Leningrad city committee elected on May 29, 1937, only 2 were reelected on June 4, 1938 (5 others were transferred); of 154 Leningrad delegates to the Seventeenth Party Congress only 2 were reelected; and all 7 Leningrad members and candidates of the Central Committee were purged. Most were probably shot.[3] Overall, between 1934 and 1939, Leningrad membership as a percentage of national Party membership was nearly halved, from 10.3 percent to 5.7 percent, and by 1946 the figure had dropped to 2 percent; it had been 16.6 percent on October 1, 1917.[4]

As Leningrad fell as a scientific center, Moscow rose in prominence. Owing to the poor quality of Soviet statistics, it is difficult to generate quantitative indicators of this shift between Leningrad and Moscow, but most observers agree that the number of institutes, scientific personnel, and graduate students increased in Moscow at the expense of Leningrad. In 1913 over 16 percent of all individuals with higher education in Russia resided in Leningrad. By 1940 the percentage was halved. As of 1955, only 5.2 percent of the country's citizens with higher education lived in Leningrad. For technical education the drop is more significant, from nearly 15 percent to under 3 percent.[5]

The major step in the decline of Leningrad as the center of Soviet physics began with the transfer of the Academy of Sciences to Moscow in 1934, followed by the establishment of the Academy's Physics Institute and the Institute of Physical Problems as major fundamental research centers. In addition, the physics associations of the Academy of Sciences and Narkomtiazhprom, both of which were located in Moscow, had assumed most of the all-union planning, professional, and supervisory functions, replacing the Leningrad-based Russian Association of Physicists, which had been disbanded in 1931.

THE PARTY AND THE PHYSICS ENTERPRISE

Stalinist policies toward science, which had been set in place in 1929–1932, became more pronounced for the remainder of the decade. Efforts to centralize administration, funding, and allocation of resources and to establish autarky in science created significant impediments to research, both in such long-established institutes as LFTI and in newly founded ones. Indeed, the Stalinist policy of centralization had the opposite effect of that intended. The creation of a bureaucracy for the administration of R and D inhibited innovation, created bottlenecks of supply, belabored hiring and firing of personnel, prevented the dispersion of resources to the provinces, precluded international contacts, and interfered with the training of creative young scholars. Finally, emphasis on industrial R and D imposed significant pressures on theoretical departments to justify their continued existence, and may have undermined the fundamental basis of scientific research institutes. The result was the establishment of a system which to this day handicaps the efforts of scientists in the largest R and D apparatus in the world to assume the lead in any field or to maintain it in areas where Soviet scholars have been first.

These policies, combined with the natural growth of the discipline, led to a decline in the preeminence of LFTI. The creation of other physico-technical institutes, in particular the Kharkov center, and of P. S. Kapitsa's Institute of Physical Problems (Institut fizicheskikh problem, or IFP) and S. I. Vavilov's Physics Institute of the Academy of Sciences (Fizicheskii institut Akademii nauk, or FIAN) helped to shift the locus of power regarding the national physics agenda from Leningrad to Moscow.

The formation of FIAN in 1934 indicates the increased importance of physics in the Academy in the 1930s, after decades during which all but seismological research was neglected. The creation of FIAN was a result of pressures within both the Academy and the government. In 1931, officials at Narkomtiazhprom desired the Academy to play a greater role in R and D, and believed that the formation of a technological division within the Academy would provide the burgeoning industry with technological innovations. New institutes were needed, and although FIAN fell within the division of mathematical and natural sciences, officials sensed its research would serve "socialist reconstruction." At the same time, party officials resolved that, given the subjugation of the Academy to Party organs, disciplines represented by the Academy ought to reflect the values of socialist reconstruction. That is, the physical and technological sciences should assume primacy over the humanities, which had long dominated the Academy. Finally, physicists, too, wished to expand the national research base far beyond the physico-technical institutes, the State Optical Institute, and P. P. Lazarev's Institute of Physics and Biophysics.

Until the formation of FIAN, physics research in the Academy labored from inattention and poor funding. After Prince B. B. Golitsyn's death in 1916, the physics laboratory of the Academy for some time had no director. Little changed when, after the Revolution, Lazarev assumed direction, as he was occupied with the formation of his Institute of Physics and Biophysics under the Commissariat of Health (Narkomzdrav). The decision at the end of 1917 to move the laboratory to Moscow in the face of the German advance and the loss of nearly all funding during the Civil War, including support for its famous seismic facilities and publication, nearly destroyed what remained. In 1921, on the initiative of Academicians V. A. Steklov, A. N. Krylov, and A. F. Ioffe, the Academy decided to create a physico-mathematics institute with these meager facilities under the direction of Steklov. In 1923–1924, with 25,000 gold rubles, the institute purchased new equipment and books for the first time since 1914, and acquired a staff of forty including, as senior physicists, P. M. Nikiforov and Iu. A. Krutkov—the latter of whom would be arrested in 1936.

When Steklov died, Ioffe became director of the new institute. But inasmuch as he was often abroad, Ioffe asked the Academy to appoint someone else. Krylov assumed responsibility and established an independent seismological institute (now the O. Iu. Smidt Institute of the Physics of the Earth) with the seismology equipment, so that the physical division of the institute remained relatively small in all respects.

At the beginning of the 1930s it became clear that a real physics institute should be created to align the research program of the Academy with national industrialization plans.[6] The first step was a new physico-mathematical institute with a physics department under S. I. Vavilov, recently elected Academician. Vavilov emphasized research in three promising areas of physics—nuclear structure and quantum mechanics, internal and external photoeffect and optics, and the structure of complex molecules—and secured Academy funding to double staff size by the end of 1933, including the theoreticians Nikol'skii, Bronshtein, Krutkov, and Fock, and the specialist in nuclear physics L. V. Mysovskii. The institute also carried out research under contract with Narkomtiazhprom.

Following the pattern of other physics institutes before it, in May of 1934 the Academy resolved to separate the institute into two organizations, the Institute of Mathematics and the Physics Institute of the Academy of Sciences, a division which actually occurred only after the transfer of the Academy of Sciences to Moscow. In the fall of 1934, Blokhintsev, Leontovich, Mandel'shtam Markov, Rumer, Papaleksi, and Tamm, all closely associated with Moscow University, were added to the staff of FIAN. For a while, a theoretical department existed in Leningrad, while departments of optics, oscillation, and the structure of matter existed in Moscow.[7] But once the transfer of the Academy to Moscow was completed in 1935, all facilities were united, and FIAN began steady growth into the broad-profile institute of theoretical, elementary particle, laser, and solid-state research that it is today. In 1939 the institute began to hold joint seminars with the Institute of Physical Problems located some two kilometers away.

The establishment of the Institute of Physical Problems (IFP) was fraught with difficulties of administration, materials and equipment, supplies, staff, and the storm surrounding Kapitsa's detention in the USSR.[8] Stalin had decided not to permit Kapitsa to return to Cavendish Laboratory in Cambridge in the fall of 1934. Having suffered the loss of Vladimir Ipatieff in 1931 and George Gamov in 1933, falling increasingly under the spell of xenophobic ideological pronouncements about the advantages of the Soviet system and the bankruptcy of bourgeois science, and needing to build up industry based on low-temperature liquefaction of such gases as oxygen and nitrogen, Stalin decided to force Kapitsa to work in Moscow. Ernest Rutherford, Peter Debye, and Niels Bohr attempted to convince the Soviet government to relent, without success.

Kapitsa at first refused to cooperate with the Soviet state. He was being held against his wishes; his wife, Anna, remained in

England; and his beloved laboratory languished in Cambridge without him. He felt isolated, and thought of abandoning low-temperature physics and moving into biophysics in Pavlov's laboratory, but ultimately grew so despondent he could not think of research. His isolation may have been a blessing in disguise, since it enabled him to write such Party figures as Molotov and Stalin—he had nothing to lose—without fear of bringing anyone else down with him, and once he ascertained their support for the new Institute of Physical Problems, he wrote often and directly to the top to put his institute in order.

Kapitsa's sense of isolation was exacerbated by the fact that most other scholars, including his former mentor, Ioffe, avoided him. After all, who would risk his own neck being associated with a rebellious young physicist in Stalin's Russia? But Kapitsa brought some of this on himself. He was extremely critical of the Academy, its top-heavy administration by septegenarians who slept through meetings, and its appointment of Vavilov as director of FIAN; he found Vavilov to be competent but not inspiring, and believed Skobel'tsyn and Fock more deserving of membership in the Academy.[9] This may have been a case of professional jealousy as much as well-placed criticism. Fortunately, Rutherford came to his rescue, working through British and Russian authorities to arrange the purchase of Kapitsa's laboratory and its shipment to Moscow.

Once Kapitsa had decided to set IFP in order, he was required to begin regular correspondence and meetings with leading government figures, including V. I. Mezhlauk, vice-chairman of the Sovnarkom and chairman of Gosplan, and one of four brothers arrested and shot between 1918 and 1938. In May 1935, Kapitsa also established contact with V. M. Molotov, chairman of the Sovnarkom, about continuing his low-temperature physics research. But he declined to publish a statement indicating his preference to remain in the USSR. "While in Cambridge science develops freely, and scholars freely travel abroad," Kapitsa explained to Molotov, "here, in the [Soviet] Union all of this is under the direct supervision of the government." True, government involvement was appropriate to ensure that "science will become not an incidental element in the life of a country but a leading and fundamental factor of [its] cultural development"; but it was also true that science, as "the highest level of intellectual work," demanded gracious and attentive support. Kapitsa, to the contrary, was deprived of his self-respect in the USSR. For four months after his detention no official paid him any notice or even authorized a bread ration card, and two agents of the NKVD conspicuously followed him. Somehow he remained devoted to his country and had great respect for the Communist party, which had accomplished such great things, "in spite of the entirely incorrect, cruel, and humiliating posture" it had adopted toward him.[10]

The government agreed to purchase his equipment from Cavendish Laboratory and set up the Institute of Physical Problems so that Kapitsa might resume research in low-temperature physics and powerful magnetic fields. The Party promised Kapitsa all the support he needed in building, equipping, and staffing the laboratory, but the institute experienced endless problems regarding acquisition of equipment and attraction of qualified young physicists, and Kapitsa was overwhelmed by bureaucratic details concerning funding and bookkeeping, salaries, and wages.[11] He was especially concerned about the fact that the treasury (the Commissariat of Finance, Narkomfin) had failed to honor its promise to provide leeway in staffing and salaries, transfers between budgetary line items without needless reporting, and the establishment of an accounting system "appropriate for a research institute."[12]

Kapitsa turned directly to the Sovnarkom and the Central Committee to get approval for his staff and budget, a clear indication that science policy decisions were made at the highest levels of the Party. Through 1937 Kapitsa wrote scores of letters to the presidium of the Academy, the Sovnarkom, and the Central Committee asking them to follow through on decisions concerning equipment and staff. On March 3, 1937, he requested the presidium to permit him to hire Landau as a senior worker at the institute, in accordance with an agreement with the former deputy chairman of the Sovnarkom, Mezhlauk, who had authorized him "to invite workers and to establish pay scale on the level of that received in the former place of work."[13] This was essential, so that his scientists would not have to take on two jobs (*"rabotat' po sovmestitel'-stvu"*) but could concentrate their efforts at IFP.[14]

The problem of cadres continued to interfere with the operation of the Institute of Physical Problems into the 1940s. Kapitsa was especially concerned about the weakness of Soviet higher-educational institutes, since faculty "value more not the student who understands the most but the one who knows the most. And science needs people who first of all understand." Kapitsa hoped to overcome this problem by developing a close working relationship with the physics department of Moscow University which, through its practica, provided him with a pool of talent from which to select the most promising students, in the same way that Ioffe used the physico-mechanical department of the polytechnical institute to provide students for LFTI.[15]

Ultimately, Kapitsa's efforts paid off. By 1946 the institute had grown to 132 people (23 Party members) including 15 scientific workers (1 Academician, 4 doctors, and 4 candidates of science, 3 of whom had won Stalin prizes, and 4 of whom were Party members); 26 support staff of whom 2 were Party members, 1 was a candidate member, and 8 were *komsomoltsy*, 9 graduate students, and 9 workers, of whom 8 were Party members.[16] But Kapitsa remained critical of the overemphasis on applied science in the USSR, the backwardness of Soviet industry and its inability to produce the equipment needed for his institute, as well as the bureaucratization of administration and finance of research.[17]

Similar problems plagued R and D even at LFTI, a well-established research center, in the late 1930s. Relations with the government reflected continuing Party efforts to improve the "science-production tie" and force the pace of innovation by administrative measures instead of relying upon such reforms as wage, patent, and copyright incentives, or even competition between research centers. Officials at the Commissariat of Machine Building (Narodnyi komissariat mashinostroeniia, or Narkommash, created upon the division of Narkomtiazhprom into several commissariats representing major branches of industry) proposed a series of different reorganizations and affiliations for LFTI, all of which unsettled the physicists and diverted their attention from research to organizational questions.

A. A. Armand, director of the scientific research sector of Narkommash, first ordered the division of some of the facilities of the institute—its library and housing—between the institute and N. N. Semenov's Institute of Chemical Physics. The physicist P. P. Kobeko complained at an academic council meeting in November 1937 that this would weaken the library's holding, if not destroy it completely. His colleague, A. P. Aleksandrov, later president of the Academy, strongly agreed. L. A. Artsimovich, one of the founders of plasma physics in the USSR, classified Armand's order as "an act of sabotage." Frenkel', Kobeko, and Aleksandrov took up the matter with the scientific-technical council of the Institute of Chemical Physics, and fortunately gained its acquiescence on this.[18] Potentially more damaging, the scientific research sector ordered the liquidation of the laboratory of plastic properties of A. V. Stepanov and M. V. Klassen-Nekliudova, as well as the entire theoretical department of LFTI. The physicists were stunned by the groundlessness of these plans; Frenkel' called them simply "unbelievable."[19]

In the face of unified opposition, Narkommash relented. The physicists seem to have temporarily persuaded officials that applied research required a firm fundamental base. The fact that physicists spoke with one voice was also important. Narkommash notified the academic council that it would bring the institute within its jurisdiction without any loss of facilities. It agreed to support the building of a cyclotron in principle, and would continue to fund physics research in such areas as synthetic rubber, insulation, rectification and mechanical properties of the solid state, metallurgy, and some defense activities.

Since the majority of these research activities fell into categories not far removed from the national electrification program, Narkommash's Bruskin then suggested that LFTI might best come under the jurisdiction not of the scientific research section but of Glavelektroprom, Narkommash's branch for the electrical industry. This reflected continuing pressure to conduct applied research. Ioffe found Bruskin's suggestion unsettling and preferred to leave the question of a close affiliation with a *glavk*, or industrial administration, open for the time being. Aleksandrov was more pessimistic, seeing a tie with Glavelektroprom as dangerous and believing it more appropriate to transfer LFTI directly to the Academy of Sciences, raising additional funds as needed through contracts with various other governmental bureaucracies as it had in the past.[20]

But the Party would not abandon its attempts to force the pace of industrial R and D. It sought greater control of scientific research through the creation of ad hoc supervisory bodies which monitored the activities of institute academic councils. Throughout its history, the academic council of LFTI, which consisted in the mid-

1930s of Ioffe, the eight other full members of the institute, two candidate senior scientists, and nineteen others including the secretaries of the institute's Party committee and Komsomol,[21] had been responsible for the day-to-day activities of the institute. It established the research agenda, hired, fired, and promoted workers, developed programs for *aspirantura*, had great freedom—except during the years of cultural revolution—in rejecting candidates for entrance, and held scientific seminars and dissertation defenses, occasionally with the assistance of such "visiting scholars" as Tamm or Vavilov.

In an effort to force scientific institutes to cede some of these responsibilities to Party organizations, the Party now insisted that academic councils share their power with scientific-technical councils (*nauchno-tekhnicheskii sovet*) and annual review commissions (*proverochnaia kommissiia*). These organizations represented an attempt to ensure by administrative fiat closer attention to applied R and D. The hope was that research advances would somehow find their way more rapidly into production. Managerial, organizational, manpower, and educational reforms which might have permitted greater institutional flexibility were avoided.

Narkommash established the scientific-technical council at LFTI to assist the academic council in administration, "to discuss the direction of [its scientific research] . . . , especially to take account of the demands of the economy for the five-year plan," and to ensure the responsiveness of research to the needs of other institutes and factory laboratories.[22] The new council met virtually every week, and usually more often, for five to seven hours at a time, to consider both annual and five-year plans. Together, the academic and scientific-technical councils were responsible for the day-to-day administration of the institute, with the latter serving as the Party's local representative in LFTI.

The Party was also involved in the supervision of research institutes through annual review commissions. As far as can be determined, these commissions were first created in the mid-1930s. The earliest mention of the work of a review commission for LFTI concerns a February 1937 report which was highly critical of the institute's research program for failing to produce "any substantial scientific results."[23] The institute established another review commission in February 1938, whose members included Ioffe; the deputy director of the institute; representatives of each department; Rusinov, the chairman of the local Party committee; its secretary, Kuprienko; and a Komsomol representative; the commission reported on its work in December 1938.[24]

Graduate education was another area in which the Party used government organs to implement Stalinist policies. After *vydvizhenie* had been abandoned and the traditional educational titles and awards restored, the Soviet government maintained control of scientific training from the top down, primarily by specifying all aspects of the program of study: curriculum, textbooks with the "proper" dialectical materialist worldview, and areas of specialization. This was a major step toward the creation of the system, so evident in the Brezhnev years, under which the national Supreme Attestation Commission (Vysshaia attestatsionnaia komissiia) established "minimum programs" for doctor and candidate degrees for hundreds of specializations. Some Western observers argue that this system encourages rote learning and narrow specialization, and stifles creativity.[25]

The scientific section of Narkommash contributed to the development of this ossified system of higher education by regulating aspirantura, as Glavnauka and Narkomtiazhprom had before it. In February 1926, M. P. Kristi and other representatives of Glavnauka had instructed all institutions under its jurisdiction to formulate programs of graduate education which specified a list of materials to be consulted, practical problems to be considered, and an outline of laboratory work, for examination by the State Academic Council (Gosudarstvennyi uchenyi sovet, or GUS).[26] GUS in turn verified the qualifications of all applicants on an all-union scale (with potential applicants required to submit a curriculum vitae, an autobiography, a list of scientific works, a diploma, and a letter of evaluation by the student's advisor).[27]

By 1930 the responsibility for all higher technical education had been transferred to Vesenkha. In the mid-1930s the All-Union Committee on Higher School Activities of the Sovnarkom assumed control over *aspirantura*. It mandated the study of dialectical materialism in all fields, including physics. In August 1937, the committee sent the directors of LFTI instructions for minimum requirements for candidate of science degrees. In addition to the usual scientific coursework and a foreign language, students were required to pass an examination in dialectical materialism.[28]

In September, A. A. Armand of Narkommash's scientific sector criticized the program of *aspirantura* at the institute, apparently for failing to admit enough students; it is true that the institute had consistently failed to meet targets from the first days of cultural revolution. A visiting committee reported that the institute "had not fulfilled any one of the main points of Order No. 31 of May 27, 1937." There was no academic planning or proper methods for evaluation. Armand scolded scientific advisers for giving students less than their due: many physicists resisted having graduate students, since it interfered with their research. In a June meeting it seems that the academic council had considered the qualifications of each *aspirant* in an open meeting in a rather brutal fashion, allowing only a few of the eight "to remain in *aspirantura*."[29] Armand concluded that the academic council of the institute had engaged in the "completely unfounded discreditation of individual *aspirants* as incapable of scientific work instead of uncovering these inadequate conditions [of *aspirantura*] and subjecting [them] to sharp criticism."

Armand singled out A. F. Ioffe and L. A. Artsimovich, who at that time was the director of *aspirantura* at LFTI, as the guilty parties, ordered the institute to establish plans for *aspirantura* by October 15, to give students in the theoretical department suitable places to "fulfill their work within the walls of the institute," and warned of another commission investigation in January.[30] Initially, Armand's dictates were ignored. At a meeting in early October, the institute accepted only four of eleven candidates for *aspirantura*, rejecting the others as being insufficiently prepared. By the end of the month, however, seven had been approved, with three in nuclear physics, two in semiconductor physics, and two in the physics of amorphic bodies. Ioffe lamented "the desire of all who passed entrance examinations to work . . . on the physics of the nucleus."[31]

THE LENINGRAD PHYSICO-TECHNICAL
INSTITUTE BEFORE WORLD WAR II

While the larger events of the Great Terror were being played out, the Leningrad Physico-Technical Institute went through a series of jurisdictional changes which reflected the continuing effort of the

Party to strengthen ties between science, technology, and production, and improve the performance of Soviet research institutes.

After eight years in Narkomtiazhprom and then Narkommash, LFTI entered the Academy of Sciences in 1939, establishing an affiliation it holds to this day. The affiliation with the Academy of Sciences was appropriate considering the general orientation of the research program of the institute in fundamental and applied physics.

Under Narkomtiazhprom and Narkommash the institute was able to ensure the stability of most of its departments, attract and hire additional staff, and embark on new research programs in such areas as nuclear physics. But the money came with strings, and the institute endured persistent interference from bureaucrats who wished to push it toward industrial development activities, and who believed that further division of the institute's facilities and staff would improve its efficiency.

In 1931 the Leningrad Physico-Technical Institute was transferred along with five other physico-technical institutes to the Commissariat of Heavy Industry, a move which reflected the government's view of science and technology as vital for domestic economic programs. In October 1937, the institute was transferred, with sixteen other institutes, to the Commissariat of Machine Building—Narkommash—as an intermediate step. Some of these institutes were subordinated to various industrial administrations, or *glavki*, of Narkommash; others, including LFTI and the Dnepropetrovsk Physico-Technical Institute, were attached to its Scientific Research Sector on Inventions (Nauchno-issledovatel'skii sektor po izobretatel'stvu, or NISIZ). The Siberian, Ukrainian, and Ural physico-technical institutes seem to have remained under Narkomtiazhprom.[32]

Having experienced the disruption and programmatic pressures associated with the jurisdictional change from Glavnauka to Narkomtiazhprom in 1931, physicists were wary of any further reorganizations. At an academic council meeting of his institute in December 1937, Ioffe raised the possibility of affiliation with the Academy of Sciences. While the council did not oppose transfer to the Academy, its members feared the preliminary plan of the Academy's presidium to divide the institute into four independent research centers. This would mean both a substantial cut in funding and a weakening of ties with Leningrad factories. The factories provided both expensive equipment and funding to the institute through production contracts. Physicists were also concerned that defense contracts which had provided the institute with income could not be continued within the Academy of Sciences; the institute was involved in anti-mine and other defense-related research.[33] When the transfer to the Academy occurred in 1939, these fears were not realized.

In 1939, LFTI finally won the approval of the scientific sector of Narkommash—with the Sovnarkom's consent—to be transferred to the Academy of Sciences. The transfer had both positive and negative consequences. Physicists were pleased by the fact that at long last their institute fell under the jurisdiction of an organization identified with fundamental research. On the other hand, the Academy was not as flush with funds as Narkommash and had a new system of accounting, bookkeeping, and hiring of workers which caused no end of problems.

Financial difficulties were nothing new. Paradoxically, the institute occasionally found it easier to secure funding for "big science" than for common reagents and equipment. The difficulties involved in getting the government to release funds in a timely fashion or in quantities sufficient to build Van de Graaff accelerators or a cyclotron notwithstanding, these funds were allocated. At the same time, were it not for contracts with factories and government organizations outside of Narkommash, the institute would have faced a 300,000-ruble shortfall in 1938. The institute's budget was restored at roughly 2,000,000 rubles per year after successful lobbying with the government.[34]

The transfer to the Academy on June 15, 1939, prolonged these financial problems. The institute was required to seek out industrial and defense contracts, while pressure to emphasize new research forced the institute to find a new institutional home for research on the problem of plastic deformation (the lab of A. V. Stepanov and M. V. Klassen-Nekliudova), and to turn over to the Ukrainian Physico-Technical Institute all activities associated with the construction of high-voltage generators.[35]

Except for the war years, when much of the institute's equipment was disassembled, loaded onto trains, and moved to Kazan to avoid German capture, and taking into account the spinning off of such other new institutes as the Institute of Semiconductors or the Kurchatov Institute of Atomic Energy (originally "Laboratory Number 2" in Moscow, formed in 1943 for I. V. Kurchatov and the Soviet atomic bomb effort using the original LFTI cyclotron), LFTI has grown rapidly to this day. As of 1990, it consisted of four departments (currently being reorganized into loosely affiliated research centers), twenty-six laboratories and twenty-three sectors and groups, and a staff of 2,500 including 500 candidates and doctors of science. In the 1930s LFTI consisted of five major departments and foci of interest, reflecting a vital and substantial research program in nuclear, solid-state and theoretical physics (see table 10). This structure changed little over the next four years.[36]

The department of theoretical physics remained a major resource of the institute before World War II, but often faced criticism, requisite in Stalin's Russia, for the absence of a "science-production tie," that is, its weak link to experimental and applied physics, as well as other "insufficiencies."[37] Another criticism centered on the fact that the department often had difficulties coordinating its activities with other laboratories. A. P. Aleksandrov noted that the work of the theoreticians had no relevance for the plans of his group on the study of amorphous bodies; he was not able to establish contact with the theoretician for the group, who was "presently on vacation."[38]

Theoreticians were determined to demonstrate that they had indeed grasped the full meaning of Stalinist science policy. In 1934 the institute reported that its theoretical work included "criticism and simplification" of Born's electrodynamics, development of a new theory of superconductivity, a series of calculations on the electronic theory of metals, and several articles on philosophy and methodology of science, including two for a volume on Engels, Ioffe's article, "The Development of Atomistic Views in the Twentieth Century,"[39] commemorating the twenty-fifth anniversary of *Materialism and Empiriocriticism*, and several discussions of the Cambridge and Copenhagen schools of quantum mechanics.[40] D. D. Ivanenko and three other physicists left the theoretical department for other institutes in 1935, allowing Frenkel', Bronshtein— who had nearly completed his candidate of science dissertation, which advanced a quantum theory of gravitation—and Pisarenko,

Table 10.   Structure of the Leningrad Physico-Technical Institute, 1935

Department	Laboratory	Director
Solid-state physics	Photoelectric properties of semiconductors	A. F. Ioffe
	Electrical conductivity of semiconductors	A. F. Ioffe
	Thermoelectrical properties of semiconductors	B. V. Kurchatov
	Amorphous bodies	B. M. Gokhberg
	Anizotropic liquids	P. P. Kobeko
	Properties of liquids	V. K. Frederiks
		A. P. Aleksandrov
Mechanical properties of the solid state	Strength	N. N. Davidenkov
	Plasticity	F. F. Vitman
		M. V. Klassen-Nekliudova
X Rays and electronic phenomena		P. I. Lukirskii
	Defraction of electrons	V. E. Lashkarev
	Photoelectric properties of metals	S. S. Prilezhaev
	X Rays	V. M. Dukel'skii
	Positrons	A. I. Alikhanov
Nuclear physics	Nuclear reactions	I. V. Kurchatov
	Hard rays	I. V. Kurchatov
	Natural radiation and cosmic rays	L. A. Artsimovich
	High-power equipment	D. V. Skobel'tsyn
	Mass spectrography	S. A. Bobkovskii
		L. M. Nemenov
Theoretical physics		Ia. I. Frenkel'

Source: *A LFTI*, f. 3, op. 1, ed. khr. 38, ll. 119–120.

who was focusing on electromagnetic phenomena, to carry on. Fock remained as a consultant, and there were two graduate students, Mamasukhlisov and Gurevich. Before leaving LFTI for the Ukrainian Physico-Technical Institute, then Kiev, Tomsk, and finally Moscow University, Ivanenko worked on the interaction of neutrons with phosphorous on the basis of Fermi's theory of beta decay.[41]

On the eve of the Third Five-Year Plan in 1937, the theoretical department advanced a five-point program designed to augment experimental research in nuclear and solid-state physics: (1) the theory of the liquid state; (2) the mechanical and thermal properties of crystals; (3) the electrical and optical properties of crystals; (4) the theory of complex nuclei; and (5) the kinetics of atomic processes. In 1939 the academic council approved the addition of the study of polymers and phase transformations (melting and crystallization) to the theoreticians, plan, thereby meeting—at least on paper—the "demands of industry."[42]

As before, the LFTI research program centered on the physics of the solid state: semiconductors; amorphous bodies and polymers; and mechanical, electrical, and photoelectric properties of the solid state (dielectrical breakdown and loss, the internal photoeffect and thermoelectrical properties of semiconductors, and the study of plastic deformation). The Third Five-Year Plan designated research on impact brittleness as a high priority, reflecting political, economic, and physics interest in the development of new, higher-quality, less brittle steels.[43] Like work on electrification, communications, and heat engineering, research in this field served the interests of government and scientists alike: it had immediate economic applications, and physicists had identified another area in the study of the solid state.

The physicists' efforts focused on surface conditions, which, more than ductile strength, influenced the magnitude of brittleness of various steels and iron. In 1934 alone, LFTI physicists published six articles on aspects of impact brittleness,[44] followed by another eight articles over the next eighteen months on the mechanical properties of the solid state, the influence of cold hardening, chromation, and so on, on the properties of various alloys.[45] The successes of these investigations led to the drafting of a five-year plan, "in agreement with directives" concerning physical questions and "closer ties with industry,"[46] to focus on the properties of several various alloys and iron, and on mechanical deformation at various temperatures in zinc, cadmium, and bismuth.[47]

On many levels LFTI managed to continue normal operation, fulfilling both the interests of its researchers and the needs of government economic programs for industrialization while going through the adjustments of new administrative arrangements. The institutional basis of LFTI remained stable throughout the 1930s. The level of budget and staff, as far as can be determined, was fairly constant, and day-to-day activities reflected little concern or awareness of the purges. In spite of the pressure to demonstrate the superiority of socialist science and to abandon international scientific contacts the breadth and focus of the research effort at the institute seems to have expanded to the limits of the institutional structure.

Yet two events in particular signaled that the sciences were about to feel the full brunt of the purges: the special session of the Academy of Sciences in March 1936 devoted to criticism of LFTI and its director, and the so-called Luzin affair in June. Ioffe himself was dressed down for inadequate attention to applied research, his faulty scientific "style," "aristocratism," and "empire building"— that is, putting personal scientific interests before national ones, and being calm in the face of an alleged crisis in the state of Soviet physics.

THE MARCH 1936 SESSION OF THE ACADEMY OF SCIENCES

The state brought its pressure to bear on the Ioffe Institute in the spirit of the 1930s. At a weeklong session of the Academy of Sciences in March 1936, physicists, Party dignitaries, and bureaucrats gathered to evaluate the performance of LFTI, its research program, and its director. The March session also addressed the achievements and failures of the State Optical Institute. But the optical institute was not the main course of the deliberations, and came in for much less criticism than LFTI.

The participants of the March session were obligated to engage in "self-criticism," to stress the alleged failings of Soviet physics under the leadership of LFTI, instead of focusing on the tremendous strides achieved under A. F. Ioffe's direction. Most of the criticism focused on the style of Ioffe's leadership, his overly optimistic view of the state of Soviet science, and in particular the failure of physicists to parlay their results into technological advances for the productive process; LFTI leaders were even accused of poor selection of research topics. The Party seems to have used the session to generate friction among scholars, turning them against each other; several physicists—Mitkevich, A. G. Gol'd-man, and F. Kvittner, in particular—used the Academy forum to air personal grudges. In the atmosphere of "self-criticism," introspection under duress, and the general siege mentality of Soviet society during the purges, it was perhaps inevitable that some physicists would use the March forum to lash out against fellow scholars.

At an academic council meeting in November 1935, Ioffe instructed physicists in his institute to prepare materials for public scrutiny and to set their house in order. He discussed his scheduled talk at the Academy of Sciences (which he then assumed would be in January). The council voted to compile materials documenting the successes of LFTI and its central role in Soviet physics: discussions of the work of each sector; a section on cadres (at the suggestion of Frenkel'); appraisals of physics publications, the ideological content of physics, and the role of conferences in contemporary science; and an evaluation of the institute's role in the domestic and international arenas since, as Bronshtein observed, "the role of Soviet physics has been exaggerated." Ioffe agreed with Bronshtein that there had been a "decrease in the specific weight of Soviet physics in the last five to eight years."[48] Within three weeks a three-hundred-page tome on the structure, personnel, and activities of LFTI and the other physico-technical institutes had been prepared; it was published on January 17, 1936.[49]

All major national and local newspapers—*Pravda, Izvestiia, Za industrializatsiiu, Tekhnika,* and many others—gave notice of the upcoming session and reported on the day-to-day activities, including accounts of the major speeches. Between March 14 and March 20 A. F. Ioffe, S. I. Vavilov, D. S. Rozhdestvenskii, I. E. Tamm, V. A. Fock, and Ia. I. Frenkel' described the research programs of LFTI and the State Optical Institute.[50] S. I. Vavilov and D. S. Rozhdestvenskii offered detailed analyses of the achievements of the optical institute;[51] I. E. Tamm discussed the problem of the atomic nucleus, V. A. Fock presented his observations on "the problem of many bodies in quantum mechanics," and Frenkel' spoke on thermal motion in solid and liquid bodies and the theory of melting.[52] An open debate followed in which virtually every major figure of prewar Soviet physics—Kurchatov, Landau, Gessen, Tamm, Lazarev—took part.

Physicists recognized the solemnity of the occasion. As Vavilov observed, it would mark a turning point in the history of the Academy and all Soviet science if only the scientific community adopted an "attentive" and "serious" attitude to its results. The March session was indeed a "new form of scientific inspection," which would demonstrate the efficacy of "self-criticism"[53] in evaluating the performance of Soviet physics with regard to such Stalinist criteria as the unity of theory and practice, the unquestioned virtue of applied science, the danger of "ivory tower" theoretical research, and the importance of the proper training of cadres.

Ioffe's three-hour address was the centerpiece of the meetings[54] and described his work and that of his institute in solid-state physics. The first and longest part of his presentation covered some forty different research topics, including several in nuclear physics and isomerism which had yet to be published; the problems of socialist technology in light of the achievements of LFTI; the organization and structure of Soviet physics, including training and personnel; and prospects for the next five-year plan.

Ioffe described how in eighteen years his institute had grown into a system of fourteen institutes and three higher technical schools with almost one thousand scientific workers, one hundred of whom were first-rate physicists, with thirty doctors of science.[55] He emphasized the accomplishments of recently begun research programs in nuclear and semiconductor physics, although he mentioned such failures as the attempt to produce thin-layer insulation and was critical of inadequate government support in nuclear physics.[56] He concluded that Soviet physics now ranked no lower than fourth in the world.

In view of the achievements of Soviet physics since the Revolution—the formation of new institutes, the establishment of fundamental research programs, the training of cadres, and Ioffe's central role in all of this—the criticism that enveloped Ioffe from all sides stunned many of the participants. But others used the occasion to grind their own axes, dredged up such failures as the case of thin-layer insulation, and called him far too optimistic in his evaluation. The criticisms of his institute and leadership were seen by many as criticisms of the state of Soviet physics as a whole, with a number of participants finding Ioffe personally responsible for the sad state of affairs.

In the second part of his presentation, the one that opened him up to the most criticism, Ioffe devoted attention to the relationship between "physics and production," and demonstrated the close tie between the two at his institute. He presented a list of the results which had found their way into the electrical, heat engineering, chemical, and telemechanical industries from the network of physico-technical institutes. These included the string method for the measurement of tension; X-ray methods for the study of imperfections and the structure of steel and alloys; magnetic methods of control; new insulating materials; high-tension power lines, transformers, and associated equipment for GOELRO; and advances in agriculture produced within the Institute of Physical Agronomy.[57]

In his positive evaluation of the role of physics for Soviet industry, Ioffe drew fire for arguing that physicists were not responsible for ensuring *vnedrenie* (assimilation) of their achievements into industry; rather, he maintained, physicists were "consultants" to industry. Ioffe based his analysis on long-term firsthand observation of Soviet and Western factory laboratories and research institutes, beginning with his visits to Siemens, Westinghouse, General Electric, and other industrial laboratories. His inspection of Leningrad factories in 1926 had resulted in a plan for the organization of 111 factory laboratories which Vesenkha had adopted for Leningrad and then applied in other cities, and which included contracts and consultancies that LFTI had worked out with the Bolshevik, Svetlana, and Krasnyi treugolnik factories.[58]

The threefold hierarchy of physics institutes, branch industrial institutes, and factory laboratories that Ioffe had envisaged in the 1920s remained the key to accelerating the assimilation of scientific advances in industry. These facilities would be staffed by physicists, technical physicists, and engineers respectively. All industrial urban centers would have physics institutes connected with branch institutes and *vuzy*. The physics association of the Academy would supervise and coordinate physics research activities from the factory laboratory to the branch institute to the physics research institute.[59] Ioffe's discussion reflected the firm belief that his institute had successfully met the challenge of an improved "science-production" tie.

Finally, while paying lip service to Stalinist policies, Ioffe criticized the net result of those policies. As he had for nearly twenty years, Ioffe pressed for physicists' continued control over both research programs and instruction. Hoping to forestall autarky in science, he urged the maintenance of international contacts and cooperation, and spoke of his instrumental role in changing the "provincial face" of Soviet physics, turning it into a major participant in the international arena. And he called for better organization and less bureaucratization of scientific research: fewer conferences or formal responsibilities, especially for academicians.

With regard to training, Ioffe suggested a greater division between research and educational activities. Ioffe did not see, however, that this would further separate the two regions, exacerbating a problem which persists in Soviet science to this day. It is unclear why he believed that teaching and research were not incompatible in such facilities as the physico-mechanical department of the polytechnical institute. Ioffe concluded his discussion of education with the usual calls for self-criticism, improved ties between theory and experiment, and greater attention to the "proper Marxist methodology and knowledge of the history of physics."[60]

In the three sessions of debate that followed, many physicists took exception to Ioffe's report on the state of the physics enterprise. Some participants accused him of being self-serving and arrogant. But Ioffe could rightly take credit for many of the accomplishments of Soviet physics, and it is likely that such attacks reflected the tenor of the age and the tradition of "criticism" and "self-criticism" in Soviet Marxism.

Nineteen physicists, many of whom had been Ioffe's students—or students of his students—examined the Leningrad Physico-Technical Institute in microscopic detail. They were concerned with three interrelated issues. First, how much was Soviet physics obligated to Ioffe for his leadership? Did the interests of Soviet power come first, or had he created a self-serving empire of physics institutes? Second, what was the correct relationship between science and technology? That is, did physics research lead directly to technological advance and impel it onward? Or, was physics limited to the more modest role of consultant, as Ioffe had suggested? Indeed, had the achievements of his institute been insufficiently linked to technology and production? Third, was Ioffe too optimistic in his predictions for the future, and had he not adopted research programs in such unpromising regions as solar energy, thin-layer insulation, and semiconductor physics, so that there was little hope of quick return for government investment? Even those critical of Ioffe prefaced their remarks by saying that it was a "great honor to have been a student of Abram Feodorovich," but the discussions were "heated and even disturbing" and "unusual from the point of view of the traditions of the Academy of Sciences."[61]

While industry and government representatives were unanimous in their belief that the work of LFTI should be more closely linked to the demands of industry, leading Soviet scholars disagreed over this matter. A. A. Armand, the government representative from the scientific sector of Narkomtiazhprom, began with a positive evaluation of Ioffe's contributions to Soviet science and industry. However, Armand focused most of his attention on the significant impediments to assimilation (*vnedrenie*) of the achievements of science by industry, all of which represented exceptions to Ioffe's views. There were too few physicists in industry, and still fewer understood that their role was more than as consultants. Armand gave chapter and verse on the fact that physicists "had nothing to do with industry." One such example was X-ray analysis for product control, which had been developed in LFTI but had yet to find its way into production, while European industry had already wholeheartedly embraced the process.[62]

D. S. Rozhdestvenskii, whose institute had a very close relationship with the Soviet optical industry, naturally shared Armand's views, but as a colleague was surprisingly unrestrained in his criticism. He argued that LFTI was "almost free of responsibility of any kind for applications in physics." It was "occupied with abstract problems. This is a step backwards. . . . I consider the organization of any physics institute exclusively on theoretical disciplines a mistake."[63] The theoretician I. E. Tamm took exception to these comments, arguing the importance of preserving a strong base of "pure" research. B. N. Finkel'shtein and A. K. Val'ter agreed with Tamm, attributing the problem of delayed applications to temporary weaknesses in the work of the theoretical department.[64]

Other scholars used the special session to attack Ioffe personally, although their criticisms seemed to focus more generally on the "insufficiencies" of Soviet physics. F. Kvittner, who had engaged Ioffe in a bitter dispute over the theory of crystal structure at the beginning of the 1930s, thanked the Soviet authorities for the opportunity to speak out against Ioffe, for only in the USSR was such criticism possible and necessary! He then castigated Ioffe for such failures as thin-layer insulation, which demonstrated lack of care or of a critical attitude in experimentation. Ioffe's theory of crystal construction was behind the failures of thin-layer insulation. Ioffe had applied wave-mechanical conceptions to Born's theory of ideal crystals which, being "undialectical and essentially formally constructed," in Kvittner's thinking, simply did not conform to real conditions. Worse still, Ioffe had lied to cover his knowledge of impending failure, and had neglected to make note of this error with sufficient modesty. Ioffe would not have wasted resources on these pursuits had he listened to Kvittner.[65]

Another physicist who exploited the proceedings for personal gain was A. G. Gol'dman, a specialist in the physics of the solid state who had worked in Petrograd during World War I before moving to Kiev where he played a leading role in the organization of Ukrainian physics. Ostensibly, Gol'dman intended to devote his attention to an evaluation of semiconductor research at LFTI. But Gol'dman's major goal was to revive a dead issue. As Vol'kenshtein reported,

> Further, A. G. Gol'dman dwelled in detail upon the discussion between himself and A. F. Ioffe several years ago. The discussion touched on the question of the form of the dependence between electromotive force which is manifested in a crystal as a result of its illumination, and the intensity of this illumination. In Gol'dman's opinion, the views of A. F. Ioffe had undergone tremendous evolution in recent years. A. F. Ioffe had finally come to the formulation which Gol'dman had first experimentally established already in 1928. Academician Gol'dman considered it necessary to mention this, since in the article of Ioffe references are given only to the work

of [physicists] of LFTI and the work of Gol'dman is entirely ignored.[66]

According to Gol'dman, the course of Ioffe's work suggested a path of "clever guesses" based on individual experiments, not systematic research. Indeed, contrary to Ioffe's sense of things, LFTI was not the leading center in semiconductor physics. Insufficient applications in technology resulted from Ioffe's view of physics as the mere "consultant" of industry.[67]

Mitkevich also took advantage of the meeting "to settle an old score" with Ioffe, Fock, Frenkel', and Shpil'rein. He pleaded once again for a "yes" or a "no" response to queries regarding his belief in the impossibility of action at a distance. He referred to Frenkel' as his "antagonist and friend," whose *Scheinproblem* "Does the devil have a tail?" still bothered him. Mitkevich suggested that if the work of LFTI were better tied to production, then such "idealist" depictions of reality as action at a distance could be avoided.[68] Having had enough of the whole affair, Tamm rebutted Mitkevich's comments and called for a special Academy of Science session to look into his work.[69]

For Ioffe, two of the most troubling comments came from two former students, L. D. Landau and I. A. Leipunskii. Landau, who was infamous for his frank, if tactless, personality, was unreserved is his attack on Ioffe's style of leadership, his work, and his views on cadres. (Elsewhere, on more than one occasion, he had disputed Ioffe's claim to preeminence as a first-rate scholar.) First of all, Landau described the catastrophic problem facing Soviet physics concerning cadres. There were hardly the number of physicists Ioffe suggested, and many of these were merely test-tube washers. At most, there were several hundred physicists, who in no way could begin to address the problems facing the USSR. The situation was even worse concerning the quality of teaching: poor instruction had resulted in incompetent engineers and had slowed *vnedrenie* in industry. Unlike Rozhdestvenskii, Landau did not believe that physicists should double as technologists, however. Such a situation where physicists with insufficient technical knowledge might replace technologists would "lead undoubtedly either to simpleminded or unrealizable or already completed projects."[70]

Landau was particularly critical of Ioffe's scientific style. His projections for future energy technologies were farfetched; the work of his physicists was messy; and he had a tendency to make claims about Soviet priorities in discovery where Western physicists had earlier achieved similar results. Ioffe was, in Landau's mind, responsible for the insufficiencies of experimental work throughout the USSR, as the example of thin-layer insulation indicated.[71] More deleterious for Soviet physics was Ioffe's overblown, haughty, and self-aggrandizing leadership. Landau offered the example of an experiment conducted by A. K. Val'ter and K. D. Sinel'nikov which essentially had repeated work of Cockcroft and Walton of some years earlier. But the two researchers had sent Stalin a telegram describing the experiment as if it were a world first (see chap. 5). Of course, Ioffe himself was not responsible for the telegram, but his bravado had encouraged his co-workers to exaggerate.[72] While there was some truth to Landau's charges, it is clear they were overstated and unfair, and it is hard to believe that he could not fathom their impact in Stalin's Russia.

Leipunskii's criticism were less personal and therefore perhaps more telling. While Leipunskii agreed with Ioffe that Soviet physics had achieved third or fourth position in the world, he urged physicists to become cognizant of just how far behind they remained. Soviet "schools" of research were hardly as productive as those in the West, and often followed the European lead. The root of the Soviet lag, he believed, was an unsystematic approach to research, the failure to identify the most promising areas of study or to carry them through to the end. Here Ioffe was responsible for orienting young physicists wrongly toward the slogan "Physics is the technology of tomorrow." Ioffe's "fantastic" ideas on energy technologies—tapping the temperature difference between arctic water and air, solar energy, and so on—had diverted physicists from imminently solvable contemporary problems. This was all tied in with a criticism like Landau's, of Ioffe's alleged "aristocratism" in physics and technology.[73]

In the environment of mistrust and attack on physicists' autonomy, only such senior figures as Tamm, Fock, and Frenkel' stood firm in their allegiance to Ioffe, and it is difficult to understand why the others directed so much vitriol toward him. They must have known the importance of their words, but perhaps were caught up in the seige mentality or, to be even less flattering, hoped to use the March session for personal gain, and end Ioffe's domination of the field.

Tamm most eloquently took exception to all of the negative remarks. He argued that it was precisely Ioffe's leadership that had secured the development of the Soviet physics enterprise and set its course for the future:

> The basic question consists not in the presence of insufficiencies, but whether the sum total of positive results during the multiyear work of A. F. Ioffe exceeds those insufficiencies which are present in any labor. For me personally, there is not the slightest doubt in the fact that the positive results of the organizational activity of A. F. Ioffe, the positive results of the assimilation of scientific and physical culture in Soviet science, the deep ideological influence which he and his school have had on the entire preceding level of development of Soviet science, that in all these relations the merits of A. F. Ioffe and his school undoubtedly in essence exceed [those] of two other Soviet schools, the school of Rozhdestvenskii and the school of Mandel'shtam.[74]

Tamm attempted to smooth the waters with Landau, Kvittner, and others, urging them to reconsider some of their accusations.

Ioffe failed to hear the few hymns of praise. He was stung by the criticism, surprised at its tone, shocked at its extent. Ioffe's leadership had been attacked, and LFTI had lost its previously unassailable position as the leading physics institute in Soviet Russia. Ioffe rebutted the personal attacks incisively, disagreeing with Landau and Leipunskii, and rejecting Gol'dman's and Kvittner's pleas that they receive praise for his work. And he carefully defended his program for the continued development of Soviet physics.[75] But the damage had been done: Soviet physicists were divided over questions of cadres, the correct relationship between science and technology, and how to fix research programs.

From this point on, individuals or entire institutes would tend to dominate whole fields of research. Some scholars used Stalinist methods to eliminate rivals. Others took advantage of an overly centralized system of R and D to establish fiefdoms, and the broad-profile central research institute which Ioffe had envisaged began its decline. By the middle of the Brezhnev period, for example, FIAN had become in the eyes of many Soviet physicists an applied research institute for the Nobel laureate N. G. Basov, who sacrificed other programs for his own.

The March 1936 Academy of Sciences session appears to have been a well-orchestrated attempt—coming from within the Academy, Narkomtiazhprom, and the Party itself—to weaken Ioffe's influence once and for all. As such, it was part and parcel of the move to eliminate centers of power or influence that rivaled the Moscow Party apparatus and Stalin. And Soviet physicists had been denied Ioffe's able stewardship.

THE PURGES IN PHYSICS

The Luzin affair was the second indication that fundamental science would soon fall under the momentum of the purges. Academician N. N. Luzin, a first-rate mathematician and educator, was the leader of Moscow theoretical mathematicians from the 1920s, and influential in a number of Academy organizations in the 1930s. In July of 1936, apparently at the instigation of Ernst Kol'man, Luzin came under attack for sabotage, wrecking, toadyism, political double-dealing, "traditions of servility," and "kowtowing [*nizkopoklonstvo*]" before Western science (this became a favorite criticism during the anticosmopolitanism campaign of the Zhdanovshchina in the 1940s).[76]

The important point was the emphasis on kowtowing and the existence of "enemies in Soviet masks" in science. At the end of August 1936, the presidium of the Academy called for the eradication of enemies of the people and of the Luzinshchina. While Luzin himself was not arrested, and ultimately returned to most of his former positions, the importance of this event was its public visibility. It represented a rise in anti-Western feeling and in hostility toward Trotskiism and other tendencies prominent among alleged enemies within the USSR, and signaled an attack on theoretical scientists.[77]

To backtrack slightly before turning to the purges, it must be recalled that the Party had for several years now tried to limit the contact of Soviet scientists with Western scholars and ideas. This was a logical outcome of "proletarian" science, its denigration of the achievements of Western science, and its roots in "socialism in one country," that is, the belief that truly patriotic Soviet scientists could surpass Western science on their own. The result was autarky in science, just like that created in the economy, where by 1937 no more than one percent of the gross national product involved foreign trade.

The pressure for autarky caught physicists between countervailing forces. On the one hand, they had expended great effort to reestablish scientific ties with the West and recognized the importance of those ties for the health of scientific enterprise. International connections promoted currency in selection of research topics and served as verification that Soviet scholars were on the right path. Indeed, they were required by the state to surpass Western achievements because of the self-proclaimed "advantages" of the Soviet social system. This was a task made easier by frequent, normal contacts with Western scientists and technologists. On the other hand, the pressure to conduct research divorced from the West grew during the Stalin years. The pressure was both political and philosophical, political because physicists had to demonstrate that Soviet science could outdistance capitalist science, and philosophical because they were required to reject a number of contemporary scientific theories advanced by such "idealists" as Born, Bohr, and Schrödinger. As Stalinist ideologues saw it, success under autarkic conditions would demonstrate the superiority of socialist science in addition to securing the motherland's borders.

Soviet physicists were able to blunt the impact of autarky to a small degree. They maintained some of the international contacts they had worked so hard to establish. Conference reports reveal that such physicists as Bohr, Langevin, Joliot-Curie, and Pauli attended meetings in the USSR, although their numbers declined precipitously. Viktor Weisskopf, Boris Podolsky, Fritz Houtermans, Martin Ruhemann, and Alexander Weissberg all worked at the Ukrainian Physico-Technical Institute; during the purges Houtermans and Weissberg were arrested and somehow survived to be extradited to Germany. Ruhemann was forced to leave. But travel to the West virtually ceased after Vladimir Ipatieff in 1931 and George Gamov in 1933 failed to return from conference travel.

George Gamov, who had originally gone abroad in 1928 at the suggestion of O. D. Khvol'son, and later accepted Rockefeller funding for his stay in Göttingen, then received an invitation from Bohr to study in Copenhagen with a Carlsberg fellowship. Gamov returned to the USSR in the spring of 1931 under instructions from the Soviet ambassador in Copenhagen. He had hoped to spend a

few more months in Europe before traveling to the first international congress on nuclear physics in October 1931. Upon arrival
in Leningrad, Gamov recalled, he immediately sensed the great
changes brought about by cultural revolution. Earlier the government had recognized the importance of contact with scientists
beyond Soviet borders, and had been proud of Soviet scholars who
were invited to conferences in the West.

Now, however, all that had changed. Gamov and his wife began
to plot to escape, first purchasing a kayak through the sportsmen's
club of Dom uchenykh for an aborted attempt to paddle to Turkey
from a TsEKUBU resort hotel in Crimea, next considering a ski
assault in a Karelian village on the Finnish border, and finally talk-
ing Molotov into approving visas for both Gamov and his wife to
attend the Solvay conference on nuclear physics in Brussels, to
which Gamov had been invited officially in 1933.[78] Gamov would
not return, and eventually settled in the United States.

Kapitsa understood Gamov's desire to work abroad. There was
little to hold him in the USSR in terms of nuclear research, and he
was clearly happier abroad. But as Kapitsa wrote Niels Bohr from
Cambridge, before his own detention in Russia after a summer
1934 visit, Gamov's failure to return made it extremely unlikely
that other young physicists who wanted to study abroad would be
given the opportunity. The only solution would be to get permission from the Soviet government to remain abroad, and perhaps
only Ioffe was capable of achieving such an outcome.[79]

The number of Soviet scientists who traveled to the West
dropped precipitously in the early 1930s, and although exact
numbers are not available, the sense is they could be counted on a
few hands. Only after Stalin's death, and especially after the
appearance of Kurchatov with Khrushchev at the British center for
nuclear research in Harwell in 1956, were controls on foreign travel
somewhat grudgingly released, but usually only for scientists who
had shown their political reliability.

The centralized receipt and censorship of scientific publications
also grew into an art at this time. Physics institutes continued to
subscribe to and receive a large number of Western scientific journals; LFTI alone subscribed to 124 journals in 1935, spending
sacred hard currency on "bourgeois" publications. But scientific
information was allowed to diffuse very slowly while journals were

cleansed of any "anti-Soviet" information. The large number of journals "received" during the 1930s is therefore no guarantee that physicists were able to read them all as needed in a timely fashion.[80]

Contacts with the West of any sort could be dangerous, and scholars had to protect themselves against unfounded charges of collusion with "enemies" of the Soviet Union during the height of the purges. This resulted in the practice of having the entire academic council of an institute vote to approve sending reprints to Western physicists. This spread the responsibility among the entire staff, saving individual scholars from the appearance of collusion; as elsewhere, it had been common practice for scholars merely to send copies of their offprints abroad.[81] LFTI sponsored exchanges of work and correspondence with a few leading European scholars, but not one physicist by the late 1930s participated in foreign congresses, conferences, or travel.[82] Upon receiving a request for a physics journal subscription from Germany in 1940, Ia. G. Dorfman carefully and pointedly indicated that the request had been unsolicited, and that he had maintained only the most proper scientific relations with Western scholars.[83] In addition, in a sign of protest over the rise of National Socialism, physicists ceased sending reprints to German physicists and halted publication of *Physikalische Zeitschrift der Sowjet Union* in 1937.

Far more damaging to science in human terms were the purges which accompanied the Great Terror. Until archives are fully open we can only guess at their extent in physics. David Joravsky provides a benchmark for comparing the loss for biologists and physicists, listing twenty-two repressed physicists and philosophers of physics (S. F. Vasil'ev, I. M. Uranovskii, S. Iu. Semkovskii, and Boris Gessen) and eighty-three biologists, philosophers of biology, and agricultural specialists in the Appendix of *The Lysenko Affair*.[84] The list, he notes, comprises "only a fraction of all [those] . . . who suffered repression . . . Rank-and-file specialists who did their work without publishing suffered repression in obscurity."[85]

On the basis of recently published memoir literature, articles by Soviet historians of science, access to archive materials, and several interviews, it seems likely that over one hundred physicists were arrested in a purge of Leningrad physicists in 1937–1938. In light of Ioffe's claim that there were altogether one thousand physicists in the USSR in 1936, this is a staggering sum. If one out of ten physicists is too high a figure, we can still judge the depth of the repression through an evaluation of the arrests of Boris Gessen, Matvei Bronshtein, Lev Landau, Alexander Weissberg, Vladimir Fock, and Iurii Rumer. One could add to the list Shubnikov, Shpil'rein, the relativists Frederiks and Krutkov, Gorskii, Rozenkevich, and Romanov, a physicist at the university in Moscow, who were arrested and shot, and Obreimov, Leipunskii, and others who were arrested but later released.

What is clear is that the processes at work in society at large—a seige mentality, a belief that enemies lay everywhere, informing on former comrades, torture and execution in the prisons, and disappearance in the camps—all of these things condemned physicists—almost always young, promising Leningrad theoreticians, Marxist or not, including future Nobel laureates—to fear, uncertainty, and inhumanity, and it is surprising that they were able to accomplish as much as they did.

Gessen does not fit the profile of most of the purged physicists. (What happened to Semen Iulevich Semkovskii, the Marxist scholar in the Ukraine whose career pattern most resembles Gessen's, is unclear.) As a Marxist physicist Gessen had spent nearly ten years in the public view defending ideas which the Mechanists and Stalinist ideologues found dangerous and offensive. He had been associated with Trotskiite, internationalist parties at the time of the Revolution, an affiliation which almost guaranteed a pass to prison. He had managed to stay active in the Communist Academy of Sciences, in spite of the identification of his position as that of a "menshevizing idealist," and was dean of the physics department at Moscow University until 1936, when Timiriazev's incessant accusations that he was an "enemy of the people," and an organizer of a center of Trotskiite, fascist physics at the university and on the editorial board of *UFN* found resonance in the secret police, or GPU.

Gessen offended a series of physicists during his decade of activity as a philosopher of science. Gamov, Landau, and Bronshtein ridiculed his belief in the need to maintain the existence of the ether, relativity theory notwithstanding, and had sent him a telegram in which they likened it to the other "fluids" once rejected in the history of physics.[86] D. D. Ivanenko's self-aggrandizing view of the development of Soviet physics pushed him to side with Timiriazev against Leningrad physicists and Gessen. Ivanenko was subsequently arrested and then released, and again took up cudgels against the Leningrad physics community in the 1940s, trying to orchestrate a session for physics in 1949 like that organized by Lysenko for biology the previous year.[87]

But the stakes were highest with Timiriazev, who stood on the outside of the physics community looking in and used denunciation for personal gain. Gessen and Vavilov angered Timiriazev with the report of a commission they chaired to investigate Timiriazev's charges about the sad state of affairs in the physics department at Moscow University. They defended the department well from typical accusations about the weakness of ties with industry and the presence of idealism in faculty teaching. They also rejected the value of Timiriazev's *Vvedenie v teoreticheskuiu fiziku* as a textbook.[88]

According to a physicist who was a student at the university at that time, Gessen was arrested in late summer or early fall of 1936. It came as a shock to many students and faculty; Gessen was widely respected, made a good impression with students, and knew physics well enough. But at a general meeting of the scientific institute of the university in August or September, without prior notification to senior physicists or administration, young Communists and Komsomol members announced that Gessen had been arrested as a wrecker and "enemy of the people," which meant he had already most likely been shot, and asked all participants to contribute to the general abuse. They accused Gessen of "a lot of nonsense," said he "cultivated bad instincts among the young workers" since they drank vodka all night—as if Russian workers drinking vodka at the university was Gessen's fault—and claimed he was a "wrecker" concerning the educational program, since few qualified students were graduating.[89] The fault here truly lay with the Party and its educational policies, but it was easier to create scapegoats than admit fallibility.

In 1937 the purges gained momentum among Leningrad physicists and those associated with the Ioffe school. Lev Landau often provoked negative reactions because of his abrasive personality. This in itself would have been enough during the Stalin years to have earned him some time in prison, but it was Landau's ardent

Trotskiism which drew more attention.[90] Landau escaped the hostile environs of the Ukrainian Physico-Technical Institute, where arrests were brewing in 1937, for Kapitsa's Institute of Physical Problems. In Moscow, Landau was ensnared by the purges and was saved only through Kapitsa's intervention.

For Iu. B. Rumer, the arrival of Landau in Moscow was like a holiday. The two had long been close friends and spent all of their free time together; indeed, Landau lived with Rumer until his apartment at IFP had been built. In 1938 Landau and Rumer were arrested on the same day and put in the same cell. "We smiled at each other," Rumer said, "and, of course, we assumed they wanted to hear our conversations, which is why they put us in one 'envelope'. . . where you could stand and sit, but not walk."[91] But soon it became clear a mistake had been made, and two days later, on April 28, Rumer's birthday, they were separated. The arbitrariness of the purges would now work its magic as Landau was in prison for one year, Rumer for ten; and Landau returned to IFP as leader of its theoretical department, whereas Rumer suffered another exile as a teacher to the Eniseisk Teachers' Institute.[92]

Iurii Borisovich Rumer[93] entered Petrograd University in the mathematics department in 1917, but then transferred to the mechanico-mathematics department of Moscow University. Upon graduation he taught mathematics in a series of technical schools and *rabfaky* (workers' faculties). In 1927 he traveled to Göttingen to study, and was taken on as Max Born's assistant through 1932. During this time he worked on relativity theory, discussed his work with Einstein in Berlin, and collaborated with Weyl and Teller on the quantum theory of chemical valence. He returned to Moscow in 1932 as a docent and in 1933 became a professor. He was awarded his doctorate of science in 1935 without a defense. In 1935 he became a senior scientist at FIAN. In 1938 Rumer was arrested, accused, without much originality, of being a German spy. This was more than two-faced, since the Soviet government soon signed a secret nonaggression pact with the Nazis. But when Rumer agreed to work as an "engineer for Soviet power," he was sent to work in the so-called Tupolevskaia sharaga—the system of special camps for scientists and engineers involved in secret defense work[94]—in a series of aviation factories in the forth special division of the NKVD. When released in 1948 he was given a teaching position at the Eniseisk Teachers' Institute in Siberia; he then, at Vavilov's request, became head of the department of technical physics in the west Siberian filial of the Academy, and later director of the Institute of Radiophysics and Electronics in Novosibirsk (1957–1964).[95] Finally, he joined the Institute of Nuclear Physics, where he remained until his death in 1981.

Landau, it turned out, was on thin ice. He brought the worst to bear on himself from the secret police and was forced to sign confessions to the most unimaginable crimes.[96] He would have perished had not Kapitsa intervened by writing Stalin on April 28:

> This morning they arrested a scientific worker of [my] institute, L. D. Landau. In spite of his twenty-nine years, he along with Fock are the most outstanding theoreticians in the Soviet Union. . . . There is no doubt that the loss of Landau as a scholar both for our institute and for Soviet and for world science will not go unnoticed and will be felt deeply. Of course, no matter how great his scholarship and talent may be, this does not give a man the right to violate the laws of his country, and if Landau is guilty, then he should pay. But I beg you in view of his supreme talent to order very close attention to his case.[97]

Kapitsa acknowledged that Landau had made a lot of enemies because of character faults—he was a squabbler, a tease, and a bully who loved to find others' mistakes, and granted that it was not always easy to have him at the institute. But he was young and had a great future in science.[98] The letter to Stalin went unanswered.

Landau was in prison for one year when he realized he would perish one way or another within six months. Kapitsa tried again to free him, suggesting that he would leave IFP if Landau was not released. Beria called Kapitsa to the Kremlin, where he signed a declaration to Molotov asking that Landau at the very least be given permission to work as an engineer in the camps. Kapitsa gained Landau's release on April 28, 1939, exactly one year after his arrest—and on Rumer's birthday.[99]

Fock also was rescued by Kapitsa after his arrest in 1937. Kapitsa, who was in Leningrad at the time, heard about the arrest and immediately sent a letter to Mezhlauk, deputy chairman of the Sovnarkom. Kapitsa informed Mezhlauk that he considered Fock "our most capable physicist-theoretician." Fock was nearly deaf, therefore completely uninterested in problems of day-to-day life and entirely devoted to science. Fock was accused of "wrecking" results of geological investigations. "I cannot imagine that such a man could commit a great crime. There must be a mistake," Kapitsa told Mezhlauk. It was possible someone had misused Fock's work.[100]

Fock was imprisoned for about a week, carried to Moscow for interrogation, interviewed by Ezhov, then head of the secret police, and returned to Leningrad intact. Fock did not initially recognize Ezhov, who informed him that the presence of so many traitors in the Soviet Union required that the innocent sometimes suffer. But Fock did not condemn anyone else to arrest.[101]

Fock's arrest was associated with the dispute with the Mechanists and Mitkevich, the journalist V. E. L'vov, and the theoretician K. V. Nikol'skii. In 1935, the division of theoretical physics at Moscow University had voted "to invite Professor K. V. Nikol'skii to read a course on . . . quantum mechanics" in the next academic year.[102] This was the same Nikol'skii who wrote "the first serious attack on the customary interpretation of quantum mechanics in a physics journal rather than a philosophical journal"—*UFN* in 1936.[103] Fock defended quantum mechanics, although not its Copenhagen interpretation.

Fock was acting to a certain extent on the instructions of Maksimov, who seemed to be playing the ends against the middle. Maksimov had read a lengthy article which Fock had penned, "Physics in the Soviet Union and the West," and urged Fock to abandon his hope of publishing it in *Pravda* but to consider something like it for *PZM* where Maksimov was on the editorial board. He called specifically for Fock to explain how physics in the West was "putrefying," in particular the work of Eddington and Jeans. This required "our leading physicists to attack idealists and clerical conclusions."[104]

After his release by Ezhov, Fock defended himself from attacks by those most likely associated with the pressure to arrest him. First, he discredited the vile attack in *Novyi mir*, in which L'vov described him as a "scholarly obscurantist who had exacted the methods of Hitler" and attacked his "openly fascist" relationship to the work of Mitkevich. Fock protested the Leningrad Regional Procuracy on July 20, 1937: "Regarding whether or not I am a Fascist, it would be best if Comrade N. I. Ezhov, member of the Politburo and Commissar of Internal Affairs who summoned me to

Moscow on 15 February 1937 for a personal interview and conversation which I will never forget, judged."[105] Fock simultaneously wrote a letter to the science department of the Central Committee of the Party concerning L'vov's attack on him and Mitkevich's anachronistic physics. The science department apparently decided to do nothing in support of Mitkevich's ideological struggle.[106]

In 1938 the Mechanists and their allies stepped up their criticism of the Leningrad "school," focusing on quantum mechanics and conservation of energy. Nikol'skii accused Fock of establishing a "filial" of the Copenhagen school in the USSR and of defending idealist and Machist ideas with the help of Tamm. This criticism took in Heisenberg and Bohr, but found Nikol'skii in the strange position of siding with Einstein, the Machist of Machists, who had quarreled with Bohr over the philosophical suppositions of quantum mechanics.[107] Fock, for his part, admirably defended the quantum theory, calling for a struggle with the idealist thesis of the contradiction between the principle of complementarity and materialism.[108] But nothing further happened, and Fock maintained a prominent role in the defense of quantum mechanics until his death in 1966.

Matvei Petrovich Bronshtein was also arrested for Trotskiite tendencies. His case, like that of Fock, probably had the assistance of the antirelativist and Einstein-hater Vladimir L'vov. Bronshtein's sympathies for Trotskii were well known. He had often been heard saying, "if Trotskii comes to power, I will call him my nephew." He also ridiculed dialectical materialism during his student days in the 1920s. There was a "jazz band" at Leningrad University in which Gamov, Ivanenko, and Landau—the "Three Musketeers," as they were known—played. Bronshtein fell into this circle as D'Artagnon! He grew close to N. A. Kozyrev and V. A. Ambartsumian, a future leader of Soviet astrophysics, which sparked his interest in relativistic cosmology. During the purges Ambartsumian was arrested but soon released, while Kozyrev served ten years in the so-called Tupolevskaia sharaga.

L'vov was also at the university at this time, having graduated in 1926. A journalist with a physico-mathematical bent, from the early days of philosophical disputes he attacked Landau for setting natural science back centuries, Ambartsumian for supporting clerical ideas about the expanding universe, Ivanenko for writing an idealist book on relativity, and Bronshtein for disseminating harmful ideas among the masses. Bronshtein was active on radio and in print as a popularizer of the new physics among the *vydvizhentsy*. He wrote a series of popular-scientific books on solar physics, X rays, inventors of the radio, and so on. These books opened Bronshtein to criticism from amateur Marxist philosophers like L'vov.

L'vov knew Bronshtein personally, but sharply attacked him for his proximity to the Copenhagen interpretation. L'vov later justified his behavior in light of the ideological struggle going on in the country at this time. In *Noyvi mir* in the mid- to late 1930s, L'vov published several articles critical of the Leningrad physicists—Fock, Frenkel', Bronshtein. He claimed their political views were suspect at best, that they worked for the "destruction of materialistic physics." The inspiration for the group, according to L'vov, was Bronshtein, who was an agent of the "fascist bourgeoisie."[109] Arrested in 1937, Bronshtein broke under the pressure of interrogation, "admitting" that he had recruited Frenkel', Ambartsumian, Fock, Lukirskii, Landau, Bursian, Frederiks, and Krutkov into a fascist-territorial organization. Torture drew other physicists to implicate N. N. Semenov and N. I. Muskhelishvili.

Iu. A. Krutkov was arrested on New Year's Eve, December 31, 1936, as the head of a counterrevolutionary organization of educational and scientific institutions centered in Leningrad, the secret police alleged, to which Bursian, Frederiks, and Bronshtein belonged. He was released ten years later and returned to work in Leningrad. In the meantime he had suffered in the Kanskie camp of the Tupolevskaia sharaga, where the aircraft engineer Tupolev himself, S. P. Korolev, and Iu. B. Rumer were interned. Lurkirskii was released in 1942. Frederiks died in 1943 on the road from one camp to another, and Bursian died in prison in 1946.

In February 1939, L. D. Chukovskaia, Bronshtein's wife and the daughter of the children's author Kurt Chukovskii, wrote the chief procurator of the USSR, Vyshinskii, with Vavilov's support, asking for permission to send Bronshtein some books so that he might continue his work. She was told that Bronshtein had been sentenced to ten years in prison and deprived of mail privileges. But this was a pure fiction. Bronshtein had been shot a year earlier, on February 8, 1938.[110]

The repression in astrophysics and astronomy was more devastating than that in physics itself, owing to the small size of the field and its concentration in Leningrad. It was part and parcel of the attack on the Leningrad intelligentsia and Party members orchestrated by Stalin. Ten senior physicists who included the leading Soviet specialists at the State Astronomical Observatory in Pulkovo just outside of Leningrad were arrested in 1936–37 for "participation in a fascist, Trotskiist terrorist organization which arose in 1932." Over one hundred individuals in all were arrested in the Leningrad region as a result of this affair, among them B. V. Numerov, N. A. Kozyrev, and four of the five authors of the two-volume *Kurs astrofiziki i zvezdnoi astronomii* (1934, 1936), the primary Soviet astrophysics text, including V. A. Ambartsumian and B. P. Gerasimovich. Once again Kapitsa tried to intervene, in this case on Gerasimovich's behalf, and that of his wife, who also was arrested and taken away from their six-year-old daughter. But on November 30, 1937, Gerasimovich was shot.[111]

In 1944 S. I. Vavilov and the astronomers G. A. Shain and A. A. Mikhailov attempted to gain the release of N. A. Kozyrev, writing Beria, head of the secret police. They described the critical state of astrophysics in the USSR, its near destruction. The authors did not mention the direct responsibility of the Party in creating the situation but instead referred to the loss of young astronomers at the front and in Kharkov, and the physical destruction of Pulkovo at the hands of the Germans. Pulkovo was in fact just about void of scientific personnel. To rebuild astronomy in the USSR and create a powerful observatory in the south of the USSR, staff were needed. Would Beria not look into "the return to astronomical work of N. A. Kozyrev, a former astronomer at Pulkovo, who was sentenced in 1936 to ten years . . . and at present works as an engineer and geophysicist in a northern expedition (Turkhanskii region)"? He was young (thirty-six) and just the sort of person required.[112]

The purges extended to the Ukrainian Physico-Technical Institute. Alexander Weissberg, who first arrived in Kharkov in 1931, had recruited a number of left-leaning European scholars to join him at the new institute. They had hopes for the construction of socialism in Soviet Russia, and were fearful already of the Right in Germany. Weissberg invited Martin Ruhemann, a British subject, to be in charge of a new low-temperature research facility to develop a nitrogen industry for Narkomtiazhprom. Ruhemann lived through the Ukrainian famine of the 1930s, in which millions perished.[113] This caused him and his family no small discomfort. But he only left UkFTI in 1938, when the institute was instructed by the GPU not to renew his contract; he alone had spoken up for Weissberg, who was denounced in the institute by the secret police after his arrest.[114]

On December 1, 1934, the day that Kirov was assassinated, a "Davidovich" had been appointed director of UkFTI. A Stalinist through and through, Davidovich orchestrated an attempt to discredit Shubnikov, Landau, and Weissburg as Trotskiites and wreckers. At the early stage of the affair, before the momentum of the Great Terror was beyond the control of the center, the physicists had the support of the Central Committee in Moscow against the local secret police office. But during the late 1930s the purges were fed by gossip, arrest, innuendo, and more arrest. "Trotskiites" and "enemies of the people" were eliminated from the theoretical department.[115] Houtermans, Leipunskii, Obreimov, Rozenkevich, Shubnikov, and Weissberg were all repressed.

The purges in physics and astronomy struck hardest at young theoreticians and Leningrad residents. The tens and hundreds of lesser-known young students, assistants, and researchers may never be identified. Senior scholars were for some reason spared, although the March 1936 session of the Academy of Sciences had demonstrated that the progenitor of Soviet physics was not immune to attack. Though battered and beaten, the Ioffe Institute, its director and its personnel, and its research programs escaped irrevocable damage. But the efforts of Ioffe, Tamm, Vavilov, and Kapitsa could not forestall the purges of physics in 1937 and 1938, and the destruction of the Leningrad physics community. And even Ioffe could now be heard at a conference on low-temperature physics in Kharkov in January 1937, expressing "the anger and indignation of Soviet scholars at the ignoble work of Trotskiite bandits which demands from the proletarian court its destruction."[116]

Epilogue

In 1952, after a series of increasingly divisive meetings at the Leningrad Physico-Technical Institute concerning alleged idealism in physics, A. F. Ioffe was removed from the directorship of the institute he had founded in 1918. Because of his loyalty to "Soviet power," the ignominy of Ioffe's removal was somewhat surprising, even for Stalin's Russia. After lifelong service to science and the Academy, Ioffe had joined a series of government and Party organs in the 1940s. He was appointed to the Leningrad Military-Technical Commission in 1941 or 1942 in connection with the evacuation of LFTI to Kazan by train in the face of the rapid German advance. He became a Communist Party member and a vice-president of the Academy in 1942, and a member of its presidium from 1945.

For some time, however, Ioffe had faced frequent criticism for the failure of his institute to produce applications in the field of semiconductor physics. More embarrassing were philosophical and other personal errors. Under pressures from Moscow physicists, ideologues at the Institute of Philosophy, and the general political environment of the late Stalin period, the academic council of LFTI had been torn apart by a series of open meetings which condemned Frenkel' in particular, but also Ioffe for his philosophical deviations and self-serving leadership. He defended himself in Stalinist fashion, by admitting mistakes but dishing out unrestrained criticism of others, including his lifelong associate Iakov Frenkel'.

The Zhdanovshchina and the rising wave of anti-Semitism ensured that he would soon be sacked. The Zhdanovshchina, a period of heightened ideological vigilance in the postwar USSR, focused particular energy on eradicating cosmopolitanism and other manifestations of "kowtowing [*nizkopoklonstvo*]" before bourgeois society among its members. In spite of the accomplishments of a generation of physicists trained entirely within the Soviet period, a generation which had embraced dialectical materialism and striven to show how modern physics and Soviet philosophy were compatible, in spite of physicists' accomplishments for the state during the war and in rebuilding the economy, they nonetheless again faced increased scrutiny from ideologues for their alleged "idealism"—which meant unquestioningly embracing the Copenhagen interpretation of quantum mechanics and mathematical formalism.[1]

In the late 1940s and early 1950s, at LFTI, the Physics Institute of the Academy of Sciences, the Institute of Physical Problems, and other institutions, scientists held general meetings on philosophical problems in physics during which some scholars admitted past mistakes and pledged vigilance against the dangers of idealism, while still others fought the unjust accusations. The unrelenting pressure culminated in the publication of a collection of articles in 1952 which subjected quantum mechanics, Einstein's scientific work, and such leading Soviet theoreticians as Mandel'shtam, Landau, Frenkel', and the experimentalist Ioffe to severe criticism. If in the 1930s the assault on Leningrad physicists had been latently anti-Semitic, in this case is was blatantly so. What is more, the so-called Doctors' Plot, the arrest of prominent Jewish physicians and no doubt thousands of others for an alleged attempt to poison Stalin and other Kremlin leaders, was about to be hatched.

At LFTI, the session for "self-criticism" occurred on March 25, 1949, when a large number of physicists from LFTI and other Leningrad research centers, including Frenkel', Ioffe, G. D. Latyshev, P. P. Kobeko, L. A. Artsimovich, and B. P. Konstantinov, met at a special session of the academic council of the institute devoted to "the ideological struggle on the front of contemporary physics." The atmosphere was tense and unfriendly. Ioffe opened the meeting by noting that the world now lay divided into two hostile camps, one led by the USSR which stood for freedom, peace, and socialism, the other under the control of the United States under the banner of aggression. While military preparedness and the need to build atomic weapons played a role in the struggle between the two camps, Ioffe urged physicists not to underestimate the role of ideological struggle in determining the outcome of the battle.[2]

The ideological struggle was based on efforts to discredit dialectical materialism, Soviet science, and Soviet culture. Ioffe spent much of his talk pointing to the compatibility of quantum mechanics, relativity theory, and dialectical materialism, and arguing that "no idealists could exist on Soviet soil" since the "objective" conditions for their development did not exist. His real purpose, however, was to call for vigilance against "cosmopolitanism" and to criticize a series of Soviet scholars for ideological failings, including the resilient A. K. Timiriazev. Ioffe reserved his harshest comments for Frenkel'.[3] The meeting ended with a call for joint efforts of theoretical physicists and philosophers in the identification and eradication of philosophical idealism and cosmopolitanism in Soviet physics.

Ioffe himself came under attack for his book *Osnovnye predstavlenii sovremennoi fiziki* (*Fundamental Concepts of Contemporary Physics*), a major text devoted to the defense of contemporary physical concepts. Here Ioffe explained the place of relativity theory and quantum mechanics in modern science. He described the revolution in physics that had occurred since the turn of the century and which had upset previous conceptions of space, time, and motion, clarified the behavior of subatomic particles, and replaced mechanistic conceptions of phenomena with electromagnetic notions of the field and of action at a distance. With the help of the proper methodology, dialectical materialism, he argued, physics had reached the point where it could observe subatomic processes directly, and discuss them with a high degree of certainty, if not absolute causality. While classical physics continued to operate in the macroworld, the behavior of electrons and light required fundamentally new understandings—which the new physics provided. Finally, while acknowledging that mathematical formalisms had limited the accessibility of physics to a narrow circle of experts, Ioffe rejected the criticism of Stalinist ideologues by pointing out that "the questions which are being examined here have great significance for the entire system of our knowledge and for many of the most important regions of technology. Progressive science and technology cannot ignore these advances, which have been called forth by the progress of physics."[4] Such "irrefutable proofs of the correctness" of relativity theory existed that it was "impossible to name even one opponent among creatively active physicists."[5]

Under different circumstances, Ioffe's straightforward discussion of the state of physics in 1949 would have done nothing to inflame the passions of opponents in a debate. But Maksimov, Timiriazev, and their stooges, Kuznetsov and Ovchinnikov, as well as Vladimir L'vov, the writer and notorious Einstein critic, attacked Ioffe's book in a series of widely published reviews.[6] Ioffe's most complete response to this criticism, and one of the first attempts to reassert physicists' control over the philosophy of physics, was published only after the death of Stalin in a 1954 letter to the editors of *UFN*.[7]

The embarrassment to Stalin himself and to Soviet physics caused by Georgii Dmitrievich Latyshev, an LFTI physicist, was perhaps the final straw that convinced an increasingly xenophobic and anti-Semitic leadership to punish Ioffe. Latyshev had apparently faked results leading to the award of a Stalin Prize. He had all the right credentials. A peasant by origin, he joined the Komsomol in 1923, studied at the *fizmekhfak* of the polytechnical institute, moved to UkFTI in the early 1930s, joined the Party in 1939, and became a corresponding member of the Academy in 1945, while returning to LFTI to do research on radioactivity. He received a Stalin Prize for his work on gamma rays in 1948 when, in fact, the work turned out to be incorrect. A special investigative commission criticized Latyshev for his work, the organization of his laboratory at LFTI, and a series of other inadequacies at a meeting of the academic council of the institute on February 17, 1950.[8] It was almost too much to bear to admit the award of a Stalin Prize to be a fake! But in the opinion of a number of physicists and policymakers, this debacle once again demonstrated failings in Ioffe's leadership style.

Finally, Ioffe himself was caught in a lie. On July 5, 1951, he wrote the personnel office of LFTI: "In view of the disagreement in my documents on the question of my nationality [which had been listed as 'Russian'], I request to define my nationality according to the origin of my parents as Jewish."[9] He was then relieved of his position as director of the institute. In 1952, months after removing him from this position, the Academy of Sciences approved the reorganization of his institute's laboratory of semiconductor physics into the independent Institute of Semiconductors, which he directed until his death in 1960. Its activities and successes in discovering applications indicate that not all of the criticism was warranted.

The case of the removal of Ioffe was emblematic of the system of administration and control of R and D created under Stalin, with its top-down, heavy-handed, coercive, and anti-intellectual methods of control. The physics enterprise itself would not crumble, but only because of the strength of its research institutes and their courageous leaders.

The courage of leading Soviet physicists was nowhere more apparent than in their response to rumors of an impending national conference sometime in 1949 to discuss philosophical problems of the "new physics." At the urging of such physicists as D. D. Ivanenko, A. S. Predvoditelev, N. S. Nozdrev, and others located primarily at Moscow State University, the Ministry of Higher Education held a series of preparatory meetings late in 1948 and early 1949 to consider rampant idealism in Soviet physics, the "kowtowing" of Soviet scholars to Western philosophical trends, in particular the subjective idealism of the Copenhagen interpretation of quantum phenomena, and the need to cleanse the discipline of these influences. The meetings served as a forum for hostile personal attacks.

Attacking such scholars as Landau, Frenkel', and Ioffe had become second nature to Stalinist ideologues. Now, with the assistance of some members of the Moscow physics community who perhaps felt slighted that they had not been recognized by their colleagues for the brilliance and insight of their contributions to physics, they hoped to hold a conference in physics like that in biology in 1948 which had declared Lysenkoism supreme.[10] The result in theoretical physics would have been the same stultification which genetics suffered until 1965.

But a number of nuclear physicists working on the Soviet atomic bomb project got wind of the idea. They gathered in I. V. Kurchatov's office, called Beria, the head of the secret police and the project, and informed him that they could not build an atomic bomb without taking note of the special theory of relativity and the equivalence of matter and energy. A conference on idealism in physics which declared relativity theory and quantum mechanics to be bourgeois pseudoscience would halt their research. After consulting briefly with Stalin, Beria announced that there would be no conference, ordering research to proceed, fully cognizant of the fact that the secret police could arrest and shoot the offending physicists at a later date if necessary.[11]

The physiognomy of LFTI also underwent significant changes during the 1940s. World War II, the blockade of Leningrad, and the resulting evacuation of the institute to Kazan ensured that it would never again hold the central position it had once occupied in physics research and science policy during the prewar years. Its physicists had served on all the major academic and governmental councils and policymaking bodies; created a professional society which assumed national and international prominence; and dominated publication of Soviet physics journals. But when Nazi armies invaded the Soviet Union on June 20, 1941, they accelerated the transfer of Russia's scientific power to Moscow and hastened the decline of the Leningrad Physico-Technical Institute. While remaining a world leader in fission, fusion, and solid-state physics, LFTI would never come close to its former preeminence.

In terms of the organization of science, the twentieth century is the century of the research institute. Scientists and governments have established a series of large-scale research centers, for purposes that often overlap: scientists have sought to insulate their disciplines from social and political events, to ensure regularity of funding, and to secure the long-term stability and growth of their research programs; governments have sought prestige, international security, and economic performance.

The physics institute in Soviet Russia served as a center for physicists' professional activities, a bridge between an illiterate society and physical ideas, and, most importantly, a place of mediation between government and science. Within years of the 1917 Revolution, Soviet physicists had built a series of government-supported scientific research institutes, such as the Physico-Technical Institute in Leningrad, which were vital to the stability of the discipline and helped to reestablish activities normal to physics research. The new institutes rapidly accumulated advanced equipment for undertaking investigations of the physics of atoms, molecules, and crystals. They created an environment in which a strong sense of community could develop, and overcame the tremendous physical and psychological hardships which had threatened their discipline during the first turbulent decades of

Soviet history. They took initiative before the Bolsheviks had turned their attention to science and technology, organized a national association, and turned their science into an active participant in the international arena. By the end of the 1930s, a network of physics institutes existed in such cities as Kharkov, Sverdlovsk, Kiev, and Dnepropetrovsk, and there were over one thousand physicists of whom surely two hundred were independent researchers as compared with only eighty on the eve of the Revolution.

From the early years of its existence, LFTI tried to steer a course between maintaining autonomy in the conduct of its research program, keeping formal relations with the Bolshevik government to a minimum, and establishing economic relations through contracts with a number of different bureaucracies. By the mid-1920s a symbiotic relationship had developed between the Soviet state, society, and physicists. Achievements in physics which supported government economic development programs were the most crucial aspect of this relationship, but they were not the only one, as the cultural and social functions of the institute indicate.

Soviet Russian physicists were unable to enjoy fully the advantages of research institutes since they underestimated or ignored the influence of social and political factors upon the development of their discipline and faced a government increasingly unwilling to cede power to any professional group. While the Communist party granted physicists some autonomy in terms of the formulation of research programs, it insisted upon accountability to the needs of "government construction" and "socialist reconstruction,"—that is, economic development and modernization. This resulted in the subordination of the scientific enterprise to state economic and political organs. The October Revolution had seen the introduction of a relatively decentralized system of administration of scientific research. With the introduction of Stalinist policies, this system was abandoned in favor of centralized organs of control, planning of research activities, a stress on the "unity of theory and practice" (in this case, the primacy of applied research over basic), subjugation of all professional organizations and institutions to Party control, and ultimately autarky in science—international isolation.

Cultural and scientific revolution also had a significant impact on the research institute in Soviet Russia. In the late 1920s Soviet physicists began to experience pressure from an increasingly class-conscious Communist party to allow more scientists of proletarian origin and the "proper" worldview into their institutes. While cultural revolution had few short-term effects on the conduct of physics research, in the long run it changed the class makeup and worldview of physicists through *vydvizhenie*, the advancement of working-class and Communist cadres into positions of responsibility, cooptation, and coercion. Cultural revolution also involved the formation of Marxist study groups in institutes which were intended to educate all scholars about the virtues of dialectical materialism. This was just one aspect of "class war," an attack on the "bourgeois" specialist and his authority, and his replacement by someone of more reliable social background.

The revolution in science that accompanied those in politics, culture, and the structure of the discipline had a profound impact on the physics enterprise, since it cut to the heart of the discipline in terms of epistemology, methodology, and the roles of theory, experiment, and mathematics in modern physics. During this scientific revolution, new conceptions of space and time, energy and mass, momentum and position in quantum mechanics and relativity theory replaced classical mechanical conceptions of "matter in motion." These advances helped physicists better understand atomic and molecular processes in the solid state, liquids, and gases. On the whole, Soviet physicists were in tune with these developments, debated them, and participated in their further refinement.

The revolution in science required even more radical adjustments among Soviet physicists, however, than among Western scholars. After a debate between two groups of Marxist scholars, the Mechanists and Deborinites, had gone beyond the walls of Marxian institutions, Soviet physicists were asked to adopt a unique philosophy of science, dialectical materialism, which their Western colleagues avoided. This involved the hoisting of dialectical materialist conceptions of matter, energy, and motion on top of recent developments in physics, and avoidance of alleged "idealism" at any cost. Stalinist ideologues and Marxist scholars, both of whom were hostile to the bourgeois specialists and caught up in the frenzy of class war in the early 1930s, insisted that the proper worldview be applied to physics research. They believed

that indeterminacy, acausality, and relativity required the acceptance of an idealist epistemology. They believed further that
during a period of class war and hostile capitalist encirclement
these epistemological leanings would lead directly to the victory
of capitalism over socialism. They therefore required the Soviet
physicist to abandon those aspects of the new physics regarding
causality, determinacy, and action at a distance which seemed to
contradict dialectical materialism.

This was the final straw for physicists. They had tolerated the
administrative excesses of political and cultural revolution but
could not willingly accept being told which concepts or epistemology to apply to their study. In many cases—quantum mechanics,
the existence of the ether, and conservation of energy in nuclear
processes—their opponents held anachronistic or completely
erroneous points of view. Physicists fought charges of idealism
and anti-Soviet activity; they published rebuttals widely in scientific, philosophical, and popular journals; and they refused to yield
authority to ideologues and philosophers. In the environment of
the Stalinist politics of science and the Great Terror, casualties were
inevitable. Tens of physicists were arrested and disappeared in the
camps. Many others were shot. Still others faced public disgrace.
And the Soviet physicist had lived through political, cultural, and
scientific revolutions which had changed the face of his discipline
forever.

Within a few years of Stalin's death, especially after the rise of
the "cult of science," many of the policies which had required
scholars to genuflect before Stalinism and its ideologues were relaxed. The cult of science, which was part of the general environment of de-Stalinization in the 1950s, was initiated by postwar successes in nuclear weapons research and development and peaked
with achievements in the peaceful uses of atomic energy (the demonstration of the world's first power-generating nuclear reactor
in 1954), space exploration (the launching of the first artificial satel-
lites), and high-energy physics (a series of increasingly powerful
particle accelerators).

The cult of science contributed to physicists' growing influence
over both their own discipline and national science policy in terms
of setting research agendas, securing financial support, and playing a broader role in Soviet politics and society. But until the late

1980s, when administrative, structural, and R and D funding reforms were initiated during the period of *perestroika* and *glasnost'* under Gorbachev, the system of physics institutes that had developed between 1900 and 1940 remained relatively unchanged, capable of significant achievements in some areas of physics, but significantly handicapped by politics and ideology.

APPENDIXES

Appendix A: The History and Politics of Soviet Physics

Appendix B: The Leningrad Physico-Technical Institute

Appendix C: Publication

Appendix A: The History and Politics of Soviet Physics

Table A.1   Membership in the RFKhO, 1890–1916

Year	No. Physicists	No. Chemists	Budget of Physics Section	
			Income	Expenses
1890	114	233	1,646	1,759
1891	122	237	2,058	1,945
1892	125	239	2,131	2,180
1893	126	245	1,587	1,543
1894	121	257	2,066	1,994
1895	114	257	1,688	1,786
1896	119	268	1,546	1,613
1897	134	267		
1898	126	273	1,733	1,708
1899	139	293	1,281	1,111
1900	127	327	1,288	1,203
1901	134	354	1,913	1,651
1902	121	366	1,895	1,721
1903	129	389	2,305	1,772
1904	154	349		
1905	176	372	2,981	2,067
1906	171	387	2,068	2,001
1907	180	355	2,558	4,454
1908	206	360	5,348	5,368
1909	213	358	4,269	4,171
1910	214	364	4,824	3,556
1911	223	370	4,961	5,579
1912	197	387	5,701	5,504
1913	174	414	8,840	9,754
1914	196	514	7,903	9,817
1915	205	535	7,093	7,017
1916	221		6,870	9,370

Source: These data are taken from a reading of the *ZhRFKhO*. There was no association activity (and hence data) for 1917–1918. Budgetary data are in rubles, rounded to the nearest ruble. The decline in membership from 1911 to 1912–1913 may be attributed to a debate in the physics division over support for Moscow physicists over the Kasso affair, as described below.

Table A.2   Scholars Receiving TsEKUBU Ration, Total Number,
1919–1923

Year	Petrograd	Moscow	Provinces	Total
1919	550		0	550
1920	2,375		630	3,005
1921	2,000	2,786	631	5,617
1922 (Nov.)	6,010	6,200	3,384	15,594
1923 (Oct.)	7,250	8,540	4,925	20,755

Sources: God raboty TsEKUBU, pp. 30–36, 48; Zaporov, *Sozdanie i deiatel'nost' tsentral'-noi kommissii po uluchsheniiu byta uchenych (1919–1925 gg.),* Candidate Dissertation, Lenin State Pedagogical Institute, Moscow, 1979, pp. 119, 131; and *Piat' let raboty TsEKUBU pri SNK PSFSR, 1921–26* (Moscow, 1927), pp. 11–13, 15–16.

Table A.3   Scholars Receiving Ration, by Discipline, 1921–1925

	Humanities	Exact Sciences	Applied Sciences	Medicine	Other	Total
1921	1,112	1,255	790	1,303	303	4,873[a]
(%)	(22.6)	(26.1)	(16.2)	(26.8)	(6.2)	
1925	3,243	3,535	2,723	3,289	1,132[b]	13,922
(%)	(23.3)	(19.6)	(25.4)	(23.6)	(8.0)	

Sources: God raboty TsEKUBU, pp. 30–36, 48; Zaporov, *Sozdanie i deiatel'nost' tsentral'-noi kommissii po uluchsheniiu byta uchenych (1919–1925 gg.),* Candidate Dissertation, Lenin State Pedagogical Institute, Moscow, 1979, pp. 119, 131; and *Piat' let raboty TsEKUBU pri SNK RSFSR, 1921–26* (Moscow, 1927), pp. 11–13, 15–16.

Table A.4   All-Union Conferences in Physics, 1931–1939

Acoustics 1931	X-Ray Analysis for
Mendeleev Congress 1932	Industry 1936
Theory of (Nonmetallic) Solid	Low-Temperature Physics 1937
State 1932	Electroinsulation 1937
Oscillation 1932	Optico-technology 1937
Atomic Nucleus 1933	Magnetic Analysis 1937
Applied Roentgenography in	Atomic Nucleus 1937
Metallurgy 1933	Gas Discharge 1938
Chemical Physics 1933–1934	Atomic and Molecular
Electrotechnology 1934	Spectroscopy 1938
Semiconductors 1934	Dielectrical Loss 1938
Ferrous Metallurgy 1934	Surface Phenomena 1938
Radioactivity 1934	Photoeffect and Secondary Emis-
Solid Rectifiers and Photo-	sion of Electrons 1938
elements 1935	Atomic Nucleus 1938
Acoustics 1936	Semiconductors and Photo-
Oscillation 1936	elements 1938

Table A.5   Nobel Prizes for Soviet Physicists

	Subfield	Period Work Completed	Year of Award
N. N. Semenov	chain reactions	1926–1934	1956
P. A. Cherenkov		1934–1936	
I. E. Tamm	Cherenkov effect	1937	1958
I. M. Frank		1937	
P. L. Kapitsa	superfluidity	1937	1978
L. D. Landau	theory of superfluidity	1941	1962
N. G. Basov			1964
A. M. Prokhorov	maser	1954	

Table A.6   Number of Scientific Workers in RSFSR, By Discipline and Party Membership, 1923–1929

Year	Total No.	% Male	% Female	% under 30 Yrs. Old	% Between 30 and 40 Years Old	% 40 and 50 Years Old	% Exact Scientists	% Medical Scientists	% Applied Scientists	% Humanistic Scientists	% Party Members
1923	8,658	–	–	–	–	–	–	–	–	–	–
1924	12,030	79.8	20.2	13.3	47.5	19.0	21.5	22.5	24.3	31.7	3.5
1925	13,696	79.9	20.1	–	–	–	–	–	–	–	4.6
1926	13,423	80.5	19.5	17.6	34.4	28.0	24.8	20.7	20.3	30.8	4.6
1927	14,822	81.5	18.5	10.7	33.2	28.7	20.8	17.8	24.0	34.2	6.9
1927–1928	14,805	–	–	–	–	–	–	–	–	–	4.6

Sources: NR, no. 1 (1925): 165–166; no. 1 (1927): 6; no. 5–6 (1928): 79–92; Itogi desiat' let, p. 90.

Appendix B: The Leningrad Physico-Technical Institute

Table B.1   Permanent Staff of LFTI (Excluding *Sverkhstatnye*), 1922–1926

	1922–23	1923–24	1924–25	1925–26
Total staff	57	56	57	57
Directors	4	3	4	4
Sr. physicists	6	7	6	6
Physicists	8	9	11	10
Sr. assistants	4	5	4	5
Assistants	8	9	8	8
Technical, administrative and workshop employees	27	23	24	24

Source: A LFTI, f. 1, op. 1, ed. khr. 3, 4, 9, 14, 23, 27, 32, 42.

Table B.2 Graduates of Physico-Mechanical Department of LPI, 1922–1930, by Specialty and Place of Work

	1922–23	1923–24	1924–25	1925–26	1926–27	1927–28	1928–29	1929–30	Total
Physics		2	4	10	4	4	2		26
Mechanics	1	1	4					1	7
Radiotechnology				1	2	1	4	1	9
Electrotechnology							3		3
Materials testing						1	1	3	5
Heat engineering			1		1	1	5		8
Other				1		1	1		3
Total	1	3	9	12	7	8	16	5	61

	Scientific Work		Pedagogical Work	Production	Other	Total
	LFTI	Other NIIs		Factory Lab		
Physics	20	1	3	2		26
Mechanics	1	2	3		1	7
Radiotechnology	5	1	2	1		9
Electrotechnology	2			1		3
Materials testing	2			3		5
Heat engineering	3	2	1		2	8
Other		1			2	3
Total	33	7	9	7	5	61

Source: Fizika i proizvodstvo, no. 3(1930): 56–59.

Table B.3   Academic Council Meetings of LFTI, 1918–1926

	1918	1919	1920	1921–22	1922–23	1923–24	1924–25	1925–26
Meetings	17	61	30	33	37	39	31	30
Papers read by LFTI staff		49	39		75	85	48	34
						74	43	

Source: A LFTI, f. 1, op. 1, ed. khr. 17, 19, 23, 27 and 42; VRiR 1, no. 1 (1919): 4–5; and Nauchnaia deiatel'nost' Ioffe, pp. 15–21.

Table B.4   Institutes Created from the Leningrad
Physico-Technical Institute

Institute of Heat Engineering   1927
Siberian Physico-Technical Institute   1928
Ukrainian Physico-Technical Institute   1929
Leningrad Electrophysical Institute   1931
Leningrad Institute of Chemical Physics 1931
Central Asian Heliotechnical Institute   1931
Institute of Musical Acoustics   1931
Ural Physico-Technical Institute   1932
Dnepropetrovsk Physico-Technical Institute   1933
Institute of Physical Agronomy   1934
Kurchatov Institute of Atomic Energy (formerly "Laboratory No. 2")
    1943
Institute of Theoretical and Experimental Physics (formerly "Laboratory
    No. 3")   1945
Institute of Semiconductors   1954
Leningrad Institute of Nuclear Physics   1972
Leningrad Scientific Computer Center   1977

Source: A. F. Ioffe Physico-Technical Institute (Leningrad, 1978), p. 85.

Table B.5   Academicians and Corresponding Members
of the Academy of Sciences Who Have Worked
at the Leningrad Physico-Technical Institute as of 1978

Academicians

N. V. Ageev	I. V. Kurchatov
A. P. Aleksandrov	G. V. Kurdiumov
Zh. I. Alferov	L. D. Landau
A. I. Alikhanov	G. S. Landsberg
N. N. Andreev	P. I. Lukirskii
L. A. Artsimovich	A. B. Migdal
A. A. Chernyshev	M. A. Mikheev
N. D. Deviatikov	I. V. Obreimov
G. N. Flerov	N. D. Papaleksi
V. A. Fock	I. Ia. Pomeranchuk
I. M. Frank	N. N. Semenov
V. E. Golant	A. V. Shubnikov
A. F. Ioffe	D. V. Skobel'tsyn
P. L. Kapitsa	M. A. Styrikovich
Iu. B. Khariton	I. E. Tamm
I. K. Kikoin	V. M. Tuchkevich
M. V. Kirpichev	S. N. Vernov
Iu. B. Kobzarev	B. M. Vul
V. N. Kondrat'ev	Ia. B. Zeldovich
B. P. Konstantinov	S. N. Zhurkov
M. P. Kostenko	

Academicians, Union Republics	Corresponding Members
V. I. Arkharov (AN UkSSR)	A. I. Alikhan'ian
N. N. Davidenkov (AN UkSSR)	Ia. I. Frenkel'
M. A. Eliashevich (AN BSSR)	G. A. Grinberg
V. M. Kel'man (AN KazSSR)	E. F. Gross
A. P. Komar (AN UkSSR)	P. P. Kobeko
M. I. Korsunskii (AN KazSSR)	M. M. Koton
G. D. Latyshev (AN KazSSR)	Iu. A. Krutkov
V. E. Lashkarev (AN UkSSR)	S. Z. Roginskii
B. G. Lazarev (AN UkSSR)	D. A. Rozhanskii
A. I. Leipunskii (AN UkSSR)	A. I. Shalnikov
L. M. Nemenov (AN KazSSR)	G. A. Smolenskii
A. F. Prikhot'ko (AN UkSSR)	P. E. Spivak
K. D. Sinel'nikov (AN UkSSR)	A. V. Stepanov
S. V. Starodubtsev (AN UzbSSR)	P. G. Strelkov
A. K. Val'ter (AN UkSSR)	D. L. Talmud
	A. F. Val'ter
	V. P. Vologin
	Iu. V. Vul'f
	B. P. Zakharchenia
	V. P. Zhelepov
	B. S. Zhelepov

Source: A. F. Ioffe Physico-technical Institute (Leningrad, 1978), pp. 81–82.

Table B.6   Party Membership of LFTI Workers, July 1926

	Nonparty	Party (%)	Total
Administrative and leading scientific personnel	10	1 (9)	11
Scientific workers	35	1 (2.8)	36
Office and clerical workers	26	0 (0)	26
Workshop and technical personnel	70	16 (18.6)	86
Total	141	18 (11.3)	159

Source: A LFTI, f. 1, op. 1, ed. khr. 32, ll. 12–13.

Table B.7 List of Members of LFTI Party *Kollektiv*, July 1926

	LFTI Position	Date of Birth	Date Entered Party	Party Position
A. I. Brovkin	Master mechanic	1889	1922	Member, mgr., *kollektiv*
V. N. Glazanov	Administrator	1899	1920	Member, mgr., *kollektiv*
V. I. Ivanov	Master mechanic	1874	1920	Member
N. V. Kolotygin	Master mechanic	1888	1922	Member
I. A. Mal'kevich	Apprentice	1906	1926	Candidate
P. M. Nikolaev	—	1898	1918	Member, chair., *kollektiv*
P. F. Ponomarenko	Metal worker	1878	1928	Member
–. Sudakov	—	1888	1924	Candidate

Source: A LFTI, f. 1, op. 1, ed. khr. 32, ll. 12–13.

Table B.8 Planned Staff of LFTI, by Position, Discipline, and Party Membership, 1930–1932

	1930					1931					1932				
	A	B	C	D	E	A	B	C	D	E	A	B	C	D	E
Mathematics	–	3	–	–	–	–	3	–	–	–	–	3	–	–	–
Physics	23	116	9	–	2	26	126	12	2	6	34	169	24	10	18
Chemistry	4	36	5	2	1	6	43	8	4	3	19	58	11	8	19
Electrotechnology	4	40	5	4	1	6	55	10	8	7	8	67	13	11	12
Metallurgy	1	4	1	–	–	1	4	1	–	–	1	4	1	–	–
Communications	3	55	2	1	2	5	67	6	3	5	7	70	9	6	9
Other	1	2	–	–	–	1	2	–	–	–	1	2	–	–	–
Total	36	256	22	7	6	45	271	37	17	21	70	390	58	35	58

Source: Piatiletnyi plan GFTIa, p. 324.
Key: A = leading scientific personnel; B = total number of personnel; C = those of working-class origin; D = Party members; E = Komsomol (Communist Youth league).

Table B.9   *Aspirantura* at the Leningrad Physico-Technical Institute,
1934–1935

	Jan. 1934	New in 1934	Finished/Left in 1934	Jan. 1935
Scientific *aspirants*	8	2	2	8
Aspirant-assistants	1	3	0	4
Total	9	5	2	12
of which Party	—	—	—	5
of which Komsomol	—	—	—	6

Source: A LFTI, f. 3, op. 1, ed. khr. 32, l. 26.

Table B.11   Staff and Salaries of Leningrad Physico-Technical Institute,
1928–1932

	1928	1929	1930	1931	1932
Total number workers	151	327	428	500[a]	600[a]
scientific	107	209	270	290[a]	375
leading			36	45	70
other	44	118	157	210[a]	240[a]
Total salary (in 1000s of rubles)	460[a]	730[a]	1100[a]	1350[a]	1700[a]

[a]Estimate based on average of two sources.

Sources: Piatiletnyi plan GFTIa, pp. 3–4, 315, 321, 324, and *Piatiletnyi plan GFTL*, p. 45. Figures from 1928 and 1929 are actual; for 1930–1932 they are estimated and approximate. It has been difficult to break down the aggregate figures provided to a more disaggregate level, to compare academic-year expenses with calendar-year numbers of workers, or determine the extent of overlap between workers of LFTI and LFTL.

Table B.10 Budget of Leningrad Physico-Technical Institute, First Five-Year Plan (1928–1932)

	1928–29	1929–30	Special Quarter 1930	1931	1932
Scientific salaries	477.1	580.1	167.4	1271.0	1763.2
Capital construction	528.9	703.9	274.2	2110.0	2689.0
Equipment	278.7	500.3	116.6		1260.0
Experimental expenses	112.2	405.2	40.9	875.5	1506.6
Other (administration, aspirantura, etc.)	420.0	188.1	66.8	564.5	
Total	1818.9	2377.6	685.9	4821.0	7219.0

Source: Piatiletnyi plan GFTIa, pp. 321–323. Figures given in thousands of rubles. Figures for 1928 through "special quarter" of 1930 are actual; 1931–32 figures are estimated.

Appendix C: Publication

Table C.1 Publication of Physics Journals in Russia,
1917–1923, by Volume and Number of Issues

Journal	1917	1918	1919	1920	1921	1922	1923
UFN		1:4			2:2	3:4	4:6
ZhRFKhO	49:3	50:3	51:3			52–54:4	55:1
VRiR			1:3				
Izv. IFBF			1:6			2:1	

Table C.2   Institutional Affiliation of *ZhRFKhO* Authors, 1923–1930, by Number of Articles and Percentage

Year and Volume	1923 55	1924[a] 56	1925 57	1926 58	1927 59	1928 60	1929 61	1930 62
Institute or City								
LFTI	8	18	6	19	13	17	16	21
(%)	(66.6)	(25.5)	(12.8)	(21.2)	(24.1)	(31.5)	(27.1)	(36.2)
LFTL	—	—	1	4	5	7	6	3
(%)	—	—	(2.1)	(4.5)	(9.2)	(13.0)	(10.2)	(5.2)
Total LFTI/L	8	18	7	23	18	24	22	24
(%)	(66.6)	(25.5)	(14.9)	(25.8)	(33.3)	(44.4)	(37.3)	(41.4)
Other FTIs	—	—	—	—	—	—	—	5
(%)	—	—	—	—	—	—	—	(8.6)
Leningrad[c]	—	16	4	11	6	6	5	4
(%)	—	(22.2)	(8.5)	(12.3)	(11.1)	(11.1)	(8.4)	(6.9)
Moscow	3	21	14	19	10	9	6	5
(%)	(25.0)	(29.6)	(29.8)	(21.3)	(18.5)	(16.7)	(10.2)	(8.6)
Provinces	1	11	4	13	15	9	12	11
(%)	(8.3)	(15.5)	(8.5)	(14.6)	(27.8)	(16.7)	(20.3)	(19.0)
Other	—	5	18	23	5	6	14	9
(%)	—	(7.0)	(38.3)	(25.8)	(9.2)	(11.1)	(23.7)	(15.5)
TOTAL[b]	12	71	47	89	54	54	59	58

[a] If vol. 56, no. 5/6 (1924), which is devoted entirely to papers read at the fourth congress of the RAF, is excluded, the number of articles written by scholars at LFTI/L is 12 of 40 (30%).
[b] The total percentage may exceed 100 because of rounding.
[c] Excluding LFTI/LFTL.

Table C.3   Institutional Affiliation of ZPF Authors, 1924–1930,
by Number of Articles and Percentage

Year and Volume	1924 1	1925 2	1926 3	1927 4	1928 5	1929 6	1930 7
Institute or City							
LFTI	3	1	—	1	—	5	2
(%)	(9.1)	(3.2)	—	(3.0)	—	(10.0)	(3.9)
LFTL	—	—	—	—	14	13	4
(%)	—	—	—	—	(28.0)	(26.0)	(7.9)
Total LFTI/L	3	1	—	1	14	18	6
(%)	(9.1)	(3.2)	—	(3.0)	(28.0)	(23.7)	(11.7)
Leningrad[a]	2	1	4	2	3	6	7
(%)	(6.1)	(3.2)	(12.9)	(6.0)	(6.0)	(11.7)	(13.7)
IFBF[b]	15	17	9	6	14	22	18
(%)	(45.4)	(54.8)	(29.0)	(18.2)	(28.0)	(28.9)	(35.4)
Moscow[c]	23	21	18	17	28	40	36
(%)	(69.9)	(67.7)	(58.8)	(51.6)	(56.0)	(52.6)	(71.0)
Provinces	2	3	5	5	2	10	5
(%)	(6.1)	(9.7)	(16.1)	(15.2)	(4.0)	(13.2)	(10.0)
Other	3	5	4	7	3	2	5
(%)	(9.1)	(16.1)	(12.9)	(21.2)	(6.0)	(2.6)	(10.0)
TOTAL[d]	33	31	31	33	50	51	51

[a] Excluding LFTI and LFTL.
[b] Institute of Physics and Biophysics.
[c] Excluding IFBF.
[d] Percentage may exceed 100 because of rounding.

Table C.4   Articles Written by Soviet Scholars in *Zeitschrift für Physik*, 1–103 (1920–1936)

Year	Vol. Nos.	No. Articles	No. Articles by Soviets	(%)	of which LFTI	(%)
1920	1–3	149	0	(0)	0	(0)
1921	4–8	228	1	(0)	0	(0)
1922	9–12	140	2	(1.4)	0	(0)
1923	13–19	223	12	(5.3)	5	(43.3)
1924	20–30	406	33	(8.1)	13	(39.4)
1925	31–34	364	38	(9.1)	14	(36.8)
1926	35–39	444	71	(16.0)	18	(25.4)
1927	40–45	492	64	(13.0)	20	(31.3)
1928	46–51	475	63	(13.3)	12	(19.0)
1929	52–58	563	70	(12.4)	12	(17.1)
1930	59–66	657	71	(10.8)	19	(26.8)
1931	67–72	429	43	(10.0)	8	(18.6)
1932	73–80	662	39	(5.9)	14	(35.9)
1933	81–86	460	32	(7.0)	8	(25.0)
1934	87–92	486	29	(6.0)	9	(31.0)
1935	93–97	423	14	(3.3)	6	(42.9)
1936	98–103	361	13	(3.6)	5	(38.5)
Total	1–103	6,962	595	(8.5)	163	(27.4)

Table C.5   Articles Published in *ZETF*, by Institute and City, 1931–1939

Year Pages/Vol.	1931 350	1932 340	1933 580	1934 1,100	1935 1,000	1936 1,100	1937 1,300	1938 1,400	1939 1,500	Total
Institute or City										
LFTI	16	23	23	20	20	20	15	15	26	178
(%)	(36.4)	(43.4)	(36.5)	(16.0)	(16.5)	(17.5)	(8.9)	(9.0)	(12.6)	(16.7)
SFTI	2	6	2	12	6	1	4	14	6	53
(%)	(4.5)	(11.3)	(3.2)	(9.6)	(5.0)	—	(2.4)	(8.4)	(2.9)	(5.0)
UkFTI	2	—	—	1	13	6	19	19	17	77
(%)	(4.5)	—	—	—	(10.7)	(5.3)	(11.2)	(11.4)	(8.2)	(7.2)
UralFTI	—	—	7	10	9	5	12	8	6	57
(%)	—	—	(11.1)	(8.0)	(7.4)	(4.4)	(7.1)	(4.8)	(2.9)	(5.4)
DFTI	—	—	—	10	3	1	4	—	6	24
(%)	—	—	—	(8.0)	(2.5)	—	(2.4)	—	(2.0)	(2.3)
Total FTI	20	29	32	53	53	33	54	56	61	391
(%)	(45.5)	(54.7)	(50.8)	(42.4)	(43.8)	(28.9)	(32.0)	(33.5)	(29.5)	(36.8)
Leningrad[a]	7	7	3	10	7	19	22	31	49	115
(%)	(22.7)	(22.6)	(12.7)	(20.8)	(18.2)	(25.4)	(23.1)	(21.6)	(26.5)	(22.3)
GOI	3	1	3	11	10	4	8	5	2	47
(%)	(6.8)	(1.9)	(4.8)	(8.8)	(8.2)	(3.5)	(4.7)	(3.0)	—	(4.4)
IKhF	—	4	2	5	5	6	9	—	4	35
(%)	—	(7.5)	(3.8)	(4.0)	(4.1)	(5.3)	(5.3)	—	(1.9)	(3.3)

Moscow[b]	2	5	7	9	10	16	11	32	19	125
(%)	(4.5)	(9.4)	(11.1)	(7.2)	(8.2)	(14.0)	(6.5)	(19.4)	(9.2)	(11.8)
MGU	2	—	13	23	26	22	26	18	25	155
(%)	(4.5)	—	(20.6)	(18.4)	(21.5)	(19.3)	(15.3)	(10.8)	(12.1)	(14.6)
Provinces	5	4	—	12	6	11	20	22	43	123
(%)	(11.6)	(7.5)	—	(9.6)	(5.0)	(9.6)	(11.8)	(13.2)	(20.8)	(11.6)
Other/?	5	3	3	2	4	3	5	3	4	32
(%)	(11.4)	(5.7)	(4.8)	(1.6)	(3.3)	(2.6)	(3.0)	(1.8)	(1.9)	(3.0)
Total	44	53	63	125	121	114	169	167	207	1,063

[a] Excluding LFTI, GOI, and IKhF
[b] Excluding MGU.

Table C.6 Articles Published in ZTF, by Institute and City, 1931–1939

Year	1931	1932	1933	1934	1935	1936	1937	1938	1939	Total
Pages/Vol.	800	900	1,300	2,000	1,800	2,200	2,400	2,200	2,300	
Institute or City										
LFTI	38	6	14	15	15	18	10	21	37	174
(%)	(52.1)	(11.5)	(11.5)	(9.9)	(8.5)	(9.2)	(4.4)	(10.2)	(14.8)	(12.0)
SFTI	1	—	4	—	2	—	4	5	5	21
(%)	(1.4)	—	(3.3)	—	(1.1)	—	(1.7)	(2.5)	(2.0)	(1.4)
UkFTI	1	1	1	3	10	9	7	6	3	41
(%)	(1.4)	(1.9)	—	(2.0)	(5.6)	(4.6)	(3.1)	(3.0)	(1.2)	(2.8)
UralFTI	—	—	3	—	2	3	5	5	4	22
(%)	—	—	(2.5)	—	(1.1)	(1.5)	(2.2)	(2.5)	(1.6)	(1.5)
Total FTI	40	7	24	20	33	40	30	48	56	298
(%)	(54.8)	(13.4)	(19.7)	(13.2)	(18.6)	(20.4)	(13.1)	(24.1)	(22.4)	(20.5)
Leningrad[a]	14	14	27	23	34	58	58	52	69	338
(%)	(19.2)	(26.9)	(22.1)	(19.0)	(19.2)	(29.6)	(25.3)	(25.6)	(27.6)	(23.3)
LEFI	—	12	22	29	18	11	4	—	—	107
(%)	—	(23.1)	(27.0)	(19.2)	(10.2)	(5.6)	(1.7)	—	—	(7.4)
GOI	2	1	2	9	6	7	9	6	4	46
(%)	(2.7)	(1.9)	(1.6)	(6.0)	(3.4)	(3.6)	(3.9)	(3.0)	(1.6)	(3.2)
Moscow[b]	10	9	33	55	42	39	70	56	72	386
(%)	(13.7)	(17.3)	(27.1)	(45.5)	(23.7)	(19.9)	(30.6)	(27.6)	(28.8)	(26.6)

	1	2	3	4	5	6	7	8	9	Total
MGU (%)	—	—	—	—	15 (8.5)	10 (5.1)	9 (3.9)	10 (4.9)	9 (3.6)	53 (3.6)
Provinces (%)	4 (5.5)	5 (9.6)	8 (6.6)	9 (6.0)	13 (7.3)	19 (9.7)	33 (14.4)	19 (9.4)	34 (13.6)	144 (9.9)
Other/? (%)	3 (4.1)	4 (7.7)	5 (4.9)	6 (4.0)	16 (9.0)	12 (6.1)	16 (7.0)	12 (5.9)	6 (2.4)	81 (5.6)
TOTAL	73	52	122	121	177	196	229	203	250	1,453

[a] Excluding LFTI, LEFI, and GOI.
[b] Excluding MGU.

Notes

1: THE POLITICS OF TSARIST PHYSICS

1. Alexander Vucinich, *Science in Russian Culture, 1861–1917* (Stanford: Stanford University Press, 1973), p. 215.

2. K. V. Ostrovitianov et al., eds., *Istoriia Akademiia nauk SSSR* (Moscow-Leningrad, 1964), 2: 491–495.

3. Vucinich, *Science in Russian Culture*, pp. 214–221.

4. See *Otchet o deiatel'nosti KEPS* (Petrograd, 1917), 8: 149–156, 161, as cited in B. I. Kozlov, *Organizatsiia i razvitie otraslevykh nauchnoi-ssledovatel'skikh institutov Leningrada, 1917–1977* (Leningrad, 1979), pp. 27–29. On the early history of the Kaiser Wilhelm Gesellschaft, see Frank Pfetsch, "Scientific Organization and Science Policy in Imperial Germany, 1871–1914: The Formation of the Imperial Institute of Physics and Technology," *Minerva*, 8 (October 1970): 557–580.

5. M. S. Bastrakova, *Stanovlenie sovetskoi sistemy organizatsii nauki (1917–1922)* (Moscow, 1973), p. 25.

6. Ibid.

7. "Avtobiograficheskii ocherk" in N. A. Umov, *Izbrannye socheneniia* (Moscow, 1960), p. 27.

8. Bastrakova, *Stanovlenie*, p. 26.

9. Umov, "Avtobiograficheskii ocherk," pp. 26–27, and *Vremennik obshchestva sodeistviia uspekham opytnykh nauk i ikh prakticheskikh primenenii im. Kh. S. Lednetsova*, 1910–1913, vols. 1–4. Between May 1909 and December 1912, the society received 697 applications of which 155 (22.2 percent) received funding, 440 (63.1 percent) were rejected, and 102 (14.7 percent) were asked for supplementary information. See also Vucinich, *Science in Russian Culture*, pp. 210–212, for a discussion of the Ledentsov Society. It remains to write an article on scientific philanthropy in prerevolutionary Russia.

10. P. P. Lazarev, *P. N. Lebedev i russkaia fizika* (Moscow, 1912). Lebedev, for example, received funding to continue his research on light pressure and electromagnetic waves.

11. P. N. Lebedev, "An Experimental Investigation of the Pressure of Light," *Annual Report of the Smithsonian Institution* (Washington, D.C.: Government Printing Office, 1903), pp. 177–178; originally published in unabridged form in *Annalen der Physik*, 6 (1901): 433–458.

12. A. F. Kononkov, "Moskovskoe fizicheskoe obshchestvo im. P. N. Lebedeva," in *IiMEN*, (Moscow, 1965), Fizika, 3: 270–273.

13. *TsGA RSFSR*, f. 237, op. 2, ed. khr. 380, ll. 2–24; and Kononkov, "Moskovskoe fizicheskoe obshchestvo."

14. K. A. Timiriazev, *Ob organizatsii i dal'neishei uchasti sostoiashei pri universitete shaniavskago fizicheskoi laboratorii im. Lebedeva* (Moscow, 1916).

15. See Daniel J. Kevles, *The Physicists* (New York: Vintage, 1979), pp. 91–116, for a discussion of the development of industrial physics R and D in the United States in the first decades of this century.

16. Theodore H. Von Laue, *Sergei Witte and the Industrialization of Russia* (New York: Columbia University Press, 1963), especially chap. 8.

17. V. N. Ipatieff, *The Life of a Chemist*, ed. X. Eudin, H. D. Fisher, and H. H. Fisher (Stanford: Stanford University Press, 1946), pp. 190–217.

18. Bastrakova, *Stanovlenie*, pp. 45–46.

19. See V. A. Mikhel'son, *Rashirenie i natsional'naia organizatsiia nauchnykh issledovanii v Rossii* (Moscow, 1916), for a discussion of the necessity of national support for the scientific-technological enterprise in prerevolutionary Russia.

20. "Fizicheskaia laboratoriia u nas i za granitsei," in A. G. Stoletov, *Izbrannye socheneniia* (Moscow, 1950), pp. 510–519.

21. The figure of one hundred physicists is somewhat higher than that given in the excellent study by Forman, Heilborn, and Weart, *Physics ca. 1900*, which provides data on academic physics for twelve countries including Russia, Germany, France, the U.S., and Great Britain, specifically for scholars occupying teaching positions explicitly devoted to physics (not technical, applied, industrial, or medical physics) in higher schools. They calculate that there were thirty-five "academic" physicists in Russia in 1900. They also include data on the growth of academic physics, posts in theoretical and mathematical physics, and distribution by field. Using their definition, and based upon an analysis of membership roles of the physical section of the RFKhO for the years 1905–1916, I arrive at a figure of sixty "academic" physicists—that is, professors, laboratory assistants such as A. F. Ioffe with higher degrees from German universities, docents, or faculty of *vuzy* (higher-educational schools). The remaining members of the physical section were secondary or primary school teachers, and individuals who pursued physics as a curiosity or hobby. But for a country of Russia's size, the number of academic physicists, no matter how calculated, was small indeed; of the countries studied, Russia exceeded only Japan, the Netherlands, and Belgium. See Paul Forman, John Heilbron, and Spencer Weart, Jr., *Physics ca. 1900* (Princeton: Princeton University Press, 1975), pp. 12, 31, 34–35.

22. Ibid., pp. 18–24, 40, 57. While it is difficult to develop consistent and comprehensive budgetary and employment data for these categories, it is clear that government support was inadequate.

23. V. V. Lermantov, "Ocherk istorii razvitiia fizicheskoi laboratorii imperatorskago S. Peterburgskago universiteta pod rukovodstvom prof. F. F. Petrushevskago, 1865–1903," [*ZhRFKhO?*] 33 (n.d.), vii–xvi.

24. Ibid. Petruschevskii was remembered in 1904 for his contributions to Russian physics as the organizer and creator of the Physics Institute of Petersburg University, the Physics Department of the RFKhO, and for other professional activities. See *Elektrichestvo*, no. 5 (1904): 1.

25. V. Ia. Frenkel', ed., *Erenfest-Ioffe: Nauchnaia perepiska* (Leningrad, 1973), pp. 78–80; and A. F. Ioffe, *Vstrechi s fizikami* (Leningrad, 1983), p. 38.

26. Forman et al., *Physics ca. 1900*, p. 31.

27. P. P. Lazarev, *Ocherki istorii russkoi nauki* (Moscow, 1950), p. 40.

28. Ioffe, *Vstrechi*, p. 100. Khvol'son wrote more than thirty popular-scientific books and is to this day widely respected in the Soviet Union although it is often asserted that his philosophical views are mistaken on a number of topics.

29. Ibid.

30. I. I. Borgman, ed., *Novye idei v fizike* (St. Petersburg, 1911), 1: i.

31. Ibid., pp. 108–109.

32. N. A. Gezekhus, "Istoricheskii ocherk desiatiletiia deiatel'nosti fizicheskogo obshchestva," *ZhRFKhO*, no. 9 (1882): 5, as cited in M. S. Sominskii, *A. F. Ioffe* (Moscow-Leningrad, 1965), p. 164. For the history of the first decades of the RFKhO, see V. V. Kozlov, *Ocherki istorii khimicheskikh obshchestv SSSR* (Moscow, 1958); Sominskii, *Ioffe*, pp. 162–171; and the recently completed dissertation by Nathan Brooks (Columbia University).

33. Erwin Hiebert, "The State of Physics at the Turn of the Century," in *Rutherford and Physics at the Turn of the Century*, ed. Mario Bunge and William R. Shea (Kent, England: Dawson; New York: Science History Publications, 1979), pp. 5–22.

34. *Voprosy fiziki*, no. 9 (1915): 136–139.

35. *ZhRFKhO*, no. 1 (1916): iv–xiii; and no. 1 (1917): vi–vii. All of the references for *ZhRFKhO* from 1907 onward are for its *Fizicheskaia chast'*.

36. *ZhRFKhO*, no. 8 (1915): 459–460. A later vote freed these physics section members from the obligations to represent it at any Military-Industrial Commission meetings because of their inability to attend on a regular basis.

37. *Vospominaniia ob A. F. Ioffe* (Leningrad, 1973), p. 35.

38. World War I undid years of efforts to organize science on an international scale. See Daniel J. Kevles, "'Into Hostile Camps': The Reorganization of Science in World War I," *ISIS*, 62, pt. 1, no. 211 (Spring 1971): 47–60.

39. *ZhRFKhO*, no. 3 (1911): 205.

40. Sominskii, *Ioffe*, p. 172.

41. Letter from Rozhdestvenskii to N. A. Morozov, in *A AN*, f. 543, op. 4, no. 15645, l. 1, as cited in Gulo et al., "Iz istorii osnovaniia gosudarstvennego opticheskogo instituta," *IiMEN*, 3: (1965): 275–276, as well as *ZhRFKhO*, no. 4 (1911): 206.

42. *ZhRFKhO*, no. 1 (1917): xxviii.

43. Perhaps the most workable definition of "research school" comes from J. B. Morrell in his comparison, "The Chemical Breeders: Liebig and

Thomson," *Ambix* 19 (1972): 1–46. I am indebted to Richard Kremer for pointing out this article to me. Gerald Holton has also looked at schools. He says the success of a research school requires the presence of "the likely center," a leader who possesses a subtle balance of experimental skill, theoretical insight, and encyclopedic knowledge, can choose research problems pragmatically, organize and distinguish between incidentals and research problems pragmatically, organize and distinguish between incidentals and essentials, and is not overly concerned with philosophical subtleties. See Gerald Holton, *The Scientific Imagination*, (Cambridge: Cambridge University Press, 1978), chap. 5, "Fermi's Group and the Recapture of Italy's Place in Physics," pp. 155–198.

44. There are a large number of biographical articles and books on Ioffe. The most useful and comprehensive is Sominskii's biography, *A. F. Ioffe*. Also helpful and interesting are Ia. I. Frenkel', *Ioffe* (Leningrad, 1968; but written in 1947); V. M. Dukel'skii, "A. F. Ioffe," in *Sbornik posviashchennyi semidesiatiletiiu akademika A. F. Ioffe* (Moscow, 1950), pp. 5–30, which also has an excellent discussion of the place of Ioffe's research in the development of Soviet solid-state physics; the popular T. M. Chernoshchekova, *A. F. Ioffe* (Moscow, 1983); and the autobiographical A. F. Ioffe, *Moia zhizn' i rabota* (Moscow-Leningrad, 1933); *O fizike i fizikakh* (Leningrad, 1977); and *Vstrechi*.

45. See Frenkel', *Ioffe*, p. 5; Ioffe, *Vstrechi*, pp. 23–30; and Ioffe, "Vil'gelm Konrad Rentgen," introduction to V. K. Roentgen, *O novom rode luchei*, ed. A. F. Ioffe (Moscow-Leningrad, 1933), pp. 17–20.

46. Ioffe, *Vstrechi*, p. 23.

47. Ioffe, *Moia zhizn'*, pp. 17–19; and *Vospominaniia ob Ioffe*, pp. 21–23.

48. V. Ia. Frenkel' writes that "the soul of the [Petersburg] seminar was P. S. Ehrenfest, who knew all new directions of theoretical physics splendidly: the theory of quanta and theory of relativity, and who shared his knowledge with new friends with great enthusiasm." V. Ia. Frenkel', "Vvedenie," in *Erenfest-Ioffe*, p. 10.

49. V. Ia. Frenkel', *Paul' Erenfest* (Moscow, 1977), pp. 20–39, and Martin J. Klein, *Paul Ehrenfest* (Amsterdan: Van Nostrand, 1970), pp. 84–92.

50. *Erenfest-Ioffe*, p. 62.

51. *Vospominaniia ob Ioffe*, p. 27. See also Ioffe, "Po povodu rabot Erengafta 'ob atomisticheskom stroenii elektrichestva,'" *ZhRFKhO*, no. 2 (1911) 40–42.

52. *Vospominaniia ob Ioffe*, p. 6.

53. Ibid., pp. 85–88.

54. Ibid.

55. Ia. I. Frenkel', as cited in V. Ia. Frenkel', *Iakov Il'ich Frenkel'* (Moscow-Leningrad, 1966), pp. 29–30.

56. See Ia. I. Frenkel', "Ob elektricheskom dvoinom sloe na poverkhnosti tverdykh i zhidkikh tel," *ZhRFKhO*, no. 1 (1917): 100–118, and no. 1/3 (1918): 5–20. Frenkel''s paper was also published in *Philosophical Magazine* 33 (1917): 297–322. Other students published papers read at these

seminars. Iu. Krutkov, who was a member of Ehrenfest's circle, published "O teorii kvantov," *Voprosy fiziki*, no. 2 (1916): 43–75, which was based on two lectures Krutkov gave at the students' seminar, and suggested by the work of Ehrenfest, with which he became familar during *komandirovka* to Leiden in 1913–14.

57. The following analysis is based on a reading of A. F. Ioffe, *Izbrannye trudy* (Leningrad, 1974); vol. 1; numerous articles from *ZhRFKhO*; and a number of secondary sources, including S. R. Mikulinskii and A. P. Iushkevich, *Razvitie estestvoznaniia v Rossii* (Moscow, 1977), p. 360; Dukel'skii, "A. F. Ioffe," pp. 14–18; and Ia. G. Dorfman, *Vsemirnaia istoriia fiziki* (Moscow, 1979), the latter two being the most complete and important to this study.

58. A. F. Ioffe, "Elastische Nachwirkung in kristallinischen Quarz," *Annalen der Physik*, 20, no. 4 (1906): 919–980.

59. A. F. Ioffe, "Retsenziia," *Voprosy fiziki*, no. 8 (1912): 324–329, as published in Ioffe, *Izbrannye trudy*, 1: 27–31.

60. A. F. Ioffe, and M. V. Kirpicheva, "Elektroprovodimost' chistykh kristallov," *ZhRFKhO*, no. 8 (1916): 261–296.

61. A. F. Ioffe, "Atomy sveta," *ZhRFKhO*, no. 2 (1912): 37–50.

62. Ioffe, *Vstrechi*, pp. 119–121, and Mikulinskii and Iushkevich, *Razvitie estestvoznaniia*, p. 358.

2: THE REVOLUTION AND SCIENCE POLICY

1. Several of the topics discussed in this chapter have been raised in Paul Josephson, "Science Policy in the Soviet Union, 1917–1927," *Minerva* 26, no. 3 (Autumn 1988): 342–369.

2. *Nauka v Rossii. Spravochnik ezhegodnik Petrograda* (Petrograd, 1920), pp. 91–117; and *Nauka i nauchnye rabotniki SSSR. Nauchnye uchrezhdeniia Leningrada* (Leningrad, 1926), pt. 2: 231–270. The same patterns hold for Moscow and the USSR as a whole as for Leningrad.

3. Svobodnaia assotsiatsiia dlia razvitiia i rasprostraneniia polozhitel'nykh nauk, *Rechi i privetstviia proiznesennye na trekh publichnykh sobraniiakh sostoiavshikhsia v 1917 g. 9-go i 16-go aprelia v Petrograde i 11-go maia v Moskve* (Petrograd, 1917), pp. 3–6.

4. See N. N. Sukhanov, *The Russian Revolution. 1917*, ed. and trans. Joel Carmichael (London and New York: Oxford University Press, 1955), pp. 23–30, for a discussion of the activities of Gorky and the editors of *Letopis'* during the first days of the Revolution.

5. *A AN LO*, f. 167, op. 1, ed. khr. 6, ll. 1–3.

6. Svobodnaia assotsiatsiia, *Rechi i privetstviia*, pp. 5–6, 11.

7. A. M. Gorky, "Nauka i demokratiia," in ibid., pp. 17–19. This speech was reprinted in *Letopis'*, no. 5–6 (1917): 223–231.

8. Ibid.

9. Ibid., p. 21.

10. Ibid., pp. 19–20.

11. On Taylorism and technocracy, see Charles S. Maier, "Between Taylorism and Technocracy: European Ideologies and the Vision of Industrial Productivity in the 1920s," *Journal of Contemporary History* 5 (1970): 27–61. See also Hans Rogger, "*Amerikanizm* and the Economic Development of Russia," *Journal for the Comparative Studies in Society and History* 23, no. 3 (July 1981): 382–420. Gorky's technocracy was more like that defined by Kendall Bailes as "any movement among technical specialists that urges them to develop a wider sense of social responsibility for the use of their technical knowledge, and particularly urges them to take an important role in policy formation. See Kendall Bailes, "The Politics of Technology: Stalin and Technocratic Thinking Among Soviet Engineers," *American Historial Review* 79 (April 1974): 449.

12. *A AN LO*, f. 167, op. 1, ed. khr. 9, l. 48; ed. khr. 11, ll. 1–19, 29–30, 49–53.

13. S. Z. Mandel', "Kultur'no-prosvetitel'naia deiatel'nost' uchenykh petrogradskogo universiteta v pervye gody sovetskoi vlasti," in *Ocherki po istorii leningradskogo universiteta*, N. G. Sladkevich et al. (Leningrad, 1962), 1: 143–148.

14. *A AN LO*, f. 167, op. 1, ed. khr. 6, ll. 14–14 ob.

15. Ibid., f. 167, op. 1, ed. khr. 6, ll. 38. See ibid., ed. khr. 11, ll. 53–63, for detailed budget breakdown. SARRPN estimated one-time costs of 1.1 million rubles and annual expenses of 600,000.

16. A. G. Slonimskii, *A. M. Gor'kii v bor'be za sozdanie sovetskoi intelligentsii v gody inostrannoi voennoi interventsii i grazhdanskoi voiny* (Stalinabad, 1956), p. 16.

17. *Novaia zhizn'* (hereafter *NZ*), no. 35 (May 30, 1917), as cited in Maxim Gorky, *Untimely Thoughts*, trans. and intro. Herman Ermolaev (New York: P. S. Ericksson, 1968), p. 47; and Ermolaev, introduction to ibid., pp. vi–x.

18. *NZ*, no. 6 (May 8, 1917), as cited in Gorky, *Untimely Thoughts*, p. 20.

19. Paul Forman, "The Financial Support and Political Alignment of Physicists in Weimar Germany," *Minerva* 12, no. 1 (1974): 39–66.

20. *God raboty tsentral'noi komissii po ulushcheniiu byta uchenykh pri SNK* (Moscow, 1922), and *Piat' let raboty tsentral'noi komissii po ulushcheniiu byta uchenykh pri sovete narodnykh komissarov RSFSR, 1921–1926* (Moscow, 1927).

21. S. Novosel'skii, "K demografii Petrograda," *Nauka i ee rabotniki* (hereafter *NieR*), no. 2 (1922): 3–5. *NieR* was founded with Gorky's assistance in 1919.

22. Ipatieff, *Life of a Chemist*, pp. 190–217.

23. *NieR*, no. 3 (1921): 34–38, and no. 2 (1922): 37–40.

24. Slonimskii, *A. M. Gor'kii*, pp. 25, 57.

25. Ibid., p. 79; *Gor'kii i nauka: Stat'i, rechi, pis'ma, vospominaniia*, ed. F. N. Petrov (Moscow, 1964), pp. 224–231; and V. D. Bonch-Bruevich, *Vospominaniia o Lenine*, 2d ed. (Moscow, 1969), pp. 233–234, 429.

26. Slonimskii, *A. M. Gor'kii*, p. 80.

27. I. P. Zaporov, *Sozdanie i deiatel'nost' tsentral'noi komissii po uluchshe-*

niiu byta uchenykh (1919–1925 gg.) (cand. diss. Lenin State Pedagogical Institute, Moscow, 1979), pp. 108–109.

28. Ibid., pp. 108–110.

29. Gorky, *Sobranie sochineniia* 29, Moscow (1955): 392.

30. Ibid., p. 394.

31. *God raboty TsEKUBU*, p. 28.

32. See Richard Stites, *Revolutionary Dreams* (New York and Oxford: Oxford University Press, 1989), chap. 6, "Republic of Equals," especially pp. 140–144, for a discussion of this issue.

33. *God raboty TsEKUBU*, p. 16.

34. Ibid., p. 19.

Initial Rating of Scholars Receiving Rations Through TsEKUBU

	Scientists			Artists		
	Moscow	*Provinces*	*Petrograd*	*Moscow*	*Petrograd*	*Total*
V	25	5	25	2	1	58
IV	118	37	83	13	5	256
III	422	266	459	57	14	1,218
II	1,218	639	1,272	131	110	3,370
I	1,101	767	637	100	17	2,622
Total	2,884	1,714	2,476	303	147	7,524

Key: V—World-class scholar; IV—of national significance; III—professional scholar; II—advanced education; I—scholar just beginning career.

35. Zaporov, *Sozdanie*, pp. 190–191.

36. *God raboty TsEKUBU*, p. 17.

37. N. N. Semenov, *Nauka i obshchestvo* (Moscow, 1981), p. 346.

38. S. E. Frish and A. I. Stazharov, eds., *Vospominaniia ob Akademike D. S. Rozhdestvenskom* (Leningrad: Nauka, 1976), p. 59.

39. *God raboty TsEKUBU*, pp. 30–36, 48; Zaporov, *Sozdanie*, pp. 119, 131; and *Piat' let TsEKUBU*, pp. 11–13, 15–16.

40. H. G. Wells, *Russia in the Shadows* (New York: George H. Doran, 1921), p. 48. During 1923–1924 the norms for each category were: V—forty rubles/month; IV—thirty rubles/month; III—twenty rubles/month; II—sixteen rubles/month; I—twelve rubles/month. Beginning in October 1924, *vuzy* and other scientific institutions took over support of the top three categories, so that only 300 of the most senior scholars continued to receive a monthly allowance of eighty-five rubles; in category IV, 250 continued to receive sixty rubles per month. *Piat let' TsEKUBU*, p. 17.

41. *Piat let' TsEKUBU*, p. 15.

42. Ibid., pp. 22–23, 38. In 1924–1925, 1,441 scientific workers received inpatient treatment at five such sanitaria; in 1925–1926, 1,215 scientists were treated. Dormitory space was also provided at nominal costs for both

temporary and long-term stays. Over 600 scientific workers stayed in TsEKUBU dormitories in 1922–1923; 1,262 in 1923–1924; 1,628 in 1924–1925; and 1,568 in 1925–1926. The use of dormitories indicates the extent to which a housing crisis plagued Russia's urban intelligentsia after the Revolution.

43. *God raboty TsEKUBU*, p. 11.

44. *Gor'kii i nauka*, pp. 131–132.

45. *Piat' let TsEKUBU*, pp. 38–42. The resolution concerned "O poriade rassmotrenii del o zhilishchnykh pravakh nauchnykh rabochikh."

46. Wells, *Russia*, pp. 48–49.

47. Paul Josephson, "Physics and Soviet-Western Relations in the 1920s and 1930s," *Physics Today*, 41, no. 9 (September 1988): 55–56.

48. Wells, *Russia*, pp. 48–51.

49. Gorky, *Sobranie sochineniia* 29: 395.

50. *NieR*, no. 2 (1921): 5–6, and *Piat' let TsEKUBU*, p. 32. In 1924 Dom Uchenykh received 86; and in 1925 and 1926 it subscribed to 142 journals, of which 61 were German, 37 French, and 74 Russian, and of which 7 were physics journals, 10 biology, 7 chemistry, 22 medicine, and 13 art.

51. See Robert V. Daniels, *The Conscience of the Revolution* (New York: Simon and Schuster, 1969), pp. 119–136, for a discussion of the Workers' Opposition.

52. *God raboty TsEKUBU*, p. 11.

53. *A LFTI*, f. 1, op. 1, no. 24, l. 13.

54. D. M. Odinetz et al., *Russian Schools and Universities in the World War* (New Haven: Yale University Press, 1929), pp. 222–223.

55. Ibid., pp. 224–229.

56. Ibid., pp. 231–233.

57. Ibid., pp. 237–239.

58. A. V. Kol'tsov, *Lenin i stanovlenie Akademii Nauk kak tsentra sovetskoi nauki* (Leningrad, 1969), p. 31.

59. Ibid., pp. 27–45; *Organizatsiia nauki v pervye gody sovetskoi vlasti (1917–1925)* (hereafter *Organizatsiia sovetskoi nauki*) (Leningrad, 1974), 1: 21–26, 104–124; and S. I. Mokshin, *Sem' shagov po zemle* (Moscow, 1972), pp. 21–22.

60. Robert Lewis, *Science and Industrialization in the USSR* (New York: Holmes and Meier, 1979), pp. 37–41.

61. Ibid.

62. Bastrakova, *Stanovlenie*, p. 99; and Mokshin, *Sem' shagov*, p. 39.

63. See *Organizatsiia sovetskoi nauki*, 1: 82–84, for Gorbunov's attitudes toward NTO.

64. Ibid., pp. 87–89.

65. Ibid., pp. 88–89.

66. Ibid.

67. V. N. Ipatieff, F. E. Dzerzhinskii et al., *Nauchnye dostizheniia v pro-myshlennosti i raboti NTO VSNKh SSSR* (Moscow, 1925), pp. 6, 34.

68. For example, the Petrograd Department, PONTO, which first met

in Small Conference Hall of the Academy of Sciences on April 16, 1919, had a presidium under the direction of N. S. Kurnakov, with members representing several branches of the natural sciences. A. F. Ioffe was chairman of PONTO's physico-mechanical section.

69. *Organizatsiia sovetskoi nauki* 1: 92–94.

70. *TsGA RSFSR*, f. 2306, op. 19, ed. khr. 114, ll. 1–1 ob.

71. Ibid., f. 2307, op. 9, ed. khr. 176, ll. 39–40.

72. *A LFTI*, f. 1, op. 1, ed. khr. 50, l. 1.

73. *NTO. Otchet o deiatel'nosti za 1922–23 gg.* (Moscow, 1923), pp. ix, 15, and 33–35.

74. *A LFTI*, f. 1, op. 1, ed. khr. 17, 19, 23, 27, and 42; *VRiR*; and *Nauchno-organizatsionnaia deiatel'nost' Akademika A. F. Ioffe* (hereafter *Nauchnaia deiatel'nost' Ioffe*) (Leningrad, 1980), pp. 15–21.

75. *Nauchnaia deiatel'nost' Ioffe*, pp. 21–23.

76. P. A. Bogdanov, in Iu. N. Flakserman, *Promyshlennost' i nauchno-tekhnicheskie instituty* (Moscow, 1925), p. 4.

77. Lewis, *Science and Industrialization*, pp. 101–103.

78. Dzerzhinskii, "NTO VSNKh i ego instituty," in Dzerzhinskii et al., *Nauchnye dostizheniia*, pp. 41–42. This talk is reproduced in Dzerzhinskii, *Izbrannye proizvedeniia*, 3d edition (Moscow, 1977), 1: 34–35.

79. R. W. Davies, "Some Soviet Economic Controllers—II," *Soviet Studies* 11, no. 4 (1959–1960): 378–386, and Ipatieff, *Life of a Chemist*, pp. 424–425.

80. Lewis, *Science and Industrialization*, pp. 41–45; and *Organizatsiia sovetskoi nauki*, 1: 98–100.

81. *A LFTI*, "Otchet fiziko-tekhnicheskogo otdela instituta za 1919 g.," f. 1, op. 1, ed. khr. 4, l. 7; and "Otchet o rabote otdela instituta za 1920 g.," f. 1, op. 1, ed. khr. 5, l. 4.

82. *Vestnik narodnogo prosveshcheniia soiuza kommun severnoi oblasti*, no. 6–8 (1918): 69.

83. B. D. Lebin, ed., *Ocherki istorii organizatsii nauki v Leningrade, 1703–1977 gg.* (Leningrad, 1980), p. 122. For the functions of the Petrograd Administration of Science, see *Organizatsiia sovetskoi nauki* 1: 40–41.

84. See *Organizatsiia sovetskoi nauki* 1: 36–37, on the functions of Akadtsentr.

85. Lewis and Bastrakova describe how Narkompros was defeated in the battle with the Academy of Sciences and Vesenkha for preeminence in the scientific research establishment in Soviet Russia (although I might add that the final outcome was determined only during the cultural revolution). See Lewis, *Science and Industrialization*, pp. 37–44, and Bastrakova, *Stanovlenie*, pp. 50–70, for a discussion of this debate.

86. Sheila Fitzpatrick, *The Commissariat of Enlightenment* (Cambridge: Cambridge University Press, 1970), pp. xiv–xvi.

87. *Biulleten' glavnauki*, no. 3–4 (1922): 8.

88. Ibid., pp. 9–11.

89. *Organizatsiia sovetskoi nauki* 1: 129–130.

90. As of 1925, there were fourteen physics or astronomy institutions within Glavnauka's Scientific Department, including the Leningrad Physico-Technical Institute, the Main Geophysical and Main Astronomical Observatories, the State Optical Institute, and the State Radium Institute; of these, ten had been founded after the Revolution. In 1927, excluding *kraevedy* (organizations which studied local customs and lore), thirty of eighty-six institutions, or thirty-five percent, were tied to physics and mathematics. See *Pervaia otchetnaia vystavka glavnauki narkomprosa* (Moscow, 1925), pp. 43–54, and *Biulleten' glavnauki*, no. 3–4 (1922): 5, 8–9.

91. *Vystavka glavnauki*, p. 5.

92. Ibid., p. 7.

93. *Biulleten' glavnauki*, no. 3–4 (1922): 39, and no. 5 (1923): 16.

94. On the development of scientific publications in the Soviet Union after the Revolution, see "Nauchnoe izdatel'stvo v SSSR," *Desiat' let sovetskoi nauki*, ed. F. N. Petrov (Moscow-Leningrad, 1927), pp. 447–463.

3: THE RUSSIAN ASSOCIATION OF PHYSICISTS

1. O. D. Khvol'son, "Shto dal oktiabr' sovetskoi fizike," *NR*, no. 12 (1927): p. 21.

2. *TsGA RSFSR*, f. 2307, op. 2, ed. khr. 377, ll. 1, 9–9 ob., 11–13.

3. *ZhRFKhO*, no. 4/6 (1919): 173–175.

4. Ibid.

5. Khvol'son, "Shto dal oktiabr'," pp. 21–22.

6. *TsGA RSFSR*, f. 2307, op. 1, ed. khr. 37, l. 4, and ed. khr. 31, l. 3.

7. S. F. Ol'denburg, *Rossiiskaia Akademiia nauk v 1922 g.* (Leningrad, 1923), p. 11, and *Rossiiskaia Akademiia nauk v 1924 g.* (Leningrad, 1925), pp. xv–xvi.

8. *ZhRFKhO*, no. 4/6 (1919): inside front cover.

9. Ibid. no. 7/9 (1919): 275–278.

10. Ioffe, *Dostizheniia fiziki* (Moscow, 1928), p. 17.

11. *TsGA RSFSR*, f. 2306, op. 19, ed. khr. 131, ll. 1–2.

12. *Trudy s"ezda fizikov v petrograde* (Petrograd, 1919), 1: v–vi; and *ZhRFKhO*, no. 4/6 (1920): 313–352.

13. *Trudy s"ezda fizikov*, pp. vi–ix.

14. Ibid., p. xxxiv.

15. Ibid., p. xxxvi. The participants also passed a resolution calling for the creation of a new organ for the publication of articles and translations of articles and monographs with *ZhRFKhO* to be the arbiter of quality.

16. Ibid. I have found no records of the deliberations or reports of either commission.

17. *TsGA RSFSR*, f. 2306, op. 19, ed. khr. 107, ll. 4–11.

18. Ibid, l. 17, and Russkaia assotsiatsiia fizikov, "S"ezd RAF," vol. 3 of *Soobshcheniia o nauchno-tekhnicheskikh rabotakh v respublikakh* (Moscow 1921), pp. 157–165. See pp. 167–206 for abstracts of the papers.

19. Ibid.

20. Russkaia assotsiatsiia fizikov, *Tretii s"ezd RAF. Otchety i protokoly* (Nizhni Novgorod, 1922), from *Telegrafiia i telefoniia bez provodov* 16: 11.

21. Ibid., *Vtoroi s"ezd RAF* (Kiev, 1921), p. 3.

22. *TsGA RSFSR*, f. 2306, op. 19, ed. khr. 107, ll. 25–31.

23. *Tretii s"ezd*, pp. 9–12.

24. *TsGA RSFSR*, f. 2306, op. 19, ed. khr. 107, ll. 25–27.

25. By the time of the conference, the Fair Committee had provided 450,000 rubles, the Commissariat of Post and Telegraph 50,000 rubles for publication and paper. Of these resources, 260,000 rubles were spent on food, 66,000 on travel, 101,000 on organization, 69,000 on housing, and the RAF pocketed 5,000 rubles.

26. Russkaia assotsiatsia fizikov, *Trudy tret'ego s"ezda RAF (v nizhnem-novgorode 17–21 sent. 1922 g.)* (Nizhni Novgorod, n.d.), pp. i–ii, and *Tretii s"ezd*, pp. 1–6.

27. *Tretii s"ezd*, pp. 14–18.

28. *Organizatsiia sovetskoi nauki* 1: 232–233.

29. *Severnaia kommuna*, May 6, 1919, as cited in *Nauchnaia deiatel'nost' Ioffe*, p. 13. See also Sominskii, *Ioffe*, pp. 195–196.

30. *A LFTI*, "Polozhenie o GRiRI," f. 1, op. 1, ed. khr. 3, l. 1.

31. Ibid., "Otchet otdela o rabote za 1921 g.," f. 1, op. 1, ed. khr. 9, l. 3.

32. Sominskii, *Ioffe*, p. 237.

33. *Nauchnaia deiatel'nost' Ioffe*, pp. 31–32.

34. *A LFTI*, "Protokoly zasedanii komissii po reorganizatsii GRiRI, 23–29 noiabria, 1921," f. 1, op. 1, ed. khr. 8, l. 3.

35. *Organizatsiia sovetskoi nauki* 1: 244.

36. *A LFTI*, "Otchet o rabote instituta za 1920 g.," f. 1, op. 1, ed. khr. 5, l. 4.

37. Ibid., ed. khr. 6, ll. 1–4.

38. N. N. Semenov, *Nauka i obshchestvo*, 2d edition (Moscow, 1981), pp. 346–349; and Sominskii, *Ioffe*, pp. 242–243.

39. *Nauchnaia deiatel'nost' Ioffe*, pp. 29–30.

40. *Biulleten' glavnauki*, no. 5 (1923): 3; Semenov, *Nauka i obshchestvo*, pp. 346–349, and Sominskii, *Ioffe*, p. 243.

41. *A LFTI*, f. 1, op. 1, ed. khr. 15, l. 2. I have been unable to locate a copy of Ioffe's talk.

42. Letter of Ioffe to his wife of February 6, 1923, as cited in Sominskii, *Ioffe*, pp. 244–246.

43. Frish and Stazharov, *Vospominaniia ob Akademike Rozhdestvenskom*, pp. 5–30; and *Nauchnaia deiatel'nost' Ioffe*, pp. 305–306.

44. Ioffe, *Vstrechi*, pp. 121.

45. *A AN*, f. 970 (V. R. Bursian), op. 2, ed. khr. 2, l. 1–2. Unfortunately, the commission published little of the transcripts of its meetings, although several papers subsequently appeared in scientific journals.

46. Mokshin, *Sem' shagov*, pp. 116–118.

47. Frish and Stazharov, *Vospominaniia ob Akademike Rozhdestvenskom*, pp. 57–59.

48. *Organizatsiia sovetskoi nauki* 1: 235–236.

49. *TsGA RSFSR*, f. 2306, op. 19, ed. khr. 160, l. 52; *VRiR*, no. 1 (1919): 4–5; and *Organizatsiia sovetskoi nauki* 1: 235–236.

50. *Nauchnaia deiatel'nost' Ioffe*, pp. 21–23.

51. A. F. Ioffe and M. V. Kirpicheva, "Roentgenograms of Stained Crystals," *Philosophical Magazine* 43 (1922): 204–206, and Ioffe, *Izbrannye trudy*, pp. 150–152. Ia. I. Frenkel' discusses this work on asterism in *Ioffe*, pp. 12–13, as does Dukel'skii in *Ioffe*, pp. 15–16.

52. A. F. Ioffe, M. V. Kirpicheva, and M. A. Levitskaia, "The Elastic Limit and Strength of Crystals," *Nature* 113 (1924): 425–426. This work is republished in Ioffe, *Izbrannye trudy*, pp. 183–185. Ioffe and Levitskaia showed that the crystal ceases to be a right monocrystal (a rhombical dodecahedron), and decays into individual crystalline blocks. Using monochromatic X rays to study crystal behavior on a screen, the researchers discovered that each crystalline block displays its system of defractional pivots, which are imposed upon each other. The displacement of individual blocks in plastically deformed crystals occurs with great uniformity. During increased load and a rise in temperature, the displacements follow after each other during similar periods of time.

53. See, for example, W. Voigt and A. Sella, "Beobachtungen über die Zerreissungsfestigkeit von Steinsalz," *Annalen der Physik* 48 (1893): 636–673.

54. A. A. Griffiths, "The Phenomenon of Rupture and Flow in Solids," *Philosophical Transactions* A221 (1921): 163–197. See Stephen P. Timoshenko, *History of the Strength of Materials* (New York: Dover, 1983), pp. 358–360, for a discussion of the work of Griffiths and Ioffe.

55. Frenkel', *Ioffe*, pp. 12–13.

56. *A LFTI*, f. 1, op. 1, ed. khr. 5, l. 5–6; ibid., ed. khr. 17, l. 15; and *Nauchnaia deiatel'nost' Ioffe*, pp. 15–21.

57. *A LFTI*, f. 1, op. 1, ed. khr. 17, l. 5–6.

58. Ibid., "Otchet o rabote instituta za 1923–1924 gg.," f. 1, op. 1, ed. khr. 23, l. 5, 16, and ed. khr. 42, l. 8 ob.

59. Ibid., "Organizatsiia proizvodstva pri TO GFTRI," f. 1., op. 1, ed. khr. 20, entire.

60. Ibid., ed. khr. 17, ll. 1–2.

61. Ibid., ed. khr. 50, l. 1.

62. For a discussion of the creation of Glavprofobr, see Fitzpatrick, *Commissariat of Enlightenment*, pp. 61–67.

63. A. P. Kupaigorodskaia, *Vysshaia shkola Leningrada v pervye gody sovetskoi vlasti (1917–1925)* (Leningrad, 1984), pp. 68–71.

64. *Gor'kii i nauka*, p. 30; Gorky, *Sobranie sochineniia* 29: 396–397.

65. See, for example, V. I. Lenin, "On the Significance of Military Materialism," in *The Lenin Anthology*, ed. Robert Tucker (New York: Norton, 1975), pp. 651–653.

66. *Vospominaniia ob Ioffe*, pp. 70–72.

67. *Vospominaniia o Ia. I. Frenkele,* ed. V. M. Tuchkevich, (Leningrad, 1976), p. 36.

68. *Vospominaniia ob Ioffe,* p. 15.

69. See Stephen P. Timoshenko, *As I Remember,* trans. Robert Addis (Princeton: Van Nostrand, 1968), for interesting comments about the life of a Russian engineer before the Revolution. Timoshenko is known in the U.S. for his famous textbook on the strength of materials. He and Ioffe were students together in Romny as youths.

70. See Harley Balzer, *Educating Engineers: Economic Politics and Technical Training in Tsarist Russia* (Ph. D. diss., University of Pennsylvania, 1980), for an in-depth discussion of prerevolutionary technical and engineering training.

71. *Nauchnaia deiatel'nost' Ioffe,* pp. 267–270; and *Fiziko-mekhanicheskii fakul'tet Leningradskogo politekhnicheskogo instituta im. M. I. Kalinina* (hereafter *FMF*) (Leningrad, 1925), pp. 11–12.

72. *Vospominaniia ob Ioffe,* pp. 70–72.

73. *FMF,* pp. 11–12.

74. Ibid., pp. 37–38.

75. *Vospominaniia ob Ioffe,* pp. 36.

76. Ibid., pp. 71–73. For a list of course offerings, programs of specialization, and the like, see *FMF*, pp. 14–17, 21–36.

77. *Vospominaniia ob Ioffe,* p. 100.

78. Ibid., p. 128.

79. Ibid., p. 100.

80. *FMF,* p. 13.

4: THE FLOWERING OF SOVIET PHYSICS

1. See A. E. Ioffe, *Mezhdunarodnye sviazi sovetskoi nauki, tekhniki i kul'tury, 1917–1932* (Moscow, 1975), for a discussion of the reestablishment of foreign contacts.

2. *TsGA RSFSR,* f. 2306, op. 19, ed. khr. 160, l. 22 ob.

3. *NR,* no. 1 (1929): 113, and *NS,* no. 8 (1928): 112–114.

4. *TsGA RSFSR,* f. 2307, op. 12, ed. khr. 8, l. 53.

5. *NTO. Otchet o deiatel'nosti 2a 1922–23 gg.* (Moscow, 1923), pp. xi, 308, and *Organizatsiia sovetskoi nauki* 1: 96–97, 377–386.

6. *Organizatsiia sovetskoi nauki* 1: 369–371.

7. V. Ia. Frenkel', *Paul' Erenfest* (Moscow, 1977), pp. 83–90.

8. Sominskii, *Ioffe,* p. 209.

9. Ibid., pp. 212–213.

10. Ibid., p. 214.

11. A. F. Ioffe Fund, Manuscript Division of the Saltykov-Shchedrin State Library, Leningrad, f. 1000, op. 2, ed. khr. 545, letters 2–4.

12. See Sominskii, *Ioffe,* pp. 216–226, for nearly complete reproduction of these letters.

13. Ioffe Fund, letters 6–8.

14. A LFTI, f. 1, op. 1, ed. khr. 17, l. 6; ed. khr. 27, l. 14 ob; and ed. khr. 42, l. 4.

15. Ioffe Fund, letters 9–10.

16. Ibid.

17. Ibid., letters 14, 16, and 17.

18. Ibid., letter 18.

19. Ibid., letters 19–22.

20. A LFTI, f. 1, op. 1, ed. khr. 9, l. 2.

21. Biulleten' glavnauki, no. 5 (1923): 3; and Ioffe Fund, letters 25 and 26.

22. Organizatssia sovetskoi naki 1: 375.

23. Frish and Stazharov, Vospominaniia ob Akademike Rozhdestvenskom, p. 95.

24. A LFTI, f. 1, op. 1, ed. khr. 24, l. 4. Later, when party officials who embraced Stalin's notion of "socialism in one country" came to power, autarkic tendencies—national self-sufficiency at any cost—were applied to the economy and science alike, depriving scholars of the international community they had struggled to establish. The debate among Soviet officials over the role of Western versus indigenous technology is discussed in Bruce Parrott, Politics and Technology in the Soviet Union (Cambridge, Mass.: MIT Press, 1983).

25. Lebin, Ocherki istorii organizatsii nauki v Leningrade, p. 136.

26. A LFTI, f. 1, op. 1, ed. khr. 23, l. 17; and ed. khr. 42, l. 11 ob.

27. Kapitsa made annual journeys to Russia during summer vacation from Cavendish. In 1934 the Soviet authorities refused to allow him to return to England. See chap. 8, below, and Lawrence Badash, Kapitza, Rutherford, and the Kremlin (New Haven: Yale University Press, 1985), for a discussion of this event.

28. See V. S. Kogan, Kirill Dmitrievich Sinel'nikov (Kiev, 1984), pp. 35–38, on Sinel'nikov's time in Cambridge with Kapitsa and Rutherford.

29. See V. Ia. Frenkel' and Paul Josephson, "Sovetskie fiziki—Stipendiaty Rokfellerovskogo fonda," UFN 26, no. 11 (November 1990): 103–134, on Soviet physicists' participation in the IEB programs.

30. International Educational Board (IEB) Archives in Rockefeller Archive Collection, series 1, box 40, folder 566.

31. Ibid., box 47, folder 688.

32. Ibid., box 52, folder 818. It appears that none of these funds was awarded.

33. Augustus Trowbridge, director of science programs in Europe in the mid-1920s, reported that Max Born rated V. A. Fock "slightly higher than Tamm," although both were "capable and original." See ibid., box 48, folder 725.

34. Ibid., box 49, folder 733.

35. Ibid.

36. Ibid.

37. These lectures appear in Max Born, Problems of Atomic Dynamics, 1st

edition (Cambridge, Mass.: MIT Press, 1926), with a focus on the structure of the atom and the lattice theory of rigid bodies.

38. For an in-depth discussion of this European trip, see V. Frenkel', *Frenkel'*, pp. 146–190. See pp. 148–149 for his meetings with Einstein. Frenkel''s letters home (many of which are published fully on pp. 144–186) provide insight into the experiences of a Soviet physicist in Europe in the 1920s.

39. IEB Archives, series 1, box 49, folder 733.

40. Ioffe Fund, letter 25; and Sominskii, *Ioffe*, pp. 468–477.

41. Ioffe, *The Physics of Crystals* (New York: McGraw Hill, 1928), p. v.

42. Loeb praised Ioffe during the honorary degree ceremony for being "above all, a man . . . whose high scientific idealism and fine personality, . . . rising above political, personal and other prejudices, has kept alive and active the physical sciences of a nation through stress of war, revolution, riot, famine and social reorganization in such a way that today the activity in the physical sciences of his country transcends many fold the activity of the period preceding the war, so that today this science of his fatherland can rear its head in pride among that of other great nations." Materials of A. F. Ioffe, CU 5, box 203, University of California, Berkeley, Archives, Bancroft Library.

43. Ibid.

44. *A LFTI*, f. 1, op. 1, ed. khr. 24, l. 17.

45. Letter, Ioffe to Loeb, February 20, 1928, Materials of Professor Leonard Loeb, University of California, Berkeley, Archives, Bancroft Library.

46. Ibid., letter, Ioffe to Loeb, June 14, 1927.

47. Ibid., letter, Ioffe to Loeb, July 8, 1928.

48. Otto Scott, *The Creative Ordeal: The Story of Raytheon* (New York: Atheneum, 1974), pp. 42–43. Scott's rendition of the story must be taken with a grain of salt, for he implies that Ioffe knew well before the early 1930s that thin-layer insulation would not work, an assertion not borne out by an examination of the record. Scott's work on the Raytheon-Soviet connection is fraught with other inaccuracies. See V. Ia. Frenkel' and A. P. Grinberg, *I. V. Kurchatov v FTIe* (Leningrad, 1984), pp. 33–36, for a discussion of Ioffe's work on thin-layer insulation.

49. See Lewis, *Science and Industrialization*, for a discussion of industrial research and development in the USSR during the 1920s and 1930s.

50. *A LFTI*, f. 1, op. 1, ed. khr. 23, l. 16; and "Otchet o rabote instituta za 1924–25 gg.," ibid., ed. khr. 27, l. 8 ob.-9.

51. Ibid., "Otchet o rabote za 1923–24 gg.," f. 1, op. 1, ed. khr. 23, l. 16.

52. Ibid., l. 5.

53. Ioffe, *Dostizheniia fiziki*, pp. 19–20; and *Moia zhizn'*, pp. 23, 26–27.

54. *Nauchnaia deiatel'nost' Ioffe*, pp. 52–54.

55. Ibid.

56. *Fizika i proizvodstvo*, no. 1 (1930): 51–56, and no. 2–3 (1930): 61–65.

57. *Nauchnaia deiatel'nost' Ioffe*, pp. 309.

58. Dzerzhinskii, *O khoziaistvennom stroitel'stve SSSR* (Moscow-Leningrad, 1926), pp. 5–6.

59. *Nauchnaia deiatel'nost' Ioffe*, pp. 52–53.

60. Ibid., pp. 50–51.

61. *A LFTI*, LFTL, d. 4, l. 8, as cited in Sominskii, *Ioffe*, p. 257.

62. *Nauchnaia deiatel'nost' Ioffe*, p. 309; and *Vospominaniia ob Ioffe*, p. 43.

63. *Vospominaniia ob Ioffe*, p. 43.

64. *Piatiletnii plan rabot gosudarstvennoi fiziko-tekhnicheskoi laboratorii* (hereafter *Piatiletnii plan GFTL*), vol. 20 of *Piatiletnii plan nauchno-eksperimental'noi raboty v sviazi s rekonstruktsiei promyshlennosti* (Moscow, 1929), p. 45.

65. *A LFTI*, f. 1, op. 1, ed. khr. 27, l. 1.

66. *FMF*, pp. 18–20.

67. *Piatiletnii plan GFTL*, p. 8.

68. *Leningradskaia fiziko-tekhnicheskaia laboratoriia. Teplotekhnicheskii otdel* (hereafter *LFTL TTO*) (Leningrad, 1928), p. 7.

69. Ibid., pp. 7–8.

70. Ibid., pp. 8–10.

71. *TLFTL* 4: 3–4. This volume contains seven articles on various aspects of heat engineering, thermal exchange and conductivity, and efficiency. See also "Raboty teplotekhnicheskogo otdela," no. 9 of ibid. (Moscow 1929), entire.

72. The following analysis of the research program of LFTI does not consider chemical physics or heat engineering. The former, under the direction of N. N. Semenov, became in 1931 the focus of an independent scientific research institute—the Leningrad Institute of Chemical Physics, and the latter was discussed briefly above.

73. A. F. Ioffe, K. D. Sinel'nikov, and B. M. Gokhberg, "Vysokovol't-naia poliarizatsiia v dielektrikakh, " *ZhRFKhO*, no. 2 (1926): 105–114.

74. K. D. Sinel'nikov and I. V. Kurchatov, "K voprosu o vysokovol'tnoi poliarizatsii v tverdykh dielektrikakh," *ZhRFKhO*, no. 3/4 (1927): 327–339.

75. A. F. Val'ter, L. D. Inge, and N. N. Semenov, "Proboi tverdykh dielektrikov," *ZPF* 2, no. 3/4 (1924): 142–160; and *TLFTL* 1: 51–63; Val'ter and Inge, "Proboi stekla," ibid., 3: 47–54; P. P. Kobeko, I. V. Kurchatov, and K. D. Sinel'nikov, "Proboi tverdykh dielektrikov," ibid., 5: 5–19; A. F. Val'ter and L. D. Inge, "Proboi stekla pri postoiannom i peremennom napriazhenii," ibid. 5: 20–39; and V. A. Fock, "K teplovoi teorii elektrichesko-go proboia," ibid. 5: 52–71, which provides a theory of high-temperature breakdown based on the work of Val'ter and Inge.

76. See, for example, the investigation of Kobeko, Kurchatov, and Sinel'nikov on the mechanism of breakdown of several kinds of resin in "Issledovanie mekhanizma proboia nekotorykh smol," *ZhRFKhO*, no. 3 (1928): 211–217.

77. Several researchers attempted to apply data on breakdown to evaluate the utility of vacuums as insulators, employing different electrodes, including steel in a vacuum tube of quartz, porcelain, and glass, but their

results were too qualitative to make final judgments. See V. N. Malyshev, N. N. Semenov, and N. V. Tomashevksii, "Vakuum kak isoliator," *ZPF*, no. 5 (1928): 93–118.

78. Val'ter and Inge, "Proboi tverdykh dielektrikov pri bol'shikh chastotakh," *ZPF* (1928 supplement): 47–65.

79. V. A. Fock, "Tochnyi raschet teplovogo soprotivleniia monozhil'nykh kabelei," *TLFTL* 3: 72–86.

80. A. F. Val'ter and G. A. Dmitriev, "Proboi propitannoi kabel'noi bumagi," *TLFTL*, 5: 48–51, and *ZPF* (1928 supplement): 67–80.

81. A. F. Val'ter, "K voprosu o mekhanizme metallicheskoi elektroprovodnosti," *ZhRFKhO*, no. 2 (1927): 219–225.

82. N. Ia. Seliakov, G. V. Kurdiumov, and N. T. Grudtsov, "Rentgenograficheskoe issledovanie," *ZPF*, no. 2 (1927): 51–68. For discussion of stress, optical, and acoustical tests, see N. Ia. Seliakov, "Khimicheskii analiz v rentgenovykh luchakh," *TLFTL* 1: 25–33; and Seliakov and E. Z. Kaminskii, "Opredelenie neodnorodnostei v metallicheskikh obraztsakh proizvol'noi formy pri pomoshchi rentgenovykh luchei," ibid., 3: 33–37.

83. N. N. Davidenkov, "Akusticheskii metod izmereniia napriazhenii," *ZPF* (1928 supplement): 37–46, and "Akusticheskii metod izmereniia napriazhenii v sooruzheniiakh," *TLFTL* 5: 122–131.

84. N. N. Davidenkov and G. N. Titov, "Repetitsionnyi sklerometr," *ZPF* (1928 supplement): 29–36.

85. N. N. Davidenkov and M. V. Iakutovich, "Opredelenie modulia uprugosti stali," *ZPF* (1928 supplement): 3–19. For Sears's work see, for example, "On the Longitudinal Impact of Metal Rods with Rounded Ends," *Proceedings of the Cambridge Philosophical Society* 14 (1908): 257–286.

86. M. V. Klassen-Nekliudova, "O prirode plastichnoii deformatsii," *ZhRFKhO*, no. 5/6 (1927): 508–515, and Ioffe, *Physics of Crystals*, pp. 50–51. Ia. I. Frenkel', T. A. Kontorova, N. Ia. Seliakov, and A. V. Stepanov later set forth the theoretical underpinnings for her work, with Stepanov advancing the idea of a parallelism between plastic deformation and local fusion in glide planes. See S. T. Konobeevskii, "Rentgenovskii analiz i rentgeno-spektroskopiia," *UFN* 33, no. 4 (1947): 535.

87. M. V. Klassen-Nekliudova, "Zakonomernosti skachkoobraznoi deformatsii," *ZhRFKhO*, no. 5 (1928): 373–381.

88. *TsGA RSFSR*, f. 2307, op. 9, ed. khr. 39, ll. 19–23; and *Nauchnaia deiatel'nost' Ioffe*, pp. 217–218.

89. *TsGA RSFSR*, f. 2307, op. 9, ed. khr. 39, l. 19; and Russkaia assotsiatsiia fizikov, *IV s"ezd russkikh fizikov v Leningrade (15–20 sentiabria 1924 godu). Perechen' dokladov*, vol. 14 of *Soobshcheniia o nauchno-tekhnicheskikh rabotakh v respublike* (Leningrad, 1924).

90. The seventeen papers presented by LFTI physicists reflect its continuing focus on the physics of crystals and methods used to study them; the theory and structure of metals; electrostatic and electrotechnical processes in gases and vapors; magnetism; and several papers in theoretical physics, including G. A. Grinberg's discussion of the derivation of equa-

tions of classical hydrodynamics from those of the special theory of relativity, a topic suggested to him by A. A. Friedmann, and Ia. I. Frenkel''s papers on the theory of cohesive forces in dielectrics, the theory of the phenomena of precipitation and absorption, and electronic theory of metals. See *IV s"ezd fizikov. Perechen' dokladov* for brief synopses of all papers given, and *ZhRFKhO*, no. 5/6 (1924) and no. 1/2 (1925) for thirty of the papers presented at the congress with English summaries.

91. *TsGA RSFSR*, f. 2307, op. 9, ed. khr. 39, ll. 24–26; ibid., op. 7, ed. khr. 176, ll. 20–20 ob., 41; and Russkaia assotsiatsiia fizikov, *IV s"ezd russkikh fizikov. Raspisanie zaniatii* (Leningrad, 1924), pp. 3–15.

92. *Nauchnaia deiatel'nost' Ioffe*, pp. 219–220.

93. Russkaia assotsiatsiia fizikov, *V s"ezd russkikh fizikov. Perechen' dokladov* (Moscow-Leningrad, 1926), pp. 5–96. LFTI scholars were well represented, reading 20 of the 167 papers for which one can determine institutional affiliation, with members of the Ioffe "school" giving 31, or almost one-fifth, of the papers.

94. S. I. Vavilov, "Shestoi s"ezd fizikov," *NS*, no. 8 (1928): 95–96.

95. T. P. Kravetz, "Shestoi s"ezd fizikov," *Priroda*, no. 10 (1928): 915.

96. *Nauchnaia deiatel'nost' Ioffe*, p. 220. More Western physicists had been expected, but August vacations apparently led to cancellations. Lewis later wrote Ioffe that he enjoyed his experience except for one event referred to cryptically: "There was one incident at the end, our visit to Minsk, which must have furnished you a good laugh when you heard about it . . . although for a few hours it did not seem laughable. But nothing could mar the general delightfulness of our Russian experience." Letter from Gilbert Lewis to Ioffe, November 23, 1928, Materials of A. F. Ioffe, University of California, Berkeley, Archives, Bancroft Library.

97. IEB Archives, op cit, Letter, Ioffe to Trowbridge, March 7, 1928, series 1, IEB box 59, Folder 979.

98. Vavilov, "Shestoi s"ezd," p. 100.

99. Kravets, "Shestoi s"ezd," pp. 915–916.

100. Ibid.

101. Vavilov, "Shestoi s"ezd," pp. 97–98.

102. Kravets, "Shestoi s"ezd," pp. 915–916.

103. Ibid.

104. *Programma zasedanii VI s"ezda rossisskoi assotsiatsii fizikov* (Leningrad, 1928), pp. 5–14.

105. Russkaia assotsiatsiia fizikov, *VI s"ezd russkikh fizikov. Perechen' dokladov* (Leningrad, 1928), pp. 3–4.

106. Vavilov, "Shestoi s"ezd," pp. 95–96.

107. For more discussion of this phenomenon, see Loren Graham, "The Formation of Soviet Research Institutes: A Combination of Revolutionary Innovation and International Borrowing," *Social Studies of Science* 5, no. 3 (August 1975): 303–330.

108. *Piatiletnii plan raboty gosudarstvennogo fiziko-tekhnicheskogo instituta, 1928/29–1932* (hereafter *Piatiletnyi plan GFTIa*) (Leningrad, n.d.; typewritten, most likely 1930), pp. 21–75.

5: PHYSICS DURING THE FIRST FIVE-YEAR PLAN

1. Some of the following discussion is covered in Paul Josephson, "Physics, Stalinist Politics of Science, and Cultural Revolution," *Soviet Studies*, no. 2 (April 1988): 245–265.

2. N.a., "Piatitetnyi perspektivnyi plan Glavnauki RSFSR na 1928/29–1932/33 gg.," *NS*, no. 6 (1929): 105–108.

3. *Organizatsiia sovetskoi nauki* 2: 141.

4. See R. W. Davies, "Some Soviet Economic Controllers—III," *Soviet Studies* 12 (1960–1961): 23–55, on Kuibyshev and Ordzhonikidze.

5. For more on the late history of Glavnauka, see *Organizatsiia sovetskoi nauki* 2: 138–140, 150; *NS*, no. 7/8 (1930): 116; and *Sorena*, no. 3 (1932): 201–204.

6. *TsGA RSFSR*, f. 2307, op. 18, khr. 18, l. 5.

7. *BNIS*, no. 2 (1930); 1.

8. *Kul'turnoe stroitel'stvo SSSR v tsifrakh (1930–1934 gg.)* (Moscow, 1934), pp. 148–150.

9. *Piatiletnyi plan GFTIa*, pp. 3, 322. See pp. 159–309 for the laboratory-by-laboratory "Problemmo-tematicheskii plan" for 1931–1932.

10. *A LFTI*, "Godovoi otchet za 1935," f. 3, op. 1, ed. khr. 38, ll. 11–12. See ll. 22–114 for a list of the fifty-five projects, budget, and personnel associated with each project.

11. See, for example, "Plan nauchno-issledovatel'skikh rabot instituta na 1972 g.," *A FIAN*, f. 532, op. 1, ed. khr. 563, ll. 1–80.

12. *Vtoroi vsesoiuznyi s"ezd nauchnykh rabotnikov SSSR (Fev. 1927). Sputnik delegata s"ezda* (Moscow, 1927), pp. 18–20.

13. *NS*, no. 3 (1929): 109–112.

14. Ioffe, "Korennye problemy NIR," in *Sotsialisticheskaia rekonstruktsiia i NIR* (Moscow, 1930), pp. 23–26.

15. *BNIS*, no. 1 (1930): 16. Ioffe's presentation is in *BNIS*, no. 3 (1930), which I have been unable to locate.

16. *A AN*, f. 351, op. 2, ed. khr. 26, l. 63.

17. *Nauchnaia deiatel'nost' Ioffe*, p. 231.

18. Ibid.

19. Ibid, pp. 240–241. Ioffe again and again identified solar, wind, ocean tide, wave, and atomic energy as the most promising technologies. See A. F. Ioffe, *Perspektivy ispol'zovaniia novykh vidov energii vo vtoroi piatiletke* (Moscow-Leningrad, 1932); "Korennye problemy," p. 25; and "Problemy novykh istochnikov energii," *Sorena*, no. 1 (1932): 23–30.

20. N.a., "O gosudarstvennom fiziko-tekhnicheskom institute," *Sorena*, no. 1 (1931): 237–238, and *Nauchnaia deiatel'nost' Ioffe*, p. 60.

21. I. A. Tugarinov, "VARNITSO i Akademiia nauk SSSR (1927–1937 gg.)," *Voprosy istorii estestvoznaniia i tekhniki*, no. 4 (1989): 46–55.

22. *Tezisy dokladov i proekty rezoliutsii II vsesoiuznaia konferentsiia po planirovaniiu nauchno-issledovatel'skoi raboty v tiazheloi promyshlennosti na vtorom piatiletie i po 1933 g.* (Moscow, 1932), pp. 89–104.

23. Ibid., pp. 91–104.

24. "Rezoliutsiia po dokladu N. I. Bukharina," *Sorena*, no. 1 (1933): 235–238.

25. See, for example, S. Frederick Starr, "OSA: The Union of Contemporary Architects," in *Russian Modernism, Culture and the Avant-Garde, 1900–1930*, ed. George Gibian and H. W. Tjalsma, (Ithaca: Cornell University Press, 1976), pp. 188–208.

26. See Loren Graham, *The Soviet Academy of Sciences and the Communist Party, 1927–1932* (Princeton: Princeton University Press, 1967).

27. Kendall Bailes, "The Politics of Technology: Stalin and Technocratic Thinking Among Soviet Engineers," *American Historical Review*, 79 (April 1979): 445–469.

28. *A MGU*, f. 201, op. 1, ed. khr., ll. 5–6.

29. *Spravochnik dlia chlenov s"ezda* (Odessa, 1930), pp. 7–10.

30. M. Savost'ianova, "I Vsesoiuznyi s"ezd fizikov," *Priroda*, no. 10 (1930): 1044.

31. Ibid., pp. 1044–1045.

32. *Nauchnaia deiatel'nost' Ioffe*, pp. 227–228.

33. Ibid. Unfortunately, three attachments to this document—a list of individuals on the organizational committee of the VAF, a proposed budget, and the resolutions of the first All-Union Congress of Physicists—have not been published.

34. Ibid.

35. *Organizatsiia sovetskoi nauki* 2: 65.

36. On the administration of physics R and D within the Academy of Sciences, see *ZSN*, March 11, 1932, p. 2.

37. *Sorena*, no. 4 (1933): 173–175.

38. Ibid.

39. *Nauchnaia deiatel'nost' Ioffe*, p. 59.

40. Sominskii, *Ioffe*, pp. 262–264.

41. Ibid., p. 264.

42. Ibid., pp. 267–268.

43. N. N. Davidenkov, "Mekhanicheskii otdel GFTIa," *ZTF*, no. 2/3 (1931): 266–276.

44. S. A. Dianin and O. M. Todes, "Otdel matematicheskoi fiziki GFTIa, *ZTF*, no. 1 (1932): 60–61.

45. A. F. Val'ter, "Elektroizoliatsionnyi otdel GFTIa," *ZTF*, no. 2/3 (1931): 276–280.

46. See *ZTF*, no. 6 (1933): 807–939, for a series of articles devoted to the history and research program of LEFI.

47. *A LFTI*, "Postanovlenie prezidiuma TsKK VKP (b) o rezul'tatikh obsledovaniia GFTI," l. 68, as cited in Sominskii, *Ioffe*, p. 270.

48. A. I. Leipunskii, "UkFTI," in *Nauchno-tekhnicheskoe obsluzhivanie tiazheloi promyshlennosti* (Moscow-Leningrad, 1934), pp. 48–55.

49. *Nauchnaia deiatel'nost' Ioffe*, p. 110.

50. Ibid., p. 111, 114, 117.

51. Alexander Weissberg, *The Accused*, trans. Edward Fitzgerald (New York: Simon and Schuster, 1951), p. 67.

52. Ibid., p. 112.

53. *Piatiletnyi plan GFTL*, p. 45.

54. IEB Archives, series 1, box 59, folder 979. Ioffe recommended Sinel'-nikov to Rutherford by saying that he was "just as able and genial as Dr. Kapitsa." Sinel'nikov returned with an English wife after a two-year tour in England. Half-joking and half-serious, Ioffe told Kurchatov that he had decided to send only married physicists abroad from that point onward.

55. Kogan, *Sinel'nikov*, pp. 59–63.

56. V. N. Kessenikh, "SFTI pri Tomskom Universitete," *Trudy SFTNI* 2, no. 3 (1934): 3–4.

57. *Nauchnaia deiatel'nost' Ioffe*, p. 121.

58. Kessenikh, "SFTI," pp. 4–5.

59. Ibid., pp. 7–12. These included laboratories of molecular physics (under V. D. Kuznetsov); electrical oscillation (A. B. Sapozhnikov); X rays (M. I. Korsunskii); physical chemistry (M. I. Usanovich and M. I. Shulgina); electronic phenomena (P. S. Tartakovskii and A. A. Vorob'ev); and theoretical physics (A. A. Sokolov). For the growth of the SFTI in terms of budget, staff, and research program, see V. Kudriavtseva, "SFTI," *ZTF*, no. 1 (1932): 51–59; *Organizatsiia sovetskoi nauki* 2: 312–314; and *Piatiletnyi plan GFTL*, p. 46.

60. Kessenikh, "SFTI," pp. 13–14. For a discussion of the SFTI research program in the 1930s, V. D. Kuznetsov, "Siberskii fiziko-tekhnicheskii institut pri tomskom gosudarstvennom universitete," *ZTF*, no. 20/21 (1937): 1968–1973.

61. Kessenikh, "SFTI," pp. 6–7.

62. For a brief discussion of DFTI, see G. V. Kurdiumov, "Dnepropetrovskii FTI," *ZTF*, no. 21/22 (1937): 1974–78. See also V. S. Savchuk and Zh. K. Efremova, "Razvitie fizicheskikh issledovanii v Dnepropetrovske v 1917–1928 gg.," *Ocherki istorii estestvoznaniia i tekhniki*, 37 (Kiev, 1989): 23–28.

63. "Ot redaktsii," *Sorena*, no. 1 (1933): 3–4.

64. Ibid., no. 3 (1936): 105.

65. Weissberg, *The Accused*, pp. 184–185.

66. Ioffe, "Vorwort," *Phys. Zeit SU* 1, no. 1 (1932): 3–4.

67. Ibid.

68. *ZTF*, no. 4 (1937): 431.

69. A. F. Ioffe, "Nasha rabota v oblasti izucheniia mekhanicheskikh i elektricheskikh svoistv tverdykh tel," in *Matematika i estestvoznanie v SSSR* (Moscow-Leningrad, 1938), pp. 206–228.

70. Kobeko and Kurchatov, "Dielektricheskie svoistva kristallov segnetovoi soli," *ZhRFKhO*, no. 3 (1930): 251–265.

71. The following discussion of work on semiconductor physics undertaken at LFTI is based largely on A. F. Ioffe, "Nasha rabota," pp. 220–225, and "Poluprovodniki—novyi material elektrotekhniki," *Sorena*, no. 2/3 (1931): 108–112; M. Sominskii, "Itogi vyezdnoi sessii fizicheskoi gruppy Akademii Nauk SSSR," *ZTF*, no. 16 (1938): 1495–1499; and V. M. Tuchkevich and V. Ia. Frenkel', eds., *Semiconductors* (Leningrad, 1979).

72. Ioffe, *Elektronnye poluprovodniki* (Leningrad-Moscow, 1933), p. 7.

73. Tuchkevich and Frenkel', *Semiconductors*, p. 4.

74. Ibid.

75. Ia. I. Frenkel', "Teoriia tverdykh tel," *Sorena*, no. 8 (1932): 168–183.

76. D. N. Nasledov, "Vsesoiuznaia konferentsiia po tverdym vypriamiteliam i fotoelementam," *ZTF*, no. 9 (1935): 1661.

77. B. M. Gokhberg, "Obshchii obzor plana rabot po fizike v institutakh NKTP na 1936 god," *ZTF*, no. 3 (1936): 571–572.

78. Kogan, *Sinel'nikov*, p. 67.

79. See Paul R. Josephson, "The Early Years of Soviet Nuclear Physics," *Bulletin of the Atomic Scientists* 43, no. 10 (December 1987): 36–39, for an abbreviated version of the section.

80. For a discussion of the international nature of nuclear physics in the 1930s, see Charles Weiner, "Institutional Settings for Scientific Change: Episodes from the History of Nuclear Physics," in *Science and Values: Patterns of Tradition and Change* (New York: Humanities Press, 1974), pp. 187–212.

81. See David Holloway, "Entering the Nuclear Arms Race: The Soviet Decision to Build the Atomic Bomb, 1939–1945," *Social Studies of Science* 11 (1981): 159–197.

82. Dmitri Skobel'tsyn, "The Early State of Cosmic Ray Particle Research," in *The Birth of Particle Physics*, ed. Brown and Hoddeson (Cambridge and New York: Cambridge University Press, 1983), pp. 111–119; and Laurie M. Brown and Lillian Hoddeson, "The Birth of Elementary Particle Physics: 1930–1950," in ibid., pp. 8–9; Frenkel' and Grinberg, *Kurchatov*, p. 61.

83. See G. Gamov, "Ocherk razvitiia uchenia o stroenii atomnogo iadra," in *UFN*, no. 4 (1930): 531–544; no. 1 (1932): 31–43; no. 4 (1932): 389–403; no. 1 (1933): 46–57; and no. 4 (1934): 389–406.

84. *A LFTI*, f. 3, op. 2, d. 4, l. 104, as cited in *Nauchnaia deiatel'nost' Ioffe*, p. 62.

85. Ibid.

86. Ibid., f. 3, op. 2, d. 24, l. 1–1 ob., as cited in *Nauchnaia deiatel'nost' Ioffe*, pp. 62–63.

87. V. M. Tuchkevich and V. Ia. Frenkel', *Vklad Akademika A. F. Ioffe v stanovlenie iadernoi fiziki v SSSR* (Leningrad, 1980), p. 26.

88. *A LFTI*, f. 3, op. 1, d. 23, l. 1, as cited in ibid., pp. 11–12.

89. For a discussion of the proceedings of the conference, see M. P. Bronshtein, "Vsesoiuznaia iadernaia konferentsiia," *Sorena*, no. 9 (1933): 155–165; "Vsesoiuznaia iadernaia konferentsiia," *UFN*, no. 5 (1933): 768–778. M. P. Bronshtein et al., *Atomnoe iadro* (Leningrad-Moscow, 1934), contains the papers from the conference.

90. M. Dobrotin, "II Vsesoiuznaia konferentsiia po atomnoi iadre," *UFN*, 18, no. 4 (1937): 583–592. Many of the conference papers are published in *Izv. AN SSSR. Seriia fizicheskaia*, no. 1/2 (1938): 7–257.

91. A. F. Ioffe, "Vstupitel'noe slovo," *Izv. AN SSSR. Seriia fizicheskaia*, no. 1/2 (1938): 8–11.

92. Ibid.

93. Ibid.

94. Tuchkevich and Frenkel', *Vklad Ioffe*, p. 24.

95. V. Levich, "III Vsesoiuznoe soveshchanie po atomnoi iadre," *UFN* 21, no. 1 (1939): 75–81. See *Izv. AN SSSR. Seriia fizicheskaia,* no. 5/6 (1938): 727–792, for some of the major papers of the third conference.

96. *Izv. AN SSSR. Seriia fizicheskaia,* no. 5/6 (1938): 792.

97. M. Stanley Livingston, *Particle Accelerators: A Brief History* (Cambridge, Mass.: MIT Press, 1969), pp. 1–16.

98. A. K. Val'ter and K. D. Sinel'nikov, "Novyi tip vysokovol'tnoi ustanovki postoiannogo napriazheniia," *ZTF* 4, no. 5 (1934): 1073–1076.

99. Kogan, *Kinel'nikov* pp. 67–79.

100. Val'ter and Sinel'nikov, "Elektrostaticheskii generatory postoiannogo vysokogo napriazheniia," *ZTF* 6, no. 1 (1936): 151–162. See A. S. Papkov, "Impul'snyi generator," *ZTF* 8, no. 2 (1938): 171–173, for critical comments on the UkFTI research program. For greater detail on the early UkFTI research program, see A. K. Val'ter, *Ataka atomnogo iadra* (Kharkov, 1933), and Sinel'nikov et al., "Vysokovol'tnaia razriadnaia trubka na 3,000,000 vol't," *ZTF* 8, no. 11 (1938): 985–993.

101. Val'ter, *Ataka atomnogo iadra,* pp. 127–129.

102. Ibid., pp. 143–144.

103. *A LFTI*, f. 3, op. 1, ed. khr. 32, l. 5–12.

104. Ibid.

105. Ibid., f. 3, op. 1, ed. khr. 38, l. 78–107. See also Frenkel' and Grinberg, *Kurchatov*, pp. 67–84.

106. *Akademik A. I. Alikhanov. Vospominaniia, Pis'ma, Dokumenty,* ed. A. P. Aleksandrov (Leningrad, 1989), Naidenov, and V. Ia. Frenkel', *Tsiklotron FTIa im. A. F. Ioffe AN SSSR (k 40-letiiu so dnia puska)* (Leningrad, 1986), pp. 5–6.

107. Frenkel' and Grinberg, *Kurchatov*, p. 87.

108. Tuchkevich and Frenkel', *Vklad Ioffe*, pp. 15–17.

109. Ibid., p. 15.

110. Ibid., pp. 15–17; *A LFTI,* f. 3, op. 1, ed. khr. 81, ll. 38–39; and Ioffe, "Elektrostaticheskii generator," *ZTF* 9, no. 2 (1939): 2071–2080.

111. See Charles Weiner, "1932—Moving into the New Physics," *Physics Today* 25, no. 5 (May 1972): 40–49; "Physics in the Great Depression," *Physics Today* 23, no. 10 (October 1970): 31–38; and "Cyclotrons and Internationalism: Japan, Denmark and the United States, 1935–1945," in *Proceedings of the XIVth International Congress of the History of Science, 1974* (Tokyo, 1975), pp. 353–365.

112. *A LFTI,* f. 3, op. 1, ed. khr. 38, l. 7.

113. Lemberg et al., *Tsiklotron,* p. 7.

114. *Pravda,* June 22, 1940, as cited in ibid., p. 10, and Tuchkevich and Frenkel', *Vklad Ioffe,* pp. 28–29. For a discussion of research conducted in the postwar years on the cyclotron, see I. Kh. Lemberg, V. O. Naidenov, and V. Ia. Frenkel', "Tsiklotron fiziko-tekhnicheskogo instituta im. A. F. Ioffe

AN SSSR," *UFN*, 153, no. 3 (November 1987): 497–519. My thanks to Dr. V. Ia. Frenkel' for sharing his research with me on the history of the LFTI cyclotron.

6: CULTURAL REVOLUTION AND
THE NATURAL SCIENCES

1. See Sheila Fitzpatrick, *Education and Social Mobility in the Soviet Union, 1921–1934* (Cambridge and New York: Cambridge University Press, 1979), especially pt. 2; Sheila Fitzpatrick, ed., *Cultural Revolution in Russia, 1928–1931* (Bloomington: Indiana University Press, 1978); Susan Gross Solomon, "Rural Scholars and the Cultural Revolution," in ibid., pp. 129–153; S. Frederick Starr, "Visionary Town Planning during the Cultural Revolution," in ibid., pp. 207–240; Kendall Bailes, *Technology and Society Under Lenin and Stalin* (Princeton: Princeton University Press, 1978); Douglas R. Weiner, *Models of Nature: Ecology, Conservation and Cultural Revolution in Soviet Russia* (Bloomington: Indiana University Press, 1988), especially chaps. 8–10; David Joravsky, *The Lysenko Affair* (Cambridge: Harvard University Press, 1970); and Gregory Guroff, "The Red-Expert Debate," in *Entrepreneurship in Imperial Russia and the Soviet Union*, ed. Gregory Guroff and Fred Carstensen (Princeton: Princeton University Press, 1983), pp. 201–222, among others.

2. Weiner, *Models of Nature*, p. 125.

3. N. M. Katuntseva, *Rol' rabochikhfakul'tetov v formirovanii intelligentsii SSSR* (Moscow, 1966), p. 70.

4. S. A. Fediukin, *Sovetskaia vlast' i burzhuaznye spetsialisty* (Moscow, 1965), pp. 199–202.

5. Fitzpatrick, Introduction to *Cultural Revolution*, p. 2.

6. Ibid. For a discussion of Lunacharskii's views on cultural revolution, see Bailes, *Technology and Society*, pp. 160–161, and A. V. Lunacharskii, *Ob intelligentsii. Sbornik stat'ei* (Moscow, 1923); *Kul'tura na zapade i u nas* (Moscow-Leningrad, 1928), especially pp. 14–121; and "Kul'turnaia revoliutsiia i nauka," *NR*, no. 4 (1928): 3–7.

7. On Bukharin and the introduction of the Five-Year Plans, see Loren Graham, "Bukharin and the Planning of Science," *Slavic Review* 23, no. 2 (April 1964): 135–148.

8. On the fascination with "Amerikanizm" see Hans Rogger, "*Amerikanizm* and the Economic Development of Russia," *Comparative Studies in Society and History* 23, no. 3 (July 1981): 382–420; and Kendall E. Bailes, "The American Connection: Ideology and the Transfer of American Technology to the Soviet Union, 1917–1941," ibid., pp. 421–448. See also A. F. Ioffe, "Vpechatleniia ot poezdki po amerikanskam laboratoriiam," *NR*, no. 4 (1926): 59–65; M. Ia. Lapirov-Skoblo, "Amerika i ee tekhnika. I. N'iu Iork," *NS*, no. 1 (1928): 76–95, "Amerika i ee tekhnika. II. V tsarstve genri forda," ibid., no. 2 (1928): 126–158, and "Amerika i ee tekhnika. III. V tsartsve elektricheskoi energii," ibid., no 4 (1928): 96–110; L. E. Koll,

"Nauka v amerikanskom sel'skom khoziastve," ibid., no. 7 (1929): 88–96; others by Ia. G. Dorfman and A. F. Ioffe; and V. A. Steklov, *V Ameriku i obratno* (Leningrad, 1925).

9. N. I. Bukharin, "Rekonstruktivnyi period i NIR," *Biulleten' NIS*, no. 2 (1930): 2–3.

10. N. I. Bukharin, "Leninizm i problema kul'turnoi revoliutsii," *Narodnoe prosveshchenie*, no. 2 (1928): 9–10.

11. Ibid., p. 10.

12. Ibid., p. 14.

13. Ibid., p. 7.

14. Ibid., p. 18.

15. Bukharin, "Sotsialisticheskaia rekonstruktsiia i estestvennye nauki," in *Sotsialisticheskaia rekonstruktsiia i nauchno-issledovatel'skaia rabota. Sbornik NIS PTEU VSNKh SSSR k XVI s"ezdu VKP (b)* (Moscow, 1930), pp. 5–26, where Bukharin discussed the need to make the scientist also an engineer ("*ob"inzhenerit' ego*") and make the engineer a scientist "*ob"nauchit' ego*"; and "Rekonstruktivnyi period i NIR," pp. 1–7.

16. See, for example, V. V. Kuibyshev, *Nauke-sotsialitscheskii plan* (Moscow, 1931).

17. Bailes, *Technology and Society*, p. 170. See ibid., pp. 162–170, for a discussion of the major actors and their positions in the debate over this transfer.

18. One explanation for the attack on the bourgeois specialist during the Industrial Party Affair was the perceived technocratic approach of some of the engineers, which the Party saw as a threat to its power. See Bailes, "The Politics of Technology," pp. 445–469. The Soviet scholar S. A. Fediukin believes that "wrecking" and anti-Soviet activity was rightly punished in the Promparty and Shakhty affairs, although he admits that Stalin may have excessively invoked class war against specialists in the following years. See Fediukin, *Sovetskaia vlast' i burzhuaznye spetsialisty*, pp. 209–211, 213–215.

19. Fitzpatrick, "Cultural Revolution as Class War," in *Cultural Revolution in Russia*, pp. 12–21.

20. *TsGA RSFSR*, f. 406, op. 12, ed. khr. 2515, l. 139, as cited in K. E. Pechkurova, *Partiinoe rukovodstvo podgotovkoi nauchnykh kadrov v gody pervoi piatiletki (1928–1932 gg.) (na materialakh Leningrada)*, cand. diss. (Leningrad: Leningrad State University, 1976), p. 44.

21. *Partrabotnik*, no. 15 (1929): 53, as cited in ibid., p. 6.

22. *TsGA RSFSR*, f. 406, op. 12, ed. khr. 2515, l. 137, as cited in Pechkurova, *Partiinoe rukovodstvo*, and *LGAOR SS* (Leningradskii gosudarstvennyi arkhiv oktiabr'skoi revoliutsii i sotsialistickeskogo stroitel'stva), f. 7420, op. 14, ed. khr. 2515, l. 227, as cited in Pechkurova, *Partiinoe rukovodstvo*, p. 44.

23. *Trudy s"ezda fizikov v petrograde* (Petrograd, 1919), p. v.

24. S. F. Ol'denburg, *Rossiiskaia akademiia nauk v 1922 g.* (Leningrad, 1923), p. 24.

25. L. V. Ivanova, *Formirovanie sovetskoi nauchnoi intelligentsii (1917–1927 gg.)* (Moscow, 1980), pp. 245–270, and S. A. Fediukin, *Sovetskaia vlast' i burzhaznye spetsialisty*, p. 243.

26. Fediukin, *Soretskaia vlast' i burzhuaznye spetsialisty*, pp. 199–201; Ivanova, *Formirovanie sovetskoi nauchnoi intelligentsii*, pp. 278–279.

27. *Biulleten' VARNITSO* 1, no. 12 (1930): 3–4. See also *Nauchnye kadry i nauchno-issledovatel'skie uchrezhdeniia SSSR*, ed. O. Iu. Smidt and B. Ia. Smulevich (Moscow, 1930).

28. *A AN*, f. 364, op. 1, ed. khr. 2, ll. 19–20; f. 364, op. 4, ed. khr. 24, ll. 130–132; and f. 351, op. 1, ed. khr. 63, l. 34.

29. Ibid., f. 364, op. 1, ed. khr. 2, ll. 14–15.

30. Ibid., ed. khr. 5, ll. 118–119.

31. Ibid., ed. khr. 2, l. 1.

32. Ibid., ed. khr. 63, ll. 34–35; f. 364, op. 4, ed. khr. 21, l. 23; and f. 364, op. 4, ed. khr. 2, l. 47.

33. Ibid., f. 364, op. 4, ed. khr. 1, l. 5; ibid., ed. khr. 24, ll. 79–80; and ibid., ed. khr. 28, l. 127.

34. Ibid., ed. khr. 1, ll. 19, 71, 931–96, and f. 354, op. 4, ed. khr. 24, ll. 79–80.

35. I. S. Samokvalov, "Chislennost' i sostav nauchnykh rabotnikov SSSR," *Sorena*, 2 (1934): 136, 140; and L. Kirsanov, "Osnovy voprosy podgotovki nauchnykh kadrov," *FNiT*, no. 7–8 (1932): 96–101.

36. K. Bukhman, *Nauchno-isssledovatel'skie instituty i vysshie uchebnye zavedeniia SSSR v 1932 godu* (Moscow, 1932), pp. 82–83.

37. A. Ziskind, "Nauchno-issledovatel'skie kadry promyshlennosti," *FNiT*, no. 7–8 (1932): 105–109.

38. *Nauchno-issledovatel'skie uchrezhdeniia i nauchnye rabotniki SSSR* 3 (Moscow, 1934): 42–49.

39. Samokvalov, "Chislennost' i sostav," pp. 139–140.

40. "Aspirantura vuzov i nii's 1933–1935 gg.," *Vysshaia tekhnicheskaia shkola*, no. 11 (1935).

41. *A AN*, f. 351, op. 2, ed. khr. 26, ll. 6–7.

42. Ioffe, "Korennye problemy NIR," p. 26.

43. *A AN*, f. 351, op. 2, ed. khr. 26, l. 38.

44. Ibid., l. 70.

45. T. H. Rigby, *Communist Party Membership in the USSR, 1917–1967* (Princeton: Princeton University Press, 1968), pp. 85, 116.

46. A. E. Beilin, *Kadry spetsialistov v SSSR: ikh formirovanie i rost* (Moscow, 1935), pp. 122, 126, 127.

47. Graham, *The Soviet Academy of Sciences and the Communist Party*, pp. 82–83, 90–93.

48. Ioffe Fund, f. 1000, op. 2, ed. khr. 545, letter 17.

49. *A AN*, f. 351, op. 1, ed. khr. 63, l. 37.

50. Graham, *The Soviet Academy of Sciences and the Communist Party*, pp. 114–115, 125, 138–143, 161.

51. *ZSN*, February 25, 1933, p. 1.

52. "Bor'ba za nauchnyi kadry," *ZSN*, September 10, 1933, p. 2.

53. *VAN*, no. 4–5 (1936): 5, and 11 (1936): 43; and *ZSN*, October 14, 1931, p. 3; February 25, 1933, p. 2; March 20, 1934, p. 3; and July 4, 1934, p. 4.

54. Pechkurova, *Partiinoe rukovodstvo*, pp. 62–69.

55. *TsGA RSFSR*, f. 406, op. 1, ed. khr. 965, May 6, 1930.

56. B. Tazulakov, "Opyt gruppy rabochikh aspirantov GFTIa," *ZTF*, no. 5 (1936): 503.

57. Letter from A. F. Ioffe to P. S. Ehrenfest, May 1, 1931, as reproduced in *Erenfest-Ioffe*, p. 289.

58. *ZSN*, July 1, 1932, p. 3.

59. Tazulakov, "Opyt gruppy rabochikh aspirantov GFTIa," p. 503.

60. Ibid.

61. K. Bukhman, *Nauchno-isssledovatel'skie instituty*, pp. 4–8.

62. *Sorena*, no. 6 (1934): 184. At UkFTI, the specializations were nuclear physics, low-temperature physics, solid-state physics, and the physics of ultra-short waves.

63. Paul Josephson, "Scientists, the Public and the Party Under Gorbachev," *Harriman Institute Forum* 3, no. 5 (May 1990): 4.

64. *ZSN*, January 15, 1935; *Vysshaia shkola*, no. 1 (1937): 15.

65. *ZSN*, March 20, 1934, p. 4.

66. *Sorena*, no. 2 (1936): 161; and *ZSN*, January 15, 1934, p. 4; November 27, 1934, p. 4; and December 2, 1934, p. 4.

67. E. Kol'man, "Sovremennye zadachi matematikov i fizikov-materialistov-dialektikov," *Estestvoznanie i marksizm*, no. 1 (1930): 128.

68. *A MGU*, f. 225, ed. khr. 12, copy of A. K. Timirazev, "Sredi nauchnykh rabotnikov," *Pervyi universitet*, no. 2 (May 30, 1927), p. 4. The goal of the circle may have been to build Communist party membership in the physics department. The Party cell had only twenty students in 1922–23, when there were nearly 4,000 students in all. *A MGU*, f. 201, op. 1, ed. khr. 339, ed. khr. 17, l. 8.

69. S. L., "Kruzhok fizikov-matematikov-materialistov," *Estestvoznanie i marksizm*, no. 1 (1929): 179–181.

70. On the Mechanist physicist Z. A. Tseitlin, see David Joravsky, *Soviet Marxism and Natural Science, 1917–1931* (New York: Columbia University Press, 1961), pp. 160–163.

71. S. L., "Kruzhok."

72. *A MGU*, f. 225, op. 1, ed. khr. 56.

73. *A AN*, f. 351, op. 1, ed. khr. 82, ll. 1–6, 23–27; and ed. khr. 161, ll. 2–3.

74. *A AN*, f. 351, op. 1, ed. khr. 83, ll. 1–5, 9, 13.

75. Pechkurova, *Partiinoe rukovodstvo*, pp. 130–162.

76. *A AN*, f. 364, op. 1, ed. khr. 2, l. 9.

77. "Otchet o rabote kursov fiziki," *Estestvoznanie i marksizm*, no. 4 (1929): 211–213.

78. Kessenikh, "SFTI," p. 14.

79. *A AN,* f. 364, op. 4, ed. khr. 28, l. 140, and f. 351, op. 1, ed. khr. 109, ll. 6–20.

80. "Marksistko-leninskoe vospitanie spetsialistov," *FNiT,* no. 12 (1934): 122–123.

81. *LPA* (Archive of Leningrad Polytechnical Institute), f. 2018, op. 2, ' ed. khr. 99, l. 32 and ed. khr. 59, l. 7; and f. 2019, op. 1, ed. khr. 175, l. 39, as cited in Pechkurova, *Partiinoe rukovodstvo,* pp. 114–117.

82. Pechkurova, *Partiinoe rukovodstvo,* pp. 164–165.

83. *A MGU,* f. 201, op. 1, ed. khr. 86, l. 7.

84. *Piatiletnyi plan GFTIa,* p. 59.

85. *Vospominaniia o Ia. I. Frenkele* (Leningrad, 1976), pp. 114–115.

86. *A LFTI,* f. 3, op. 1, ed. khr. 39, l. 17, 19; and ed. khr. 32, l. 25.

87. S. F. Vasil'ev, *Iz istorii nauchnykh mirovozrenii (sbornik statei)* (Moscow-Leningrad, 1935).

88. *A LFTI,* f. 3, op. 1, ed. khr. 81, l. 12.

89. V. Egorshin, "O polozhenii na fronte fiziki i zadachi obshchestva fizikov-materialistov pri komakademii," *Za Marksistsko-leninskoe estestvoznanie,* no. 1 (1931): 112. Egorshin's criticism in this case referred to I. I. Sokolov's *Fizika dlia pedtekhnikumov,* which I was unable to locate.

90. Ibid.

91. *A MGU,* f. 225, op. 1, ed. khr. 40, l. 3.

92. V. Egorshin, "O nekotorykh osnovykh printsipakh marksistsko-leninskogo uchebnika po fizike," *Za marksistsko-leninskoe estestvoznanie,* no. 3–4 (1931): 44–75. Egorshin criticized the following textbooks: N. V. Kashin, *Metodika fiziki* and *Kurs fiziki dlia pedvuzov;* P. D. Zernov, *Konspekt lektsii po fizike;* V. A. Mikhel'son, *Fizika;* and E. Grimzel', *Kurs fiziki.*

93. V. Berestnev, "O proekte programmy po dialekticheskomu i istoricheskomu materializmu," *Vysshaia shkola,* no. 8–9 (1937): 56–89.

94. *Nauka proizvodstvu,* no. 1/2 (1932): 17–20.

95. *Sorena,* no. 1 (1932): 241.

96. See Anne Rassweiler, *The Generation of Power: The History of Dneprostroi* (New York and Oxford: Oxford University Press, 1988), on the relationship between the manager, worker, and technology at a modern industrial site.

97. M. P. Bronshtein, "Fizika," *Nauka proizvodstvu,* no. 11 (1932): 631–632. These remarks would not serve Bronshtein well in a few years.

7: THEORETICAL PHYSICS

1. For a discussion of the history of relativity theory in Soviet Russia, see V. P. Vizgin and G. E. Gorelik, "The Reception of the Theory of Relativity in Russia and the USSR," trans. Paul Josephson, in *The Comparative Reception of Relativity,* ed. Thomas Glick (Dordrecht: D. Reidel, 1987), pp. 265–326. For a discussion of philosophical debates over quantum mechanics, see K. Kh. Delokarov, *Metodologicheskie problemy kvantovoi mekhaniki v sovetskoi filosofskoi nauke* (Moscow, 1982).

2. Tetu Herosige, "The Ether Problem, the Mechanistic Worldview, and the Origins of the Theory of Relativity," *Historical Studies in the Physical Sciences* 7 (Princeton: Princeton University Press, 1976): 3–82.

3. A. Einstein, *Efir i printsip otnositel'nosti* (Petrograd, 1921), pp. 18–27.

4. See Dayton Miller, "Efirnyi vetr," *UFN* 5, no. 3 (1925): 177–185. Similar articles were published by Dayton Miller in *Science, Nature,* and *Annalen der Physik* in 1925 and 1926. Vavilov criticized supporters of Dayton Miller in "Eksperimental'nye podtverzhdeniia sledstvii obshchie teorii otnositel'nosti," *UFN* 5, no. 6 (1925): 457–460; "Novye poiski efirnogo vetra," *UFN* 6, no. 3 (1926): 242–254; "Novoe povtorenie opyta Maikel'sona," *UFN* 6, no. 4 (1926): 421–425. See also T. P. Kravets, "Fizika v 1927 g.," for a speech published in *Priroda*, no. 3 (1928): 207–228, in which the author describes Kennedy's repetition of the Michelson-Morley experiment, and its blow to Dayton Miller.

5. V. Ia. Frenkel', *Na zare novoi fiziki* (Leningrad, 1970), p. 136. On the early period of Bolshevik history and "god-building," see Stites, *Revolutionary Dreams*, pp. 102–105, and Zenovia A. Sochor, *Revolution and Culture: The Bogdanov-Lenin Controversy* (Ithaca: Cornell University Press, 1988). See also Bertram Wolfe, *Three Who Made a Revolution*, 2d edition (Boston: Beacon Press, 1960), pp. 505–507.

6. Ia. I. Frenkel', *Teoriia otnositel'nosti* (Petrograd, 1923), p. 227.

7. Frenkel', *Elektrichestvo i materiia* (Moscow-Leningrad, 1925), pp. 88–89, and "Osnovy elektrodinamiki tochechnykh elektronov," *ZhRFKhO*, no. 3/4 (1925): 393–412.

8. A. A. Friedmann, "Über die Krümmung des Raumes," *Zeitschrift für Physik* 10, no. 6 (1922): 377–387.

9. See E. A. Tropp, V. Ia. Frenkel', and A. D. Chernin, *Aleksandr Aleksandrovich Fridman* (Moscow, 1988), especially chap. 9, pp. 166–200.

10. V. K. Frederiks, "Obshchii printsip otnositel'nosti Einshteina," *UFN* 2, no. 2 (1920–1921): 162–188. This article was followed by one by G. S. Landsberg on Eddington's expedition to view a solar eclipse to verify the bending of rays of light predicted by relativity theory, "Otklonenie sveta v gravitatsionnom pole solntsa (resultaty angliiskikh ekspeditsii po nabliudeniiu solnechnogo zatmeniia 1919 g.)," *UFN* 2, no. 2 (1920–1921): 189–193.

11. V. K. Frederiks and A. A. Friedmann, *Osnovy teorii otnositel'nosti* (Leningrad, 1924), p. 1.

12. Joravsky, *Soviet Marxism and Natural Science*, p. 278. In a special 1927 edition of *UFN* devoted to the 200th anniversary of Newton's death, which included articles by Vavilov, Frenkel', and Frederiks, the latter again asked if the axiomatization of physics was possible (that is, to make physics like geometry and proceed from definitions to axioms). Frederiks concluded that such an approach was conceivable in spite of the necessity of rigorous proof independent of axioms, the requirement that definitions and axioms not contain other hidden axioms and definitions. See Frederiks, "Nachala mekhaniki N'iutona i printsip otnositel'nosti," *UFN* 7, no. 2 (1927): 75–86.

13. *Vospominaniia o Frenkele*, pp. 78–79.

14. See, for example, N. Bohr, "Atomnaia teoriia i mekhanika," *UFN* 6, no. 2 (1926): 93–111 (originally published in *Nature*); W. Heisenberg, "Kvantovaia mekhanika," *UFN* 6, no. 4 (1926): 425–434 (originally published in *Die Naturwissenschaften*); A. Sommerfeld, "Sovremennoe sostoianie atomnoi fiziki," *UFN* 7, no. 3–4 (1927): 165–175 (originally published in *Physikalische Zeitschrift*); E. Schrödinger, "Volnovaia teoriia mekhaniki atomov i molekul," ibid., 176–201 (originally published in *Physical Review*); P. Jordan, "Prichinnost' i statistika v sovremennoi fizike," *UFN* 7, no. 5 (1927): 318–328 (originally published in *Die Naturwissenschaften*); and N. Bohr, "Kvantovoi postulat i novoe razvitie atomistiki," *UFN* 8, no. 3 (1928): 306–337 (originally published in *Nature*).

15. On Bursian's life and major works, see E. V. Bursian, "Viktor Robertovich Bursian," *Chteniia pamiati A. F. Ioffe* (Leningrad, 1988), pp. 30–50.

16. *A AN*, V. R. Bursian, f. 920, op. 2, ed. khr. 23, ll. 1, 2, 3, 8, 12, 36, 51, and 74.

17. Ioffe, "Predslovie," in *Osnovaniia novoi kvantovoi mekhaniki*, ed. A. F. Ioffe (Moscow-Leningrad, 1927), p. 3.

18. *Vospominaniia o Frenkele*, p. 104.

19. See Loren Graham, *Science, Philosophy, and Human Behavior in the Soviet Union* (New York: Columbia University Press, 1987), pp. 321–323, 337–343, for an evaluation of Fock's philosophical views.

20. *Vospominaniia o Frenkele*, p. 79.

21. Ibid.

22. For Frenkel''s evaluation of the early history of quantum mechanics, see "Proiskhozhdenie i razvitie volnovoi mekhaniki," *Priroda*, no. 1 (1930): 5–32.

23. On L. I. Mandel'shtam, see N. D. Papaleksi, "Leonid Isaakovich Mandel'shtam," in L. I. Mandel'shtam, *Polnoe sobranie trudov* 1 (Moscow, 1948): 7–66; S. M. Rytov et al., *Akademik L. I. Mandel'shtam. K 100-letiiu so dnia rozhdeniia* (Moscow, 1979); and L. I. Mandel'shtam, *Lektsii po optike, teorii otnositel'nosti i kvantovoi mekhanike* (Moscow, 1972).

24. V. A. Fock, "Kvantovaia mekhanika," in *Matematika i estestvoznanie*, p. 165.

25. Ibid., pp. 166–170.

26. I. M. Khalatnikov et al., *Vospominaniia o L. D. Landau* (Moscow, 1988), and the popular and uneven Maiia Bessarab, *Landau. Stranitsy zhizni* (Moscow, 1978).

27. R. E. Peierls, "Note on Landau," in *The Beginnings of Solid State Physics*, ed. N. F. Mott (London: Royal Society, 1980), pp. 6–7.

28. L. Hoddeson and G. Baym, "The Development of the Quantum Mechanical Electron Theory of Metals: 1900–1928," in ibid., pp. 8–23.

29. Fock, "Kvantovaia mekhanika," pp. 173–175. See also Frenkel', "Teoriia metallicheskoi provodimosti," *UFN* 8, no. 2 (1928): 155–183.

30. M. P. Bronshtein, "Kvantovanie gravitatsionnykh voln," *ZETF*, no. 3 (1936): 195–236.

31. See, for example, V. Bazarov, "Obzor nauchno-popular'noi literatury po teorii otnositel'nosti," *VKA*, no. 3 (1923): 324–341.

32. David Joravsky's classic study of the debate between the Mechanists and the Deborinites, *Soviet Marxism and Natural Science*, describes thoroughly the epistemological and philosophical concerns raised by both groups and the outcome of their vindictive and often hostile discussions. This discussion focuses solely on physics, particularly the debates involving the Leningrad physics community, and uses archival materials not available to Professor Joravsky.

33. "Ot redaktsii," *Dialektika v prirode* 2 (Vologda, 1926): i.

34. Ibid., pp. i–iii.

35. A. Var'iash, "Nekotorye problemy sovremennoi fiziki i dialekticheskii materializm," in ibid., pp. 49–50.

36. A. K. Timiriazev, "Dialektika prirody Engel'sa i sovremennaia fizika," in ibid., p. 217.

37. See note 4, above.

38. I have drawn this picture of A. K. Timiriazev from materials in *A MGU*, f. 201, op. 1, ed. khr. various. See also G. E. Gorelik, "Naturfilosofskie problemy fiziki v 1937 gody," *Priroda*, no. 2 (1990): 93–102, for discussion of Timirazev's role in philosophical disputes in 1937.

39. A. K. Timiriazev, "Filosofskie raboty tov. I. V. Stalina i ikh znachenie dlia metodologii estestvoznaniia," *A MGU*, f. 201, op. 1, ed. khr. 89.

40. *Mekhanisticheskoe estestvoznanie i dialektieheskii materializm* (Vologda, 1925), p. 53.

41. A. K. Timiriazev, "Teoriia otnositel'nosti Einshteina i makhizm," *VKA*, no. 7 (1924): 337–378, and "Voskreshaet li sovremennoe estestvoznanie mekhanicheskii materializm xviii stoletiia," *VKA*, no. 17 (1926): 116–168, and A. A. Bogdanov, "Ob"ektivnoe ponimanie printsipa otnositel'nosti," *VKA*, no. 8 (1924): 332–347.

42. N.a., "6. Legenda o mekhanistakh i redaktsiia estestvenno-nauchnykh izdanii GIZa," *Dialektika v prirode* 5 (Moscow, 1929): 52.

43. Timiriazev, "Dialektika prirody Engel'sa," pp. 208–210.

44. "6. Legenda o mekhanistakh," pp. 51, 58. See Frenkel'′s "Osnovy elektrodinamiki tochechnykh elektronov," which opened Frenkel' to the Mechanists' wrath.

45. "6. Legenda o mekhanistakh," p. 64, and O. Iu. Smidt, "Rol' matematiki v stroitel'stvo sotsializma," *Estestvoznanie i marksizm*, no. 2/3 (1930): 9.

46. "6. Legenda o mekhanistakh," p. 51.

47. "Deborintsy i klassovaia bor'ba v nauke," *Dialektika v prirode* 5: 23–24.

48. *A MGU*, f. 225, op. 1, ed. khr. 40, ll. 1–3.

49. *A MGU*, f. 225, op. 1, ed. khr. 23, ll. 1–2. The party report concluded with a list, "Characteristics of Basic Professorial Groups of the Physics Division of Fizfak." Timiriazev's group included Iakovlev, whose scientific works were insignificant and who was a "Black Hundred and henchman of Kasso," Kasterin, who was "well known" but part of a

whole group of "Black Hundreds who had been expelled in 1922 from the USSR by the GPU [secret police]." The group of Mandel'shtam, led by a "great physicist" who was "loyal and a marvelous teacher" included the "well-known" S. I. Vavilov, who was inclined to the right "but recently [is] trying to work with us. Read papers in circle of physicist-materialists at the [Communist Academy]," I. E. Tamm, a "good young physicist" whose loyalty was unquestioned, and G. S. Landsberg.

50. "6. Legenda, o mekhanistakh," pp. 49–51.

51. *TsGAOR*, f. 5221, op. 4, ed. xr. 24, l. 15 and op. 4, ed. xr. 43, l. 1, 5, and f. 5205, op 1, ed. xr. 588, l. 2, as cited in R. S. Iangutov, *Stanovlenie sovetskoi fiziki i bor'ba za dialektiko-materialisticheskoe mirovozzrenie v nei (1917–1925)*, Avtoreferat (Moscow, 1971), p. 14.

52. The Communist Academy grew throughout the 1920s, in spite of budgetary pressures which forced a cutback in staff. Seminar activity increased fourfold in three years, from 1925, when 38 papers were delivered, to 1926, when 87 papers were delivered, to 1927, when 124 papers were delivered. See *NS*, no. 3 (1928): 167–170.

53. *A AN*, f. 364, op. 3, ed. khr. 43, ll. 3, 7, 11, 13.

54. Ibid., f. 364, op. 1, ed. khr. 2, l. 35; f. 364, op. 1, ed. khr. 20. As of 1932, the academy had forty-four full-time staff with a budget of 116,000 rubles, with 25,000 rubles for translations, subsidies, consultancies, *komandirovki*, and conferences.

55. Ibid., f. 364, op. 4, ed. Khr. 1, ll. 114–116.

56. Ibid., f. 364, op. 1, ed. khr. 109, ll. 6–16.

57. Ibid., f. 364, op. 4, ed. khr. 28, l. 3.

58. Ibid., f. 364, op. 1, ed. khr. 5, ll. 2, 5. Lectures included "Hegel and Natural Science," "Neo-Kantianism and Social Fascism," and "Vulgar Materialism and the Contemporary Mechanists."

59. Ibid., f. 364, op. 4, ed. khr. 28, l. 50.

60. See Joravsky, *Soviet Marxism and Natural Science, 1917–1931*, pp. 279–287, for a discussion of the reception of relativity among Timiriazev, Semkovskii, Gessen, and others.

61. *A AN*, f. 364, op. 3, ed. khr. 24, ll. 2, 6, 8, 11; and *A MGU*, f. 225, op. 1, ed. khr. 40, l. 3.

62. V. P. Egorshin, *Estestvoznanie, filosofiia i marksizm* (Moscow 1930), pp. 144, 166–175, 199.

63. Joravsky, *Soviet Marxism and Natural Science, 1917–1931*, pp. 73, 83–84, 120–122.

64. Semkovskii, *Teoriia otnositel'nosti i materializm* (Kharkov, 1924), pp. 87–88.

65. Semkovskii, *Dialekticheskii materializm i printsip otnositel'nosti* (Moscow-Leningrad, 1926), pp. 122–123.

66. Ibid., p. 171.

67. Semkovskii continued to defend relativity throughout the 1930s. See his "Dialektika prirody Engel'sa i teoriia otnositel'nosti," *FNiT*, no. 9 (1935): 8–15.

68. MIT Press will publish many of Gessen's major essays in translation in 1991, along with an article by Wolf Schäfer, editor of the volume, Loren Graham's "The Social-Political Roots of Boris Gessen," originally published in *Social Studies of Science* 15 (1985): 705–722, which offers an explanation of the approach Gessen adopted in his paper presented at the 1931 London International Congress of the History of Science, and Paul Josephson's "Boris Gessen and Theoretical Physics in the Soviet Union in the 1920s and 1930s."

69. V. P. Egorshin, "O polozhenii na fronte fiziki i zadachi obshchestva fizikov-materialistov pri Komakademii," *Za marksistsko-leninskoe estestvoznanie*, no. 1 (1931): 122–125.

70. B. M. Gessen, "Teoretiko-veroiatnostnoe obosnovanie ergodicheskoi gypotezy," *UFN* 9, no. 5 (1929): 600–620.

71. See R. Mizes, "O prichinnoi i statisticheskoi zakonomernosti v fizike," *UFN* 10, no. 4 (1930): 437–462, with an introduction by Gessen; and Gessen, "Statisticheskii metod v fizike i obosnovanie teorii veroiatnostei R. Mizesa. Stat'ia 1. Kollektiv i ego raspredelenie," *Estestvoznanie i marksizm*, no. 1, (1929): 33–58.

72. Egorshin, "O polozhenii na fronte fiziki," pp. 122–125.

73. "Lichnoe delo B. M. Gessen," *A AN*, f. 364, op. 3a, ed. khr. 17, ll. 1, 3, 4–6, 9–10; f. 154, op. 4, ed. khr. 30, entire; f. 351, op. 1, ed. khr. 63, ll. 34–35; f. 355, op. 2, ed. khr. 71, entire; and f. 364, op. 4, ed. khr. 24, ll. 130–132; and *A MGU*, f. 225, op. 1, ed. khr. 40, ll. 1–3.

74. *A AN*, f. 1515, op. 2, ed. khr. 17, ll. 1–8.

75. B. M. Gessen, *Osnovnye idei teorii otnositel'nosti* (Moscow, 1928), pp. 64–65. Gessen wrote two earlier articles in which he questioned Timiriazev's uninformed attack on relativity theory, "Ob otnoshenii A. Timiriazeva k sovremennoi nauke," *PZM*, no. 2–3 (1927): 188–189; and "Mekhanicheskii materializm i sovremennaia fizika," *PZM*, no. 7–8 (1928): 5–47.

76. Gessen, *Osnovnye idei*, p. 64.

77. Ibid., pp. 64–66, 69.

78. Ibid., p. 35.

79. Ibid., p. 163.

80. Ibid., pp. 165–166.

81. Ibid., pp. 161–163.

82. V. P. Egorshin, "Za marksistsko-leninskaia traktovka osnovykh fizicheskikh poniatiiakh," *Za marksistsko-leninskoe estestvoznanie*, no. 1 (1932): 63–64.

83. See A. A. Maksimov, "O printsipe otnositel'nosti A. Einshteina," *PZM*, no. 9–10 (1922): 180–208; and "Teoriia otnositel'nosti i materializm," ibid., no. 4–5 (1923): 140–156; A. K. Timiriazev, "Einshtein, materializm i tov. Gol'tsman," ibid., no. 1 (1924): 127–136; and "Teoriia otnositel'nosti i dialekticheskii materializm," ibid., no. 8–9 (1924): 142–157.

84. Gessen, *Osnovnye idei*, p. 105.

85. Ibid., p. 108.
86. Ibid., p. 106.
87. Ibid., pp. 113–114.

8: DIALECTICAL MATERIALISM

1. "Nashi zadachi," *Estestvoznanie i marksizm*, no. 1 (1929): 7–8.
2. See *Za povorote na fronte estestvoznaniia* (Moscow-Leningrad, 1931) for the session at the Communists Academy and the resolutions passed on "menshevizing idealism."
3. "Za partiinost' v filosofii i estestvoznanii," *Estestvoznanie i marksizm*, no. 2–3 (1930): iv.
4. Ibid., p. v.
5. Graham, *Science, Philosophy, and Human Behavior*, p. 24. See pp. 24–67 for an in-depth discussion of the development of dialectical materialism in the Soviet Union.
6. Ernst Mach, *The Analysis of Sensations and the Relation of the Physical to the Psychical* (New York: Dover, 1959), p. 12, as cited in ibid., p. 40.
7. Engels, *The Dialectics of Nature* (New York: International Publishers, 1940), pp. 26–34, and 152–164.
8. D. I. Blokhintsev and F. Galperin, "Gipoteza neitrino i zakon sokhraneniia energii," *PZM*, no. 6 (1934): 147–157.
9. *A LFTI*, f. 3, op. 3, ed. khr. 287, ll. 1, 5, 8, 12, 13–15, 17, 19–22. See G. E. Gorelik and V. Ia. Frenkel', *Matvei Petrovich Bronshtein* (Moscow, 1990).
10. Bronshtein, "Sovremennoe sostoianie relatavistskoi kosmologii," *UFN* (1931): 126–184, and "Sokhraniatsia li energiia," *Sorena*, no. 1 (1935): 7–10.
11. F. G[alperin?], "Bor'ba materializma i idealizma vokrug osnovykh voprosov sovremennoi fiziki," *FNiT*, no. 4 (1934): 23–25.
12. F. Galperin, "K voprosu o zakone sokhraneniia energii v iadernoi fizike," *FNiT*, no. 10–11 (1934): 23–26. Other articles written on the question of conservation of energy include D. I. Blokhintsev and F. Galperin, "Atomistika v sovremennoi fizike," *PZM*, no. 5 (1936): 102–123; G. Pokrovskii and A. Nekrasov, "O vtorom nachale termodinamiki," *PZM*, no. 3 (1935): 96–100, and V. Fridman, "Protiv otritsaniia zakona sokhraneniia i prevrashcheniia energii," *PZM*, no. 11–12 (1937): 192–200.
13. *Doklady Akademii Nauk SSSR* (Leningrad, 1929), pp. 131, 259, 287.
14. G. E. Gorelik, "Naturfilosofskie problemy fiziki v 1937 godu," *Priroda*, no. 2 (1990): 95.
15. N.a. "Priroda elektricheskogo toka," *Elektrichestvo*, no. 3 (February 1930): 128–138.
16. Ibid., pp. 128–131, 133. Most of Mitkevich's major articles on his theory of Faraday-Maxwell "force lines," written between 1930 and 1939, are collected in *Osnovnye fizicheskie vozzreniia*, 3d edition (Moscow-Leningrad, 1939).

17. "Priroda elektricheskogo toka," p. 132.

18. Ibid., pp. 132–133.

19. Ibid., pp. 137–138.

20. "Priroda elektricheskogo toka," *Elektrichestvo*, no. 8 (April 1930): 337–350.

21. Ibid., pp. 337–338. Emphasis in the original.

22. Ibid., pp. 347, 350.

23. "Priroda elektricheskogo toka," *Elektrichestvo*, no. 10 (May 1930): 425–435.

24. On the history of the theoretical department, see "Otdel teoreticheskoi fiziki im. I. E. Tamma fizicheskogo instituta im. P. N. Lebedeva AN SSSR (k 50-letiiu so dnia osnovaniia)," FIAN preprint 34 (Moscow, 1985).

25. S. M. Rytov, "The Old Days," in *Reminiscences About I. E. Tamm*, ed. E. L. Feinberg (Moscow, 1987), p. 254.

26. I. E. Tamm, "Rukovodiashchie idei v tvorchestve Faradeia," *UFN* 12, no. 1 (1932): 1–30, and "Osnovnye idei Faradeia i ikh rol' v razvitii nauk ob elektrichestvo," *Elektrichestvo*, no. 23–24 (1931): 1324–1325.

27. Gogoberidze, "K voprosu ob uslovnosti matematicheskoi traktovki fizicheskikh iavlenii," *Elektrichestvo*, no. 1 (1934): 20–21.

28. Gogoberidze, "K voprosu ob uslovnosti matematicheskoi traktovki fizicheskikh iavlenii," *Elektrichestvo*, no. 5 (1935): 49–50.

29. Mitkevich, "K voprosu ob uslovnosti matematicheskoi traktovki fizicheskikh iavlenii," *Elektrichestvo*, no. 7 (1934): 40. See Ia. N. Shpil'rein, "O geometricheskikh svoistvakh silovykh linii," *Sorena*, no. 9–10 (1932): 55–61, and "Sushchestvuiut li elektricheskie silovye linii?" *Sorena*, no. 3 (1936): 13–15.

30. Mitkevich, "K voprosu ob uslovnosti matematicheskoi traktovki fizicheskikh iavlenii," *Elektrichestvo*, no. 12 (1933): 1–3. See also his "O 'fizicheskom' deistvii no rastoianii," *Elektrichestvo*, no. 1 (1934): 15–19.

31. V. A. Fock, "Za podlinno nauchnuiu sovetskuiu knigu," *Sorena*, no. 3 (1934): 132–136.

32. V. L. Ginzburg, "About Igor Tamm," in *Reminiscences About Tamm*, p. 194.

33. E. L. Feinberg, "The Man and His Time," in *Reminiscences About Tamm*, pp. 126–127.

34. Gorelik, "Naturfilosofskie," p. 95.

35. N. P. Kasterin, *Obobshchenie osnovykh uravnenii aerodinamiki i elektrodinamiki* (Moscow, 1937), p. 3.

36. Ibid., p. 4.

37. Ibid., pp. 12–13.

38. D. I. Blokhintsev et al., "O stat'e N. P. Kasterina 'Obobshchenie osnovnykh uravnenii aerodinamiki i elektrodinamiki'," *Izv. AN SSSR. Seriia fizicheskaia*, no. 3 (1937): 425.

39. Ibid., pp. 426–428. Tamm and Ioffe responded to Kasterin's work independently. Ioffe saw in the nonrelativistic generalization of these equations "mistakes, bewilderment, and nothing believable." They were

built on "a nonrealizable curvolinear system of coordinates which mask with a mathematic fog . . . the preconditions and derivations of the theory." See Ioffe, "O polozhenii na filosofskom fronte sovetskoi fiziki," *PZM*, no. 11–12 (1937):133–143, especially p. 134; and Tamm, "O rabote N. P. Kasterina po elektrodinamike i smezhnym voprosam," *Izv. AN SSSR. Seriia fizickeskaia*, no. 3 (1937): 437–448.

40. *A MGU*, f. 219, op. 1, ed. khr. 386, ll. 2–3.

41. Ibid., ed. khr. 383.

42. Ibid., ed. khr. 388, l. 85; ed. khr. 387, ll. 1–7.

43. Ibid., ed. khr. 385 (letter from V. F. Mitkevich to V. M. Molotov, Chairman of Sovnarkom, December 9, 1936). Emphasis in the original.

44. Vavilov, "Po povodu knigi akad. V. F. Mitkevicha 'Osnovnye fizicheskie vozzreniia'," *PZM*, no. 7 (1937): 56–63.

45. Ioffe, "O polozhenii na filosofskom fronte sovetskoi fiziki."

46. Gorelik, "Naturfilosofskie," pp. 97–101.

47. Maksimov, "O filosofskikh vozzreniakh Akad. V. F. Mitkevicha i o putiakh razvitiia sovetskoi fiziki," *PZM*, no. 7 (1937): 25–55.

48. Letter from A. A. Maksimov to V. F. Mitkevich, February 17, 1937, *A MGU*, f. 225, ed. khr. 63, ll. 1–3.

49. Ibid., f. 225, ed. khr. 63, letter Maksimov to Mitkevich, March 13, 1937.

50. *A AN*, f. 364, op. 4, ed. khr. 1, ll. 93–96, 98–99.

51. Ibid., ll. 101–107.

52. *A LFTI*, f. 3, op. 1, ed. khr. 171, ll. 63–67.

53. Ibid.

54. V. Egorshin, "O polozhenii na fronte fiziki," pp. 107–108.

55. Ibid., pp. 108–109.

56. Ibid., p. 112.

57. Ibid., pp. 111–112, 122–125.

58. A. A. Maksimov, "O filosofskikh vozzreniakh Akad. V. F. Mitkevicha"; "Klassovaia bor'ba v sovremennom estestvoznanii," *FNiT*, no. 9 (1932): 21–33; "Lenin o estestvoispytateliakh," *FNiT*, no. 4–5 (1933): 28–36, and "Lenin o estestvoispytateliakh. Okonchanie," *FNiT*, no. 6 (1933): 30–39; "O mekhanitsizme i marksizme v estestvoznanii," *PZM*, no. 5 (1933): 24–72; "Filosofiia i estestvoznanie za piat' let," *PZM*, no. 1 (1936): 46–65; "O Makhizme v vozzreniakh nekotorykh sovremennykh fizikov," *PZM*, no. 6 (1938): 172–205; "Materiia i massa," *PZM*, no. 8 (1939): 127–136; "Sovremennoe fizicheskoe uchenie o materii i dvizhenii i dialekticheskii materializm," *PZM*, no. 10 (1939): 86–111. See also Maksimov's *Lenin i estestvoznanie* (Moscow-Leningrad, 1933), especially pp. 110–127, and "*Materializm i empirio-krititsizm* Lenina—materialistichskoe obobshchenie dannykh estestvoznaniia," in *Materializm i empirio-krititsizm Lenina i sovremennaia fizika* (Moscow, 1939), pp. 31–43.

59. For Kol'man's early views on mathematics and physics, see "Khod zadom v filosofii A. Einshteina," *NS*, no. 1 (1931): 11–15; "Problema prichinnosti v sovremmenoi fiziki," *PZM*, no. 4 (1934): 80–109; and *FNiT*,

no. 9 (1934): 7–27; "Uzlovye problemy sovremennoi atomnoi fiziki," *FNiT*, no. 2 (1936): 24–34, where Frenkel' and Shpil'rein met criticism for their failure to posit the existence of a medium to explain how light and energy waves propagate through a vacuum; "O zlobodnevnom znachenii teorii veroiatnostei," *PZM*, no. 2 (1934): 71–76; "Problema prichinnosti v sovremennoi fiziki," *PZM*, no. 4 (1934): 80–109; "Novoe vystuplenii za i protiv indeterminizma v fizike," *PZM*, no. 6 (1934): 187–190; "Novoe vystuplenie E. Shredingera," *PZM*, no. 5 (1936): 124–125; "Fizika i filosofiia," *PZM*, no. 7 (1937): 64–80; "Teoriia otnositel'nosti' i dialekticheskii materializm," *PZM*, no. 6 (1939): 106–120; "Teoriia kvant i dialekticheskii materializm," *PZM*, no. 10 (1939): 129–145; "Protiv lzhenauki," *Sovetskaia nauka*, no. 1 (1938): 39–50; and *Noveishie otkrytiia sovremennoi fiziki v svete dialekticheskogo materializma* (Moscow, 1943).

60. Kol'man, "Novoe vystuplenie E. Shredingera," pp. 124–125.

61. Kol'man, "Teoriia otnositel'nosti i dialekticheskii materializm," pp. 106–120.

62. Kol'man, "Teoriia kvant i dialekticheskii materializm"; "Materializm i empirio-krititsizm V. I. Lenina i sovremennaia fizika," *Sovetskaia nauka*, no. 2 (1939): 57–70.

63. See, for example, D. I. Blokhintsev, "V chem zakliuchaiutsia osnovnye osobennosti kvantovoi mekhaniki," *Sovetskaia nauka*, no. 4 (1938): 47–61.

64. Kol'man, "Protiv formalizma i abstraktnosti," *Vyshhaia shkola*, no. 12 (1937): 73–77.

65. See A. M. Deborin, *Lenin i krizis noveishei fiziki*, 2d edition (Leningrad, 1930), a speech at the annual meeting of the Academy of Sciences in 1929; and *Dialektika i estestvoznanie* (Moscow-Leningrad, 1929), a collection of his essays from *PZM*, and *VKA*, where he attacks mechanism and defends physicists' right to determine the physical content of their work.

66. S. I. Vavilov, *UFN* 8, no. 6 (1928): 266–267.

67. B. M. Gessen, "K voprosu o probleme prichinnosti v kvantovoi mekhanike," introduction to Artur Gass, *Volny materii i kvantovaia mekhanika*, trans. P. S. Tartakovskii (Moscow-Leningrad, 1930), p. 22.

68. B. M. Gessen "Statisticheskii metod v fizike i obosnovanie teorii veroiatnostei R. Mizesa. Stat'ia 1. Kollektiv i ego raspredelenie," *Estestvoznanie i marksizm*, no. 1 (1929): 33–34.

69. Ibid., p. 25.

70. Ibid., p. 29.

71. Ibid., pp. 31–33, 37.

72. Ibid., p. 35.

73. The major papers from the 1934 session are published in *PZM*, no. 4 (1934): entire.

74. A. F. Ioffe, "Razvitie atomisticheskikh vozzrenii v XX v.," *PZM*, no. 4 (1934): 52–68; and *FNiT*, no. 9 (1934): 28–37. See also S. I. Vavilov, "Dialektika svetovykh iavlenii," *FNiT*, no. 9 (1934): 38–45; and *PZM*, no. 4 (1934): 69–79.

75. V. P. Egorshin, "Lenin i sovremennoe estestvoznanie," in 25 let 'Materializma i empiriokrititsizma.' Sbornik statei (Moscow, 1934), pp. 142–168.

76. E. Kol'man, "Problema prichinnosti v sovremennoi fizike," FNiT, no. 9 (1934): 8, 26–28; and "Problema prichinnosti v sovremennoi fizike," Tezisy dokladov (Leningrad, 1934), pp. 8–11.

77. Ioffe, "O polozhenii na filosofskom fronte," pp. 132–133.

78. Ibid., pp. 134, 136. For a discussion of the politics of "Aryan" physics and the views of Lenard and Stark, see Alan Beyerchen, Scientists Under Hitler (New Haven: Yale University Press, 1977).

79. Ioffe, "O polozhenii na filosofskom fronte," pp. 134–135.

80. Ibid., p. 140.

81. Ibid., pp. 138–139.

82. Ibid., p. 134.

83. See Paul Forman, "Weimar Culture, Causality, and the Quantum Theory, 1918–1927: Adaptation by German Physicists and Mathematicians to a Hostile Intellectual Environment," Historical Studies in the Physical Sciences, ed. R. McCormmach, 3 (1971): 1–115; Alan Beyerchen, Scientists Under Hitler; Stanley Coben, "The Scientific Establishment and the Transmission of Quantum Mechanics to the United States, 1919–1932," American Historical Review 76 (April 1971): 442–466; Kevles, The Physicists, pp. 155–170; and Graham, Science, Philosophy, and Human Behavior, pp. 320–353.

84. P. N. Fedoseev, ed., Filosofskie problemy sovremennogo estestvoznaniia (Moscow, 1959).

9: ASSAULT ON THE LENINGRAD PHYSICS
COMMUNITY

1. Edward Bubis and Blair Ruble, "The Impact of World War II on Leningrad," in The Impact of World War II on the Soviet Union, ed. Susan J. Linz (Totawa, N.J. 1985), pp. 191–206.

2. Robert Conquest, The Great Terror (New York: Collier Books, 1968), pp. 42–53.

3. Ibid., pp. 236–241.

4. Bubis and Ruble, "The Impact of World War II," p. 193.

5. Ibid., p. 201.

6. A commission under Ioffe, Krylov, Lazarev, Mandel'shtam, Rozhdestvenskii, and others met in February 1932, but could not initially agree on the form the new institute would take, its relationship to other such institutes as LFTI and GOI, and whether it would focus on theoretical or experimental endeavors.

7. On the early history of the Physics Institute, see I. R. Gekker, A. N. Starodub, and S. A. Fridman, Ot fizicheskoi laboratorii akademii nauk v petrograde do fizicheskogo instituta akademii nauk SSSR (Moscow, 1985).

8. For a brief early history of the Institute of Physical Problems (IFP)

and its role in Soviet physics in the 1930s, see A. I. Shalnikov, "Institut fizicheskikh problem," *UFN* 18, no. 3 (1937): 323–336. This discussion of the formation of IFP is based on a reading of Kapitsa's correspondence in Kapitsa, *Pis'ma o nauke* (Moscow, 1989), pp. 29–48, 59–61, 72–85, 91–92, 120–124, 129–132. See also Badash, *Kapitza, Rutherford, and the Kremlin,* pp. 20–36, and 51–95, for many of the letters Kapitsa wrote his wife, Anna, in England.

9. Kapitsa, *Pis'ma o nauke,* pp. 62–69.

10. Ibid., pp. 39–43.

11. Ibid., pp. 59–61, 72–85, 91–92, 120–124, 129–132.

12. Ibid., pp. 105–112.

13. *A IFP,* op. 1, ed. khr. 55, ll. 3, 5.

14. Ibid., ll. 6–8.

15. P. L. Kapitsa, "Ob organizatsii nauchnoi raboty instituta fizicheskikh problem AN SSSR" (Moscow, 1943), pp. 5–6. This essay has been translated as "The Institute of Problems of Physics [sic]," in *Peter Kapitsa on Life and Science,* trans. and ed. Albert Parry (New York: Macmillan Co. 1968), pp. 160–184.

16. *A IFP,* op. 1, ed. khr. 17, ll. 1–16.

17. Kapitsa, "Ob organizatsii," pp. 8–9. Kapitsa was removed as director of IFP in 1946 because of his refusal to be involved with the atomic bomb project, but he returned to the institute in 1955 and remained its director until his death in 1984. In the interim he was replaced by A. P. Aleksandrov, a physicist who had worked at LFTI, the developer of RBMK reactor (used at Chernobyl), and the future president of the Academy of Sciences.

18. *A LFTI,* f. 3, op. 1, khr. 63, l. 145.

19. Ibid., ed. khr. 81, ll. 26–ob. 27.

20. Ibid., ed. khr. 63, l. 153.

21. Ibid., ed. khr. 63, l. 1 ob.

22. Ibid., ed. khr. 63, l. 84.

23. Ibid., l. 3.

24. *A LFTI,* f. 3, op. 1, ed. khr. 81, l. 6–7, and f. 3, op. 1, ed. khr. 94, l. 5. I could not find the report in the LFTI archive.

25. For example, the "minimum program" for the candidate in theoretical and mathematical physics approved in 1984 includes eight subspecialties and is contained in a thirty-two-page document. See *Programma-minimum kandidatskogo ekzamena po spetsial'nosti 01.04.02-teoreticheskaia i matematicheskaia fizida* (Moscow, 1983).

26. *A LFTI,* f. 1, op. 1, ed. khr. 43 l. 4. For the LFTI plan for *aspirantura,* including texts to be used and courses to be taught, see ibid., ll. 9–10 ob.

27. Ibid.

28. Ibid., f. 3, op. 1, ed. khr. 63, l. 85.

29. Ibid., ll. 75–77.

30. Ibid., ll. 126–127.

31. Ibid., ll. 132, 136–137.

32. *A LFTI*, f. 3, op. 1, ed. khr. 53.

33. Ibid., ed. khr. 63, l. 153.

34. Ibid., ed. khr. 81, l. 5–5 ob., and 11. The 1938 budget of 2,055,000 included 500,000 for the LFTI cyclotron. The 1937 budget was somewhere between 1.8 and 2.0 million rubles.

35. Ibid., ed. khr. 94, ll. 4, 19.

36. Ibid., ll. 5–9.

37. Ibid., ed. khr. 63, l. 1; ed. khr. 81, l. 7 ob.

38. Ibid., ed. khr. 38, l. 39.

39. Ioffe, "Razvitie atomisticheskikh vozzrenii v XX v.," *PZM*, no. 4 (1934): 52–68, and *FNiT*, no. 9 (1934): 28–38.

40. *A LFTI*, f. 3, op. 1, ed. khr. 32, ll. 20–21.

41. Ibid., ed. khr. 38, ll. 110–114.

42. Ibid., l. 39.

43. *A LFTI*, f. 3, op. 1, ed. khr. 32, ll. 12–18, and ed. khr. 63, l. 117.

44. Ibid., ll. 19–20. See, for example, N. N. Davidenkov and D. Golovachev, "Strunnyi teletensometr," *ZTF*, no. 3 (1934): 587–601; and N. N. Davidenkov and E. M. Shevandin, "O sravnitel'noi prochnosti rastianutykh i szhatykh obraztsov," *ZTF*, no. 5 (1934): 925–941.

45. *A LFTI*, f. 3, op. 1, ed. khr. 38, l. 9. See, for example, A. V. Stepanov, "Prakticheskaia prochnost'," *ZTF*, no. 2 (1935): 348–361; F. F. Vitman, N. N. Davidenkov, and P. Sakharov, "O vlianii poverkhnosti na prochnost' metallicheskikh obraztsov," ibid. no. 3 (1935): 418–424; E. Shevandin and E. Sverdlov, "Primenenie strunnogo metoda k opredeleniiu uprugovo posledeistviia stali," ibid., no. 5 (1935): 776–783; N. N. Davidenkov, "Metod kuntze i ego kritika," ibid., no. 6 (1935): 915–927; F. F. Vitman, "K voprosu o raschete ostatochnykh napriazhenii v tolstostennykh trubakh," ibid., no. 9 (1935): 1589–1597; F. F. Vitman and N. N. Davidenkov, "Novyi metod tsentrirovaniia obraztsov," ibid., no. 10 (1935): 1760–1767; F. F. Vitman and Ia. B. Salitra, "Vlianie khromirovaniia na khrupkuiu prochnost' stali," ibid., no. 4, (1936): 608–613; N. N. Davidenkov et al., "Izmerenie ostotachnykh napriazhenii bez razredki izdelii," ibid., no. 11 (1936): 1964–1974; and many others in 1937 and 1938.

46. *A LFTI*, f. 3, op. 1, ed. khr. 63, l. 117.

47. Ibid., ll. 121–121 ob.

48. *A LFTI*, f. 3, op. 1, ed. khr. 39, ll. 29–30.

49. *Fiziko-tekhnicheskii institut v rabotakh Akademika Ioffe* (Moscow-Leningrad, 1936).

50. The sources consulted for this discussion of the March 1936 session are *Izv. AN SSSR. Seriia fizicheskaia*, no. 1–2 (1936): entire; V. L. Levshin, "Novye puti sovetskoi fiziki," *VAN*, no. 4–5 (1936): 62–75; S. I. Vavilov, "Sovetskaia fizika na martovskoi sessii Akademii nauk SSSR," *Priroda*, no. 5 (1936): 3–4; F. F. Vol'kenshtein, "Raboty Akademika A. F. Ioffe i Leningradskogo fiziko-tekhnicheskogo instituta," ibid., pp. 5–19; I. A. Khvostikov, "Puti razvitiia opticheskogo instituta," ibid., pp. 20–27; F. P. Faerman, "Analiz spektrov i spetral'nyi analiz," ibid., pp. 27–37;

and shorter articles in *Sorena, ZTF, UFN,* and *ZETF.* See also V. Ia. Frenkel', "K 50-letiiu martovskoi sessii akademii nauk SSSR (1936 g.)," *Chteniia pamiati A. F. Ioffe—1985* (Leningrad, 1987), pp. 63–86.

51. S. I. Vavilov, "Puti razvitiia opticheskogo instituta," *Izv. AN SSSR. Seriia fizicheskaia,* no. 1–2 (1936): 163–188; D. S. Rozhdestvenskii, "Analiz spektrov i spektral'nyi analiz," ibid., pp. 189–214.

52. I. E. Tamm, "Problema atomnogo iadro," ibid., pp. 301–323; V. A. Fock, "Problema mnogikh tel v kvantovoi mekhanike," ibid., pp. 351–362; and Ia. I. Frenkel', "Teplovoe dvizhenie v tverdykh i zhidkikh telakh i teoriia plavleniia," ibid., pp. 371–393.

53. Vavilov, "Sovetskaia fizika," pp. 3–4.

54. Ioffe, "Usloviia moei nauchnoi raboty," *Izv. AN SSSR. Seriia fizicheskaia,* no. 1–2 (1936): 7–33.

55. Ibid., p. 26.

56. Ibid., pp. 10–18, 27.

57. Ibid., pp. 21–23. Several shorter reports on the physics of agriculture (F. E. Koliasev), heat engineering (M. V. Kirpichev), and biophysics (G. M. Frank) followed Ioffe's speech. For transcripts of these talks, see ibid., pp. 34–58.

58. Ibid., pp. 23–27.

59. Ibid., pp. 27–28.

60. Ibid., pp. 25–27.

61. Vol'kenshtein, "Raboty Akademika Ioffe," p. 11.

62. *Izv. AN SSSR. Seriia fizicheskaia,* no. 1–2 (1936): 127–130.

63. Ibid., p. 61.

64. Ibid., pp. 89, 92, 96–97.

65. Ibid., pp. 98–99. For more on the Ioffe-Kvittner debate on the behavior of crystals, see F. Kvittner, "O teorii stroeniia krystallov," *Sorena,* no. 5 (1932): 50–63; Ioffe, "O stat'e Kvittnera i o fizike tverdogo tela," ibid., pp. 64–73; Kvittner, "Po teorii stroeniia kristallov," ibid., no. 6 (1932): 48–53; and Kvittner, "Tri raboty po elektroprovodnosti izoliruiushikh kristallov," ibid., no. 6 (1934): 122–125.

66. Vol'kenshtein, "Raboty Akademika Ioffe," p. 13.

67. *Izv. AN SSSR. Seriia fizicheskaia,* no. 1–2 (1936): 63–72.

68. Ibid., pp. 108–109.

69. Ibid., p. 118.

70. Ibid., pp. 83–85.

71. Ibid., pp. 85–86.

72. Ibid., p. 86.

73. Ibid., p. 76–80.

74. Ibid., pp. 87–88.

75. Ibid., pp. 145–162.

76. Aleksey E. Levin, "Anatomy of a Public Campaign: 'Academician Luzin's Case' in Soviet Political History," *Slavic Review* 49, no. 1 (Spring 1990): 90–108. Kapitsa was stunned by the publication in *Pravda* of the condemnation of Luzin, and wrote Molotov with great incredulity. How

could the affair possibly serve Soviet science, he asked. Kapitsa, *Pis'ma o nauke*, pp. 86–89.

77. On the Luzin affair, see also Margarita Riutova-Kemoklidze, "Priezzhaite. Einshtein vas primet . . . Okonchanie," *Siberskie ogni*, no. 2 (1989): 111–129, especially 112–113.

78. George Gamov, *My World Line* (New York: Viking Press, 1970), pp. 110–123.

79. Kapitsa, *Pis'ma o nauke*, pp. 25–26.

80. See Bruce Parrott, *Information Transfer in Soviet Science and Engineering* (Santa Monica: Rand Corporation, 1981), for a discussion of the problems facing Soviet science in terms of diffusion of scientific results.

81. *A LFTI*, f. 3, op. 1, 1939, ll. 3, 18.

82. Ibid., ed. khr. 38, l. 13.

83. Ia. G. Dorfman, f. 261, letter 37, Saltykov-Shchedrin State Library Manuscript Division.

84. Joravsky, *The Lysenko Affair*, (app. A, pp. 318–328). See also Roy Medvedev, *Let History Judge* (New York: Vintage, 1971), p. 226.

85. Joravsky, *The Lysenko Affair*, p. 317.

86. Gamov, *My World Line*, p. 96. The telegram read: "Being inspired by your article on the light-ether, we are enthusiastically pushing forward to prove its material existence. Old Albert is an idealistic idiot! We call for your leadership in the search for caloric, flogiston, and electric fluids."

87. *A AN*, f. 596, op. 2, ed. khr. 173–175.

88. *A MGU*, f. 201, op. 1, ed. khr. 366, ll. 1–4.

89. Interview with Immanuil Lazarevich Fabelinskii, FIAN, November 1, 1989. Fabelinskii, who graduated from Moscow University in 1936, is a specialist in optics and a corresponding member of the Academy of Sciences since 1979.

90. Gamov says that in Cambridge and Copenhagen, Landau always wore a red blazer as a symbol of his Marxist views. *My World Line*, p. 119.

91. Riutova-Kemoklidze, "Okonchanie," p. 114.

92. Ibid.

93. On the life of Iu. B. Rumer, see Riutova-Kemoklidze, "Einshtein vas primet . . ." *Siberskie ogni*, no. 1 (1989): 116–129, and "Okonchanie," pp. 111–129. See also Riutova-Kemoklidze's *Kvantovyi vozrast* (Moscow, 1989). My thanks to Dr. Riutova-Kemoklidze for her hospitality during my visit to the theoretical department of the Institute of Nuclear Physics, Novosibirsk, and for sharing with me her research on the history of physics.

94. See G. A. Ozerov, *Tupolevskaia sharaga* (Munich, 1977). For a discussion of the role of *sharashki* in defense R and D, see David Holloway, "Innovation in the Defense Sector," in *Industrial Innovation in the Soviet Union*, ed. R. Amann and J. M. Cooper (New Haven: Yale University Press, 1982), pp. 334–341.

95. Riutova-Kemoklidze, "Okonchanie," p. 115.

96. Ibid.

97. Kapitsa, *Pis'ma o nauke*, pp. 174–175.

98. Ibid., p. 175.

99. Ibid., pp. 178–179.

100. Ibid., pp. 125–126. Kapitsa also referred to the arrest of a large number of young theoreticians in the same affair, which had emptied many classes of students; and he could not believe that in twenty years of Soviet power the government had not learned "how to win scholars to its side . . . or how to leave them in a neutral frame of mind, but even turned them against it."

101. Ibid., p. 127. According to another source, Fock was asked to provide details on a whole series of enemies of the people. When the conversation became difficult he turned off his hearing aid. See Riutova-Kemoklidze, "Okonchanie," p. 115.

This was just one of many occasions when Fock would use his hearing aid to advantage. In 1952, at a special session at FIAN devoted to the "green book," after ridiculing the comments of Maksimov, Blokhintsev, Omel'ianovskii, Ovchinnikov, and Khinchin, he left the large auditorium of the institute, turning off his hearing aid and ignoring the calls of the Stalinist physicists and philosophers to respond further. Interviews with V. L. Ginzburg and E. L. Feinberg, November 1 and 2, 1989, FIAN, and *A AN*, f. 532, op. 1, ed. khr. 232.

102. *A MGU*, f. 3, op. 3, ed. khr. 5, l. 38.

103. Loren Graham, "Quantum Mechanics and Dialectical Materialism," *Slavic Review* 25, no. 3 (September 1966): 383–384.

104. Letter from Maksimov to Fock, July 1, 1937, *A AN LO*, f. 1034, op. 3, ed. khr. 532.

105. Gorelik, "Naturfilosofskie," p. 101.

106. Ibid.

107. K. V. Nikolskii, "O putiakh razvitiia teoreticheskoi fiziki v SSSR," *PZM*, no. 1 (1938): 106–172.

108. V. A. Fock, "K diskussii po voprosam fiziki," *PZM*, no. 1 (1938): 149–159.

109. G. E. Gorelik, "Dva portreta," *Neva*, no. 8 (1989): 168–172, and N. V. Uspenskaia, "Vreditel'stvo . . . v dele izucheniia solnechnogo zatmeniia," *Priroda*, no. 8 (1989): 94–101.

110. Uspenskaia, "Vreditel'stvo," pp. 96–98.

111. Ibid., pp. 86–96. Gerasimovich studied in the United States with Rockefeller funds in the late 1920s. Robert McCutheon has written an article on the purges in the Soviet astrophysical community to be published in *Slavic Review* in 1991 or 1992.

112. *A AN*, f. 595, op. 2, ed. khr. 17, ll. 1–2.

113. Robert Conquest, *The Harvest of Sorrow: Soviet Collectivization and the Terror-Famine* (New York and Oxford: Oxford University Press, 1986).

114. Weissberg, *The Accused*, pp. 67–71.

115. Ibid., pp. 162–169.

116. *ZTF*, no. 8 (1937): 884.

EPILOGUE

1. On the Zhdanovshchina, see Werner Hahn, *Postwar Soviet Politics: The Fall of Zhdanov and the Defeat of Moderation, 1946–53* (Ithaca: Cornell University Press, 1982); Graham, *Science, Philosophy, and Human Behavior,* chap. 10, "Quantum Mechanics," and chap. 11, "Relativity Physics," and Vucinich, *Empire of Knowledge* (Berkeley, Los Angeles, Oxford: University of California Press, 1984), pp. 210–256.

2. *A LFTI,* f. 3, op. 1, ed. khr. 195, ll. 4, 6, 8.

3. Ibid., ll. 9–33. Frenkel' defended himself by arguing that as a physicist, "I was never and cannot be an idealist. The external world which is studied by the physicist always presented itself to me as objective reality, and not a figment of my imagination." He admitted having doubted the applicability of dialectical materialism in 1931, however. Ibid., l. 54.

4. Ioffe, *Osnovnye predstavleniia sovremennoi fiziki* (Moscow-Leningrad, 1949), p. 7.

5. Ibid., p. 42.

6. See N. F. Ovchinnikov, "Massa i energiia," *Priroda,* no. 11 (1951): 7–16, and "Fizicheskii idealizm—vrag nauki," *Nauka i zhizn',* no. 3 (1952): 42–46; I. V. Kuznetsov, "Lenin i estestvoznanie," *Nauka i zhizn',* no. 1 (1951): 1–6; and M. E. Omel'ianovskii's review of Ioffe's book in *Voprosy filosofii,* no. 2 (1951): 203–207. See also Vucinich, *Empire of Knowledge,* pp. 222–224; and Graham, *Science, Philosophy, and Human Behavior,* pp. 355–366.

7. A. F. Ioffe, "K voprosu o filosofskikh oshibkakh moei knigi 'Osnovnye problemy sovremennoi fiziki'," *UFN* 52, 4 (August 1954): 589–598.

8. *A LFTI,* f. 3, op. 3, ed. khr. 1291, ll. 9–11, 31–36.

9. Ibid., f. 3, op. 3, "A. F. Ioffe," l. 57.

10. On Lysenkoism and the 1948 VASKhNIL (Lenin All-Union Academy of Agricultural Sciences) conference which proclaimed Lysenkoism the true and only acceptable form of Soviet biology, see Joravsky, *The Lysenko Affair;* Graham, *Science, Philosophy and Human Behavior,* pp. 102–156; Mark Adams, "Science, Ideology and Structure: The Kol'tsov Institute, 1900–1970," in *The Social Context of Soviet Science,* ed. Linda Lubrano, Susan Solomon (Boulder: Westview Press, 1980), pp. 173–205; and Zhores Medvedev, *The Rise and Fall of T. D. Lysenko* (New York: Columbia University Press, 1969).

11. In another book on the history of postwar Soviet physics, I intend to write about the preparation for the 1949 physics conference, based on the materials I have found in the personal fund of S. I. Vavilov (*A AN,* f. 596, op. 2, ed. khr. 172–174). I have found no written confirmation of the episode concerning Beria, so it may be apocryphal. On the other hand, six leading physicists, including I. N. Golovin, the deputy of I. V. Kurchatov, told me this history, so there must be some truth to it. See also the recent work by A. S. Sonin in *Moscow News* and *Priroda,* as well as an unpublished manuscript, *Physical Idealism: The History of One Campaign* (in Russian), which describes the ideological struggle in postwar Soviet physics.

Selected Bibliography

I. PRIMARY SOURCES

A. ARCHIVES. ABBREVIATIONS THAT HAVE BEEN USED THROUGHOUT THE TEXT ARE LISTED HERE.

Arkhiv Akademii Nauk SSSR	*A AN*
Arkhiv Akademii Nauk SSSR, Leningrad Division	*A AN LO*
Arkhiv Fizicheskogo Instituta Akademii Nauk SSSR	*A FIAN*
Arkhiv Instituta Fizicheskikh Problem	*A IFP*
Arkhiv Leningradskogo Fiziko-Tekhnicheskogo Instituta	*A LFTI*
Arkhiv Moskovskogo Gosudarstvennogo Universiteta	*A MGU*
Rockefeller Archive Collections	
Saltykov-Shchedrin State Library Manuscript Division	
Tsentral'nyi Gosudarstvennyi Arkhiv RSFSR	*TsGA RSFSR*
University of California, Berkeley, Archives, Bancroft Library	

B. ARCHIVAL DOCUMENT COLLECTIONS

Nauchno-organizatsionnaia deiatel'nosti Akademika A. F. Ioffe. Sbornik dokumentov (Leningrad, 1980). Referred to as *Organizatsionnaia deiatel'- nosti Ioffe* throughout text.

Organizatsiia nauki v pervye gody sovetskoi vlasti (1917–1925) (Leningrad: Nauka, 1968). Referred to as *Organizatsiia sovetskoi nauki* 1 throughout notes.

Organizatsiia nauki v pervye gody sovetskoi vlasti v 1926–1932 gg. (Leningrad: Nauka, 1974). Referred to as *Organizatsiia sovetskoi nauki* 2 throughout text.

C. JOURNALS, NEWSPAPERS, AND SERIES. ABBREVIATIONS WHICH HAVE BEEN USED THROUGHOUT THE TEXT ARE LISTED HERE.

Biulleten' Glavnauki	
Biulleten' NIS PTEU VSNKh SSSR	*BNIS*

Biulleten' VARNITSO
Chteniia pamiati A. F. Ioffe
Dialektika v prirode
Doklady Akademii Nauk SSSR
Elektrichestvo
Estestvoznanie i marksizm
Fiziki i proizvodstvo
Front nauki i tekhniki   FNiT
Istoriia i metodologiia estestvennykh nauk   IiMEN
Izvestiia
Izvestiia Akademii nauk SSSR
    Seriia fizicheskaia   Izv. AN SSSR. Seriia fizicheskaia
Letopis'
Narodnoe prosveshchenie
Nature
Nauchnoe slovo   NS
Nauchnyi rabotnik   NR
Nauka i ee rabotniki   NieR
Physical Review
Physikalische Zeitschrift der Sowjet Union   Phys. Zeit. SU
Pod znamenem marksizma   PZM
Pravda
Priroda
Proceedings of the Royal Society
Soobshcheniia o nauchno-tekhnicheskikh rabotakh v
    respublikakh
Sotsialisticheskaia rekonstruktsiia i nauka   Sorena
Sovetskaia nauka
Trudy Gosudarstvennogo Opticheskogo Instituta
Trudy Leningradskoi fiziko-tekhnicheskoi laboratorii   TLFTL
Trudy sibirskogo fiziko-tekhnicheskogo nauchnogo instituta   SFTNI
Uspekhi fizicheskikh nauk   UFN
Vestnik akademii nauk SSSR   VAN
Vestnik kommunisticheskoi akademii   VKA
Vestnik narodnogo prosveshcheniia soiuza kommun severnoi
    oblasti
Vestnik rentgenologii i radiologii   VRiR
Voprosy fiziki
Vremennik obshchestva im. Ledentsova
Vysshaia shkola
Vysshaia tekhnicheskaia shkola
Za sotsialisticheskuiu nauku   ZSN
Zavodskaia laboratoriia
Zeitschrift für Physik   Zeit. für Phys.
Zhurnal eksperimental'noi i teoreticheskoi fiziki   ZETF
Zhurnal prikladnoi fiziki   ZPF

Zhurnal russkogo fiziko-khimicheskogo obshchestva   ZhRFKhO
Zhurnal tekhnicheskoi fiziki   ZTF

D. Books and Pamphlets

Adoratskii, V. V. *Dialectical Materialism*. New York, 1934.
Agol, I. I. *Vitalizm, mekhanisticheskii materializm, i marksizm*. Moscow-Leningrad, 1928.
Akademik, A. I. Alikhanov. *Vospominaniia, pis'ma, dokumenty*. Edited by A. P. Aleksandrov. Leningrad, 1989.
Adademiia nauk SSSR: ee zadachi, razdelenie i sostav. Leningrad, 1925.
Beilin, A. E. *Kadry spetsialistov v SSSR, ikh formirovanie i rost*. Moscow, 1935.
Belinskii, P. *Prostranstvo, vremia i dvizhenie*. Moscow, 1937.
Borgman, I. I. *Efir i materiia*. St. Petersburg, 1906.
———. *P. N. Lebedev*. St. Petersburg, 1912.
———, ed. *Novye idei v fizike*. Vols. 1–6. St. Petersburg, 1911–1913.
Bronshtein, M. P., et al. *Atomnoe iadro*. Moscow-Leningrad, 1934.
Bruevich, N. G., ed. *220 let AN SSSR. Spravochnaia kniga*. Moscow-Leningrad, 1945.
Bukharin, N. I. "Leninizm i problema kul'turnoi revoliutsii." *Narodnoe prosveshchenie*, no. 2 (1928): 3–20.
———. "Rekonstruktivnyi period i NIR." *Biulleten' NIS*, no. 2 (1930): 1–7.
Deborin, A. M. *Vvedenie v filosofiiu dialekticheskogo materializm*. Moscow, 1922.
———. *Dialektika i estestvoznanie*. Moscow-Leningrad, 1929.
———. *Lenin i krizis noveishei fiziki*. 2d edition. Leningrad, 1930.
Dzerzhinskii, F. E. *Osnovnye zadachi promyshlennosti i rol' Leningrada*. Leningrad, 1925.
———. *O khoziaistvennom stroitel'stve SSSR*. Moscow-Leningrad, 1926.
———. *Izbrannye proizvedeniia*. 3d edition, 2 vols. Moscow, 1977.
Egorshin, V. P. *Sovremennoe uchenie o stroenie materii*. Moscow-Leningrad, 1928.
———. *Estestvoznanie, filosofiia i marksizm*. Moscow, 1930.
Ehrenfest-Ioffe: Nauchnaia perepiska. Edited by V. Ia. Frenkel', Leningrad, 1978.
Einstein, A. *Efir i printsip otnositel'nosti*. Petrograd, 1921.
———. *O spetsial'noi i obshchei teorii otnositel'nosti*. Moscow, 1921.
———. *Osnovy teorii otnositel'nosti*. Petrograd, 1923.
Fersman, A. E. *Puti k nauke budushchego*. Petrograd, 1922.
Fiziko-mekhanicheskii fakul'tet Leningradskogo politekhnicheskogo instituta im. M. I. Kalinina. Leningrad, 1925.
Flakserman, Iu. N. *Promyshlennost' i nauchno-tekhnicheskie instituty*. Moscow, 1925.

————. *Znachenie nauchno-tekhnicheskikh uchrezhdenii dlia promyshlennosti*. Moscow, 1926.

Florinskii, Pavel. *Mnimosti v geometrii*. Moscow, 1922.

Fock, V. A. *Osnovy kvantovoi mekhaniki i granitsy ee prilozhimosti*. Moscow, 1936.

————. "Chto vnesla teoriia kvantov v osnovnye predstavleniia fiziki." *Uchenye zapiski Leningradskogo gosudarstvennogo universiteta. Seriia fizicheskikh nauk*, no. 2 (1936): 77–84.

Frederiks, V. K., and A. A. Friedmann. *Osnovy teorii otnositel'nosti*. Leningrad, 1924.

Frederiks, V. K., and D. D. Ivanenko, eds. *Printsip otnositel'nosti. Sbornik rabot klassikov reliativizma*. Moscow, 1935.

Frenkel, Ia. I. *Teoriia otnositel'nosti*. Petrograd, 1923.

————. *Elektrichestvo i materiia*. Moscow-Leningrad, 1925.

————. *Sobranie izbrannykh trudov*. 2 vols. Moscow-Leningrad, 1958.

————. *A. F. Ioffe*. Leningrad, 1968; written in 1948.

————. *Na zare novoi fiziki*. Leningrad, 1970.

Friedmann, A. A. *Izbrannye trudy*. Moscow, 1966.

Gaas, Artur (Arthur Haas). *Volny materii i kvantovaia mekhanika*. Translated by P. S. Tartakovskii. Moscow-Leningrad, 1930.

Gamov, George. *My World Line*. New York: Viking Press, 1970.

Gessen, B. M. *Osnovnye idei teorii otnositel'nosti*. Moscow-Leningrad, 1928.

————. "K voprosu o probleme prichinnosti v kvantovoi mekhanike." Introduction to *Volny materii i kvantovaia mekhanika*, by A. Gaas. Moscow-Leningrad, 1930.

God raboty tsentral'noi komissii po ulushcheniiu byta uchenykh pri SNK. Moscow, 1922.

Gol'dgammer, D. A. *Protsessy zhizni v mertvoi prirode*. Moscow, 1903.

————. *Novye idei v sovremennoi fizike*. Kazan, 1910.

Gor'kii i nauka: Stat'i rechi, pis'ma, vospominaniia. Edited by F. N. Petrov. Moscow, 1964.

Gorky, Maxim. *Untimely Thoughts*. Translated and with an introduction by H. Ermolaev. New York: P. S. Eriksson, 1968.

Ioffe, A. F. *Stroenie veshchestva*. Petrograd, 1919.

————, ed. *Osnovaniia novoi kvantovoi mekhaniki*. Moscow-Leningrad, 1927.

————. *The Physics of Crystals*. New York: McGraw Hill, 1928.

————. *Dostizheniia fiziki*. Moscow, 1928.

————. *Fizika za desiat' let*. Moscow, 1928.

————. "Korennye problemy NIR." In *Sotsialisticheskaia rekonstruktsiia i NIR*, 23–26. Moscow, 1930.

————. *Novye problemy nauchno-issledovatel'skoi raboty v fizike*. Moscow-Leningrad, 1931.

————. *Perspektivy ispol'zovaniia novykh vidov energii vo vtoroi piatiletke*. Moscow-Leningrad, 1932.

————. *Moia zhizn' i rabota*. Moscow-Leningrad, 1933.

————. *Elektronnye poluprovodniki*. Leningrad-Moscow, 1933.

————. *Atomnoe iadro segodnia*. Moscow, 1934.

————, ed. *K otchetnomu dokladu Akademika A. F. Ioffe*. Moscow-Leningrad, 1936.

————. *Nekotorye problemy sovremennoi fiziki*. Moscow, 1941.

————. *Izbrannye trudy*. 2 vols. Leningrad, 1974.

————. *O fizike i fizikakh*. Leningrad, 1977.

————. *Vstrechi s fizikami*. Leningrad, 1983.

Ioffe, A. F., and A. K. Val'ter. *Nad chem rabotaiut sovetskie fiziki*. Moscow, 1930.

Ipatieff, V. N. *Nauka i promyshlennost' na zapade i v Rossii*. Petrograd, 1923.

————. *The Life of a Chemist*. Edited by X. Eudin, H. D. Fisher, and H. H. Fisher. Stanford, 1946.

Ipatieff, V. N., F. E. Dzerzhinskii, et al. *Nauchnye dostizhenii v promyshlennosti i raboty NTO*. Moscow, 1925.

Itogi desiat' let. Moscow-Leningrad, 1927.

K otchetnym dokladam Akademika S. I. Vavilova i D. S. Rozhdestvenskogo. Ob ikh nauchnoi rabote i rabotakh GOI na sessii AN SSSR 14–20 marta 1936 g. Moscow-Leningrad, 1936.

Kasterin, K. P. *Obobshchenie osnovykh uravnenii aerodinamiki i elektrodinamiki*. Moscow, 1937.

Kessenikh, V. N. "Nauchno-tekhnicheskie itogi 5 let raboty SFTI." *Trudy Siberskogo fizikogo nauchno-issledovatel'skogo instituta pri Tomskom gosudarstvennom universitete*, no. 3 (1934): 3–15.

Khvol'son, O. D. *Fizika nashikh dnei*. 2d edition. Moscow-Leningrad, 1929.

Kol'man, Ernst. *Noveishie otkrytiia sovremennoi fizike v svete dialekticheskogo materializma*. Moscow, 1943.

Kravets, T. P. *P. N. Lebedev i sozdannaia im fizicheskaia shkola*. Moscow, 1913.

Krzhizhanovskii, G. M., et al., eds. *Nauka i tekhnika SSSR, 1917–1927*. Moscow, 1927.

————. *Nauchnye rabotniki i sotsialisticheskoe stroitel'stvo*. Moscow, 1928.

————. *Izbrannoe*. Moscow, 1957.

Kuibyshev, V. V. *Kto ne snami, to protiv nas*. Moscow-Leningrad, 1930.

————. *Nauke-sotsialistcheskii plan*. Moscow, 1931.

————. *Izbrannye proizvedeniia*. Moscow, 1958.

Kul'turnoe stroitel'stvo SSSR v tsifrakh (1930–1934 gg.). Moscow, 1934.

Kurchatov, I. V. *Rassheplenie atomnogo iadra*. Moscow-Leningrad, 1935.

————. *Elektricheskaia prochnost'*. Moscow, 1936.

————. *Rassheplenie iader neitronami*. Moscow, 1936.

Lapirov-Skoblo, M. Ia. *Puti i dostizhenie russkoi nauki i tekhniki za 1918–23 g.* Petrograd, 1923.

————. *Perspektivnye plany nauchno-issledovatel'skikh uchrezhdenii promyshlennosti*. Moscow-Leningrad, 1931.

Lazarev, P. P. *P. N. Lebedev i russkaia fizika*. Moscow, 1912.

————. *Chelovecheskii organizm kak fizicheskaia mashina*. Moscow, 1922.

————, ed. *Sovremennye problemy estestvoznaniia*. Moscow, 1922.

Lazarev, P. P., and I. P. Pavlov. *Biofizika: sbornik statei po istorii biofiziki v SSSR* Moscow, 1940.

Leipunskii, A. I. "UkFTI." In *Nauchno-tekhnicheskoe obsluzhivanie tiazheloi promyshlennosti*. Moscow-Leningrad, 1934, pp. 48–55.

Lenin i Akademiia nauk. Sbornik doumentov. Edited by P. N. Pospelov. Moscow, 1969.

Leningrad i Leningradskaia oblast' v tsifrakh. Leningrad, 1974.

Leningrad v tsifrakh. Leningrad, 1936.

Leningrad za 50 let. Leningrad, 1967.

Leningradskaia fiziko-tekhnicheskaia laboratoriia. Teplotekhnicheskii otdel. Leningrad, 1928.

Lermantov, V. V. "Ocherk istorii razvitiia fizicheskoi laboratorii imperatorskago S. Peterburgskago universiteta pod rukovodstvom prof. F. F. Petrushevskago, 1865–1903." N.p., n.d.

Lukirskii, P. I. *O fotoeffekte.* Leningrad, 1933.

————. *Neitron.* Leningrad, 1935.

Maksimov, A. A. *Lenin i estestvoznanie.* Moscow-Leningrad, 1933.

————. *Ocherki po istorii bor'by za materializm v russkom estestvoznanii.* Moscow, 1947.

Maksimov, A. A., V. F. Mitkevich, and S. I. Vavilov. *"Materializm i empiriokrititsizm" Lenina i sovremennaia fizika.* Moscow, 1939.

Maksvell, D. K. (James Clerk Maxwell). *Rechi i stat'i.* Edited by V. F. Mitkevich. Moscow-Leningrad, 1940.

Mandel'shtam, L. I. *Lektsii po optike, teorii otnositel'nosti i kvantovoi mekhanike.* Moscow, 1972.

Matematika i estestvoznanie v SSSR. Ocherki razvitiia matematicheskikh i estestvennykh nauk za dvadtsat' let. Moscow-Leningrad, 1938.

Mekhanisticheskoe estestvoznanie i dialekticheskii materializm. Vologda, 1925.

Mikhel'son, V. A. *Rashirenie i natsional'naia organizatsiia nauchnykh issledovanii v Rossii.* Moscow, 1916.

Mitkevich, V. F. *Osnovnye fizicheskie vozzreniia.* 3d edition. Moscow-Leningrad, 1939.

Morozov, N. *Printsip otnositel'nosti i absoliutnoe.* Moscow, 1920.

Mysovskii, L. V. *Novye idei v fizike atomnogo iadra.* Moscow-Leningrad, 1935.

Nauchno-issledovatel'skie instituty i vysshie uchebnye zavedeniia SSSR v 1932 godu. Moscow, 1932.

Nauchno-issledovatel'skie instituty NKTP. Edited by A. A. Armand. Moscow-Leningrad, 1935.

Nauchno-issledovatel'skie uchrezhdeniia i nauchnye rabotniki SSSR. Vol. 3. Moscow, 1934.

Nauchnye kadry i nauchno-issledovatel'skie uchrezhdeniia SSSR. Edited by O. Iu. Smidt and B. Ia. Smulevich. Moscow, 1930.

Nauka i nauchnye rabotniki SSSR. Nauchnye uchrezhdeniia Leningrada. Part 2. Leningrad, 1926.

Nauka i nauchnye rabotniki SSSR, Petrograd. Edited by S. F. Ol'denburg. Vol. 1. Petrograd, 1920.

Nauka i nauchnye rabotniki SSSR, Moskva. Vol. 4. Moscow, 1925.

Nauka i nauchnye rabotniki SSSR, Nauchnye uchrezhdeniia Leningrada. Part 2. Leningrad, 1926.

Nauka i nauchnye rabotniki SSSR, Nauchnye rabotniki SSSR bez Moskvy i Leningrada. Leningrad, 1928.

Nauka, tekhnika, propaganda. Sbornik NISa i tekhpropa k XVII vsesoiuznoi konferentsii VKP (b). Moscow, 1932.

Nauka XX veka. Edited by I. E. Tamm. 2 vols. Moscow-Leningrad, 1928–1929.

Nauka v Rossii. Spravochnik ezhegodnik Petrograda. Petrograd, 1920.

NKTP NIS SSSR. *Planirovanie i operativnyi uchet v nauchno-issledovatel'skikh institutakh promyshlennosti.* Moscow, 1932.

NTO. *Otchet o deiatel'nosti za 1922–23 gg.* Moscow, 1923.

Obreimov, I. V. *Sostoianie veshchestva.* Petrograd, 1922.

Ol'denburg, S. F. *Rossiiskaia akademiia nauk v 1922 g.* Leningrad, 1923.

―――. *Rossiiskaia akademiia nauk v 1923 g.* Leningrad, 1924.

―――. *Rossiiskaia akademiia nauk v 1924 g.* Leningrad, 1925.

―――. *Akademiia nauk SSSR v 1925 g.* Leningrad, 1926.

―――. *Akademiia nauk SSSR v 1926 g.* Leningrad, 1927.

―――. *Akademiia nauk SSSR v 1928 g.* Leningrad, 1929.

―――. *Polozhenie nashei nauki sredi nauki mirovoi.* Leningrad, 1928.

Ordzhonikidze, G. K. *Izbrannye stat'i i rechi.* Moscow, 1939.

―――. *Stat'i i rechi.* 2 vols. Moscow, 1956.

Ostrovskii, B., and S. Romm. *Nauka i sotsialisticheskoe stroitel'stvo.* Leningrad, 1934.

Pamiati Karla Marksa. Sbornik statei k piatidesiatiletiiu so gnia smerti. Leningrad, 1933.

Papaleksi, N. D. "Leonid Isaakovich Mandel'shtam." In L. I. Mandel'shtam, *Polnoe sobranie trudov,* 1: 7–66. Moscow, 1948.

Pervaia otchetnaia vystavka glavnauki narkomprosa. Moscow, 1925.

Petrov, F. N., ed. *Desiat' let sovetskoi nauki.* Moscow-Leningrad, 1927.

―――. *65 let v riadakh leninskoi partii.* Moscow, 1962.

Piat' let raboty tsentral'noi komissii po uluchsheniiu byta uchenykh pri sovete narodnykh kommissarov RSFSR (TsEKUBU), 1917–1926. Moscow, 1927.

Piatiletnii plan rabot gosudarstvennoi fiziko-tekhnicheskoi laboratorii. Vol. 20 of *Piatiletnii plan nauchno-eksperimental'noi raboty v sviazi s rekonstruktsiei promyshlennosti SSSR.* Moscow, 1929.

Piatiletnii plan raboty gosudarsvennogo fiziko-tekhnicheskogo instituta, 1928/29–1932. Leningrad, n.d.; typewritten, most likely 1930.

Problemy sovremennoi fiziki v rabotakh FTIa Akademika A. F. Ioffe. Moscow-Leningrad, 1936.

Rais, Dzhems (James Rice). *Printsip otnositel'nosti.* Translated and with an introduction by Ia. I. Frenkel'. Moscow-Leningrad, 1928.

Reports of the Physico-Technical Roentgen Institute and the Leningrad Physico-Technical Laboratory, 1918–1926. Leningrad, 1926.

Roentgen, V. K. *O novom rode luchei.* Edited and with an introduction by A. F. Ioffe. Moscow-Leningrad, 1933.

Russkaia (Rossiiskaia) assotsiatsiia fizikov. *Trudy s"ezda fizikov v petrograde (s 4-go po 7-oe fevralia 1919 goda).* Petrograd, 1919.

Russkaia assotsiatsiia fizikov. *S"ezd RAF.* In *Soobshcheniia o nauchno-tekhnicheskikh rabotakh v respublike,* 3: 158–206. Moscow, 1921.

———. *Vtoroi s"ezd RAF.* Kiev, 1921.

———. *Tretii s"ezd RAF. Otchety i protokoly.* Nizhni Novgorod, 1922. Reprinted from *Telegrafiia i telefoniia bez provodov,* no. 16: n.d.

———. *Trudy s"ezda fizikov v petrograde.* Vol. 1. Petrograd, 1919.

———. *Trudy tret'ego s"ezda RAF (v nizhnem-novgorode 17–21 sent. 1922 g.).* Nizhni Novgorod, n.d.

———. *IV s"ezd russkikh fizikov v Leningrade (15–20 sentiabria 1924 godu). Perechen' dokladov.* Vol. 14 of *Soobshcheniia o nauchno-tekhnicheskikh rabotakh v respublike.* Leningrad, 1924.

———. *IV s"ezd russkikh fizikov. Raspisanie zaniatii.* Leningrad, 1924.

———. *V s"ezd russkikh fizikov. Perechen' dokladov.* Moscow-Leningrad, 1926.

———. *Programma zasedanii VI s"ezda RAF.* Leningrad, 1928.

———. *VI s"ezd russkikh fizikov. Perechen' dokladov.* Leningrad, 1928.

———. *Pervyi vsesoiuznyi fizicheskii s"ezd. Spravochnik dlia chlenov s"ezda.* Odessa, 1930.

Rudaev, Boris. *Na putiakh k materializmu xx veka.* Kharkov, 1927.

Rytov, S. M., et al. *Akademik L. I. Mandel'shtam. K 100-letiiu so dnia rozhdeniia.* Moscow, 1979.

Sbornik po prikladnoi fiziki. Vols. 1 and 2. Leningrad, 1925 and 1926.

Semenov, N. N. *Nauka i obshchestvo.* 2d edition. Moscow: Nauka, 1981.

Semkovskii, S. Iu. *Teoriia otnositel'nosti i materializm.* Kharkov, 1924.

———. *Dialekticheskii materializm i printsip otnositel'nosti.* Moscow-Leningrad, 1926.

Skvortsov-Stepanov, I. I. *Elektrifikatsiia SSSR.* Moscow, 1922.

———. *Istoricheskii materializm i sovremennoe estestvoznanie.* Moscow, 1925.

———. *Dialekticheskii materializm i deborinskaia shkola.* Moscow-Leningrad, 1928.

Sotsialisticheskaia rekonstruktsiia i nauchno-issledovatel'skaia rabota. Sbornik NIS PTEU VSNKh SSSR k XVI s"ezdu VKP (b). Moscow, 1930.

Sovremennaia kvantovaia mekhanika. Tri nobelevskikh doklada. Translated by D. D. Ivanenko. Moscow-Leningrad, 1934.

Spravochnik po nauchnym uchrezhdeniiam Leningrada. Leningrad, 1927.

Steklov, V. A. *V Ameriku i obratno.* Leningrad, 1925.

Svobodnaia assotsiatsiia dlia razvitiia i rasprostraneniia polozhitel'nykh nauk. *Rechi i privetstviia proiznesennye na trex publichnykh sobraniiakh sostoiavshikhsia v 1917 g. 9-go i 16-go aprelia v Petrograde i 11-go Maia v Moskve v 1917 g.* Petrograd, 1918.

Svodnyi katalog innostrannykh zhurnalov vypisannykh bibliotekami seti NKTP v 1935 g. Moscow-Leningrad, 1935.

Tamm, I. E. *Rentgenovskie luchi.* Moscow-Leningrad, 1927.

Tartakovskii, P. S. *Kvanty sveta*. Moscow-Leningrad, 1928.

——. *Noveishie techeniia v oblasti ucheniia ob atome i elektrone*. Moscow, 1929.

Tartakovskii, P. S., and M. Usanovich. *Evoliutsiia ucheniia o stroenii materii*. Moscow, 1930.

Tezisy dokladov i proekty resoliutsii II vsesoiuznaia konferentsiia po planirovaniiu nauchno-issledovatel'skoi raboty v tiazheloi promyshlennosti na vtorom piatiletie i na 1933 g. Moscow, 1932.

Timiriazev, A. K. *Kineticheskaia teoriia materii*. Moscow-Petrograd, 1923.

——. *Lenin i sovremennoe estestvoznanie*. Leningrad, 1924.

——. *Metodologiia estestvoznaniia u nas i na zapade*. Moscow, 1929.

——. *Dialektika v nauke*. Moscow, 1929.

——. *Vvedenie v teoreticheskuiu fiziku*. Moscow-Leningrad, 1933.

Timiriazev, K. A. *Nauka i demokratiia*. Leningrad, 1920.

Timoshenko, Stephen P. *As I Remember*. Translated by Robert Addis. Princeton: Van Nostrand, 1968.

Tseitlin, Z. A. *Nauka i gipoteza*. Moscow-Leningrad, 1926.

Val'ter, A. K. *Ataka atomnogo iadra*. Kharkov, 1933.

——. *Atomnoe iadro*. Moscow-Leningrad, 1935.

Vasil'ev, A. V. *Space, Time, Motion*. Translated by H. M. Lucas and C. P. Sanger, with an introduction by Bertrand Russell. London, 1924.

Vavilov, S. I. *Sobranie sochinenii*. 3 vols. Moscow, 1956.

——. *Lenin i fizika*. Moscow, 1960.

——. "Staria i novaia fizika." *IiMEN, Fizika* 3 (1965): 3–13.

Volgin, V. P. *AN SSSR za chetyre goda 1930–1933. Rechi i stat'i*. Leningrad, 1934.

Volgin, V. P., et al. *Planirovanie nauki i zadachi Komademii i Akademii Nauk*. Moscow-Leningrad, 1931.

Vospominaniia ob A. F. Ioffe. Edited by V. P. Zhuze. Leningrad: 1973.

Vospominaniia ob Akademike D. S. Rozhdestvenskom. Edited by S. E. Frish and A. I. Stazharov. Leningrad: 1976.

Vospominaniia o Ia. I. Frenkele. Edited by V. M. Tuchkevich. Leningrad, 1976.

Vospominaniia o I. E. Tamme. Edited by E. L. Feinberg et al. Moscow, 1981.

Vospominaniia o L. D. Landau. Edited by I. M. Khalatnikov et al. Moscow, 1988.

Wells, H. G. *Russia in the Shadows*. New York: George H. Doran, 1921.

II. SECONDARY SOURCES

A. DISSERTATIONS

1. Soviet

Fataliev, Kh. M. *K istorii bor'by za dialekticheskii materializm v sovetskoi fiziki*. Doctoral diss., Moscow State University, 1950.

Iangutov, R. S. *Stanovlenie sovetskoi fiziki i bor'ba za dialektiko-materia-*

listicheskoe mirovozzrenie v nei (1917–1925 gg.). Avtoreferat. Moscow, 1971.

Osinovskii, A. N. *Iz istorii sviazi sovetskoi fiziki s proizvodstvom (po materialam deiatel'nosti Akademika D. S. Rozhdestvenskogo i ego shkoly, 1918–1940 gg.).* Candidate's diss., Krupskaia Pedagogical Institute. Moscow, 1965.

Pechkurova, K. E. *Partiinoe rukovodstvo podgotovkoi nauchnykh kadrov v gody pervoi piatiletki (1928–1932 gg.) (na materialakh Leningrada).* Candidate's diss., Leningrad State University, 1976.

Zaparov, I. P. *Sozdanie i deiatel'nost' tsentral'noi komissii po uluchsheniiu byta uchenykh (1919–1925 gg.).* Candidate's diss., Lenin State Pedagogical Institute. Moscow, 1979.

2. Other

Balzer, Harley D. *Educating Engineers: Economic Politics and Technical Training in Tsarist Russia.* Ph.D. diss., University of Pennsylvania, 1980.

Goldberg, Stanley. *The Early Response to Einstein's Special Theory of Relativity, 1905–1911: A Case Study of National Differences.* Ph. D. diss., Harvard University, 1968.

Sopka, Katherine. *Quantum Physics in America, 1920–1935.* New York: Ayre Publishing Co., 1980. Originally published as author's Ph.D. diss., Harvard University, 1976.

B. Books, Pamphlets, Articles

Artsimovich, L. A., et al., eds. *Razvitie fiziki v SSSR.* 2 vols. Moscow, 1967.

Bodash, Lawrence. *Kapitza, Rutherford, and the Kremlin.* New Haven: Yale University Press, 1985.

Bailes, Kendall. "The Politics of Technology: Stalin and Technocratic Thinking Among Soviet Engineers." *American Historical Review* 79 (April 1974): 445–469.

———. *Technology and Society Under Lenin and Stalin.* Princeton: Princeton University Press, 1978.

———. "The American Connection: Ideology and the Transfer of American Technology to the Soviet Union, 1917–1941." *Comparative Studies in Society and History,* 23, no. 3 (July 1981): 421–448.

Bastrakova, M. S. *Stanovlenie sovetskoi systemy organizatsii nauki (1917–1922).* Moscow, 1973.

———. "Organizatsionnye tendentsii russkoi nauki v nachale XX v." In *Organizatsiia nauchnoi deiatel'nosti,* edited by E. A. Beliaev et al. Moscow, 1968.

Bessarab, Maiia. *Landau. Stranitsy zhizni.* Moscow, 1978.

Beyerchen, Alan. *Scientists Under Hitler.* New Haven: Yale University Press, 1977.

Brown, Laurie M., and Lillian Hoddeson, eds., *The Birth of Particle Physics.*

Cambridge and New York: Cambridge University Press, 1983.

Bunge, Mario, and William Shea, eds. *Rutherford and Physics at the Turn of the Century*. Kent, England: Dawson; and New York: Science History Publications, 1979.

Coben, Stanley. "The Scientific Establishment and the Transmission of Quantum Mechanics to the United States, 1919–1932." *American Historical Review*, 76 (April 1971): 442–466.

Conquest, Robert. *The Great Terror*. New York: Collier Books, 1968.

————. *The Harvest of Sorrow: Soviet Collectivization and the Terror-Famine*. New York and Oxford: Oxford University Press, 1986.

Delokarov, K. Kh. *Metodologicheskie problemy kvantovoi mekhaniki v sovetskoi filosofskoi nauke*. Moscow, 1982.

Dorfman, Ia. G. *Istoriia vozniknoveniia sovremennoi iadernoi fiziki*. Moscow, 1954.

————. *Vsemirnaia istoriia fiziki*. Moscow, 1979.

Dukel'skii, V. M. "A. F. Ioffe." In *Sbornik posviashchennyi semidesiatiletiiu Adademika A. F. Ioffe*. 5–30. Moscow, 1950.

Edge, David, and Michael Mulkay. *Astronomy Transformed: The Emergence of Radioastronomy in Britain*. New York: Wiley, 1976.

Esakov, V. D. *Sovetskaia nauka v gody pervoi piatiletki*. Moscow, 1971.

Fediukin, S. A. *Privlechnie burzhuaznoi tekhnicheskoi intelligentsii k sotsialisticheskomu stroitel'stvu v SSSR*. Moscow, 1960.

————. *Sovetskaia vlast' i burzhuaznye spetsialisty*. Moscow, 1965.

Fitzpatrick, Sheila. *The Commissariat of Enlightenment*. Cambridge, England: Cambridge University Press, 1970.

————, ed. *Cultural Revolution in Russia, 1928–1931*. Bloomington: Indiana University Press, 1978.

————. *Education and Social Mobility in the Soviet Union, 1921–1934*. Cambridge and New York: Cambridge University Press, 1979.

Forman, Paul. "Weimar Culture, Causality and The Quantum Theory, 1918–1927: Adaptation by German Physicists and Mathemathicians to a Hostile Intellectual Environment." In *Historical Studies in the Physical Sciences*, edited by Russell McCormmach, 3 (1971): 1–115.

————. "The Financial Support and Political Alignment of Physicists in Weimar Germany." *Minerva* 12, no. 1 (1974): 39–66.

Forman, Paul, John Heilbron, Spencer Weart. *Physics circa 1900. Vol. 5 of Historical Studies in the Physical Sciences*, edited by Russell McCormmach. Princeton: Princeton University Press, 1975.

Frenkel', V. Ia. *Iakov Il'ich Frenkel'*. Moscow-Leningrad, 1966.

————. *Paul' Erenfest*. Moscow, 1977.

Frenkel, V. Ia., and A. P. Grinberg. *I. V. Kurchatov v FTIe*. Leningrad, 1984.

Frieden, Nancy. *Russian Physicians in an Era of Reform and Revolution*. Princeton: Princeton University Press, 1981.

Frish, S. E. "Vospominanie o Leningradskom Universitete v pervye gody posle revoliutsii (1918–1924 gg.)." In *Ocherki po istorii Leningradskogo Universiteta*, edited by V. V. Makarova, 3: 85–117. Leningrad, 1976.

Gorbunov, N. P. "Velikie plany razvitiia nauki i tekhniki." In *V. I. Lenin vo glave velikogo stroitel'stva*, edited by V. Z. Drobizhev, 172–186. Moscow, 1960.

———. *Kak rabotal Lenin*. Moscow, 1959.

Gorelik, G. E. "Dva portreta." *Neva*, no. 8 (1989): 167–173.

———. "Naturfilosofskie problemy fiziki v 1937 gody." *Priroda*, no. 2 (1990): 93–102.

Graham, Loren. "Bukharin and the Planning of Science." *Slavic Review* 23, no. 2 (April 1964): 135–148.

———. *The Soviet Academy of Sciences and the Communist Party, 1927–1932*. Princeton: Princeton University Press, 1967.

———. *Science, Philosophy, and Human Behavior in the Soviet Union*. New York: Columbia University Press, 1987.

———. "The Formation of Soviet Research Institutes: A Combination of Revolutionary Innovation and International Borrowing." *Social Studies of Science* 5, no. 3 (August 1975): 303–330.

———. "The Socio-political Roots of Boris Gessen: Soviet Marxism and the History of Science." *Social Studies of Science* 15, no. 4 (November 1985): 705–722.

Guroff, Gregory. "The Red-Expert Debate." In *Entrepreneurship in Imperial Russia and the Soviet Union*, edited by Gregory Guroff and Fred Carstensen, 201–222. Princeton: Princeton University Press, 1983.

Gustafson, Thane. "Why Doesn't Soviet Science Do Better Than It Does?" In *The Social Context of Soviet Science*, edited by Linda Lubrano and Susan Solomon, 31–67. Boulder: Westview Press, 1980.

Hanson, Norwood Russell. *The Concept of the Positron*. Cambridge: Cambridge University Press, 1963.

Herosige, Tetu. "The Ether Problem, the Mechanistic Worldview, and the Origins of the Theory of Relativity." Vol. 7 of *Historical Studies in the Physical Sciences*, 3–82. Princeton: Princeton University Press, 1976.

Hiebert, Erwin. "The State of Physics at the Turn of the Century." In *Rutherford and Physics at the Turn of the Century*, edited by Mario Bunge and William Shea, 3–22. Kent, England: Dawson; and New York: Science History Publications, 1979.

Holton, Gerald. *The Scientific Imagination: Case Studies*. Cambridge and New York: Cambridge University Press, 1978.

Istoriia Akademiia Nauk SSSR. Edited by K. V. Ostrovitianov et al., Vol. 2. Moscow-Leningrad, 1964.

Ivanov, N. I. *Issledovanie russkikh fizikov v elektromagnitnoi teorii sveta (II polivina XIX veka)*. Ulan Ude, 1968.

Ivanova, L. V. *Formirovanie sovetskoi nauchnoi intelligentsii (1917–1927 gg.)*. Moscow, 1980.

Johnston-Nicholson, Heather. "Autonomy and Accountability of Basic Research." *Minerva* 15, no. 1 (Spring 1977): 32–61.

Joravsky, David. *Soviet Marxism and Natural Science, 1917–1931*. New York: Columbia University Press, 1961.

————. *The Lysenko Affair*. Cambridge, Mass.: Harvard University Press, 1970.

Josephson, Paul. "Science Policy in the Soviet Union, 1917–1927." *Minerva* 26, no. 3 (Autumn 1988): 342–369.

————. "Physics and Soviet-Western Relations in the 1920s and 1930s." *Physics Today* 41, no. 9 (September 1988): 55–56.

————. "The Early Years of Soviet Nuclear Physics." *Bulletin of the Atomic Scientists* 43, no. 10 (December 1987): 36–39.

Kevles, Daniel J. " 'Into Hostile Camps': The Reorganization of Science in World War I." *ISIS* 62, pt. 1, no. 211 (Spring 1971): 47–60.

————. *The Physicists*. New York: Vintage, 1979.

Klein, Martin J. *Paul Ehrenfest*. Amsterdam: North-Holland Publishing Co., 1970.

Kliaus, E. M., U. I. Frankfurt, and A. M. Frenk. *Nil's Bor*. Moscow, 1977.

Kogan, V. S. *Kirill Dmitrievich Sinel'nikov*. Kiev, 1984.

Kohlstedt, Sally Gregory. "Institutional History." *Osiris*. 2d series, no. 1 (1985): 17–36.

Kokin, Lev. *Iunost' akademikov*. Moscow, 1970.

Kol'tsov, A. V. *Lenin i stanovlenie Akademii Nauk kak tsentra sovetskoi nauki*. Leningrad, 1969.

Komkov, G. D., et al. *Akademiia Nauk SSSR—shtab sovetskoi nauki*. Moscow, 1968.

Kononkov, A. F. "Moskovskoe fizicheskoe obshchestvo im. P. N. Lebedeva." *IiMEN, Fizika*, 3: 270–273. Moscow, 1965.

Kononkov, A. F., and A. N. Osinovskii. "Sozdanie sovetskoi optiki i deiatel'nost' D. S. Rozhdestvenskogo." In *Ocherki po istorii sovetskoi nauki i kul'tury*, edited by V. V. Mikheeva, 113–135. Moscow, 1969.

Kozlov, B. I. *Organizatsiia i razvitie otraslevykh nauchno-issledovatel'skikh institutov Leningrada, 1917–1977*. Leningrad, 1979.

Kozlov, V. V. *Ocherki istorii khimicheskikh obshchestv SSSR*. Moscow, 1958.

Kudriavtsev, P. S. *Kurs istorii fiziki*. Moscow, 1974.

————. "A. K. Timiriazev kak populiarizator i istorik nauki." *IiMEN, Fizika* 15: 248–259. Moscow, 1974.

Kupaigorodskaia, A. P. *Vysshaia shkola Leningrada v pervye gody sovetskoi vlasti (1917–1925)*. Leningrad, 1984.

Lampert, Nicholas. *The Technical Intelligentsia and the Soviet State*. New York: Holmes and Meier Publisher, 1979.

Lazarev, P. P. *Ocherki istorii russkoi nauki*. Moscow-Leningrad, 1950.

Lebin, B. D., ed. *Ocherki istorii organizatsii nauki v Leningrade, 1703–1977 gg.* Leningrad, 1980.

Lemberg, I. Kh., V. O. Naidenov, and V. Ia. Frenkel'. "Tsiklotron fiziko-tekhnicheskogo instituta im. A. F. Ioffe AN SSSR," *UFN* 153, no. 3 (November 1987): 497–519.

Levshin, B. V. *Akademiia Nauk SSSR v gody velikoi otechestvennoi voiny (1941–1945 gg.)* Moscow, 1966.

Lewis, Robert. *Science and Industrialization in the USSR*. New York: Holmes

& Meier Publishers, 1979.

Lubrano, Linda, and Susan Solomon, eds. *The Social Context of Soviet Science*. Boulder: Westview Press, 1980.

Maier, Charles S. "Between Taylorism and Technocracy: European Ideologies and the Vision of Industrial Productivity in the 1920s." *Journal of Contemporary History* 5 (1970): 27–41.

Mendelsohn, Everett. "The Emergence of Science as a Profession in 19th Century Europe. In *The Management of Scientists*, edited by Karl Hill, 3–48. Boston, 1963.

Mikulinskii, S. R., and A. P. Iushkevich, eds. *Razvitie estestvoznania v Rossii*. Moscow, 1977.

Mokshin, S. I. *Sem' shagov po zemle*. Moscow, 1972.

Morrell, J. B. "The Chemical Breeders: Leibig and Thomson." *Ambix* 19 (1972): 1–46.

Mott, N. F., ed. *The Beginnings of Solid State Physics*. London, 1980.

Odinetz, D. M., P. J. Novgorotsev, and P. N. Ignatiev. *Russian Schools and Universities in the World War*. New Haven: Yale University Press, 1929.

Parrott, Bruce. *Information Transfer in Soviet Science and Engineering*. Santa Monica: Rand Corporation, 1981.

————. *Politics and Technology in the Soviet Union*. Cambridge, Mass.: MIT Press, 1983.

Pfetsch, Frank. "Scientific Organization and Science Policy in Imperial Germany, 1871–1914: The Formation of the Imperial Institute of Physics and Technology." *Minerva* 8 (October 1970): 557–580.

Piatidesiatiletnii iubelei FTI (Materialy sessii). Leningrad, 1971.

Piksanov, N. K. *Gor'kii i nauka*. Moscow-Leningrad, 1948.

Predvoditelev, A. S., and B. I. Spasskii, eds. *Razvitie fiziki v Rossii*. 2 vols. Moscow, 1970.

Priadkin, K. K. *Khar'kovskii fiziko-tekhnicheskii institut*. Kiev, 1978.

Rassweiler, Anne. *The Generation of Power: The History of Dneprostroi*. New York and Oxford: Oxford University Press, 1988.

Reingold, Nathan, ed. *The Sciences in the American Context: New Perspectives*. Washington, D.C.: Smithsonian Institute Press, 1979.

Riutova-Kemoklidze, Margarita. "Priezzhaite. Einshtein vas primet. . . " *Siberskie ogni* no. 1 (1989): 116–129; and "Einshtein vas primet. . . Okonchanie," *Siberskie Ogni*, no. 2 (1989): 111–129.

Rogger, Hans. "*Amerikanizm* and the Economic Development of Russia." *Comparative Studies in Society and History* 23, no. 3 (July 1981): 382–420.

Rostotskaia, N. M., ed. *Uchenie o radioaktivnosti*. Moscow, 1973.

Scott, Otto. *The Creative Ordeal: The Story of Raytheon*. New York: Atheneum, 1974.

Shpol'skii, E. V. *Ocherki po istorii sovetskoi fiziki, 1917–1967*. Moscow, 1969.

Shuleikin, V. V., et al., eds. *Sbornik posviashchennyi pamiati Akademika P. P. Lazareva*. Moscow, 1956.

Sladkevich, N. G., et al. *Ocherki po istorii Leningradskogo universiteta*. Vol. 1. Leningrad, 1962.

Slonimkii, A. G. A. M. *Gor'kii v bor'be za sozdanie sovetskoi intelligentsii v gody inostrannoi voennoi interventsii i grazhdanskoi voiny*. Stalinabad, 1956.

Sobolev, G. L. *Uchenye Leningrada v gody velikoi otechestvennoi voiny, 1941– 1945*. Moscow-Leningrad, 1966.

Sochor, Zenovia, A. *Revolution and Culture: The Bogdanov-Lenin Controversy*. Ithaca: Cornell University Press, 1988.

Sominskii, M. A. *F. Ioffe*. Moscow-Leningrad, 1965.

Stites, Richard. *Revolutionary Dreams*. New York and Oxford: Oxford University Press, 1989.

Stogov, V. V., ed. *Ukazatel' soderzhanie 'ZhRFKhO' 1873–1930. Chast' fizicheskaia*. Moscow, 1960.

Timiriazev, A. K., et al., ed. *Ocherki po istorii fiziki v Rossii*. Moscow, 1949.

Tuchkevich, V. M., and V. Ia. Frenkel', eds. *A. F. Ioffe Physico-Technical Institute*.

————, eds. *Fizika tverdogo tela*. Leningrad, 1978.

————, eds. *Vklad Akademika A. F. Ioffe v stanovlenie iadernoi fiziki v SSSR*. Leningrad, 1980.

Tugarinov, I. A. "VARNITSO: Akademiia nauk SSSR (1927–1937 gg.)." *Voprosy istorii estestvoznaniia i tekhniki*, no. 4 (1989): 46–55.

Ul'ianovskaia, V. A. *Formirovanie nauchnoi intelligentsii v SSSR, 1917–1937*. Moscow, 1966.

Uspenskaia, N. V. "Vreditel'stvo . . . v dele izucheniia solnechnogo zatmeniia." *Priroda*, no. 8 (1989): 86–101.

Vizgin, V. P., and G. E. Gorelik. "The Reception of the Theory of Relativity in Russia and the USSR." Translated by Paul Josephson. In *The Comparative Reception of Relativity*, edited by Thomas Glick, 265–326. Dordrecht: Reidel, 1987.

Vucinich, Alexander. *Science in Russian Culture: A History to 1861*. Stanford: Stanford University Press, 1963.

————. *Science in Russian Culture, 1861–1917*. Stanford: Stanford University Press, 1973.

————. *Social Thought in Tsarist Russia: The Quest for a General Science of Society, 1861–17*. Chicago: University of Chicago Press, 1976.

————. *Empire of Knowledge*. Berkeley, Los Angeles, Oxford: University of California Press: 1984.

Weart, Spencer. "The Physics Business in America, 1919–1940: A Statistical Reconnaissance." In *The Sciences in the American Context: New Perspectives*, edited by Nathan Reingold, 295–358. Washington, D.C.: Smithsonian Institution Press, 1979.

Weiner, Charles. "Physics in the Great Depression." *Physics Today* 23, no. 10 (October 1970): 31–38.

————. "1932—Moving into the New Physics." *Physics Today*, 25, no. 5 (May 1972): 40–49.

————. "Institutional Settings for Scientific Change: Episodes from the History of Nuclear Physics." In *Science and Values*, edited by Arnold Thackray and Everett Mendelsohn, 187–212. Boston, 1974.

————. "Cyclotrons and Internationalism: Japan, Denmark and the United States, 1935–1945." In *Proceedings of the XIVth International Congress of the History of Science, 1974*, 353–365. Tokyo, 1975.

Weiner, Douglas R. *Models of Nature: Ecology, Conservation and Cultural Revolution in Soviet Russia*. Bloomington: Indiana University Press, 1988.

Weissberg, Alexander. *The Accused*. Translated by Edward Fitzgerald. New York: Simon and Schuster, 1951.

Wetter, Gustav. *Dialectical Materialism*. Translated by Peter Heath. New York: F. A. Praeger, 1959.

Wolfe, Bertram. *Three Who Made a Revolution*. 2d edition. Boston: Beacon Press, 1960.

Index

Academy of Sciences (Imperial, Russian, and Soviet), 10, 11–13, 19, 34, 37, 48, 59, 60, 67, 68, 79, 108, 135, 150, 152–153, 169, 176, 187, 190, 191, 192, 198–199, 236, 248, 258, 259–260, 261–262, 274, 276, 279, 290–291, 318; Association of Physico-Mathematical Sciences of, 155; and early Soviet science policy, 56–57; and Marxist study circles, 206; move from Leningrad to Moscow, 276–277; and physics research, 280–286; and special March 1936 session, 269, 295–305, 317
Ageev, N. V., 338
Agol, I. I., 210
Akadtsentr, 67
Akulov, N. S., 156
Aleksandrov, A. P., 4, 160, 286, 292, 293, 338
Aleksandrov, P., 113
Alferov, Zh. I., 338
Alikhan'ian, A. I., 180, 339
Alikhanov, A. I., 180, 293, 338
Alikhanov, brothers, 178, 180
Ambartsumian, V. A., 314–315
American Relief Administration (ARA), 54
Andreev, N. N., 125, 156, 203, 265, 270, 338
Andronov, A. A., 223
Anri, V. A., 78, 79, 80
Arkad'ev, V. K., 16, 203
Arkharov, V. I., 339
Armand, A. A., 285–286, 289, 300
Arsen'eva, A. N., 160, 173
Artsimovich, L. A., 1, 208, 286, 289, 293, 319, 338
aspirantura (graduate education), 194–198, 202; and Academy of Sciences, 198–199; and LFTI, 200–202, 287–289, 342
Atomic Commission, 52, 60, 89, 90–91, 220

Basov, N. G., 1, 304, 333
Baumgart, K. K., 45
Beria, L. P., 210, 312, 322–323
Berg, L. S., 53
Berkeley, University of California at, 83, 117, 180
Bernstein, S. N., 33
BINT (Biuro inostrannoi nauki i tekhniki), 107, 111
Blackett, P. M. S., 176
Blokhintsev, D. I., 250–51, 256, 259, 261, 266, 282
Bobkovskii, S. A., 293
Bogdanov, A. A., 46, 186, 226, 229, 249
Bogdanov, P. A., 65
Bogoslovskii, M. M., 92
Bohr, Niels, 76, 80, 83, 89, 112, 136, 164, 175, 181, 219, 223, 246, 269, 270, 282, 306, 307, 314
Boltzman, L., 32, 260
Bonch-Bruevich, V. D., 49–50
Borgman, I. I., 20, 21, 22, 32, 33
Born, Max, 93, 114, 115, 135, 136, 221, 238, 260, 269, 292, 306, 311
Borodin, I. I., 13, 42
Boskovich, R., 218, 230
Bothe, W., 176
Bragg, W. H., and W. L., 92, 110
Braun, C. F., 222
Brillouin, L., 115, 135, 136
Broglie, Louis de, 135, 137, 195, 221, 223, 266
Bronshtein, M. P., 116, 173, 176, 210, 212, 215, 251, 266, 270, 282, 292, 296, 309, 314–315
Bruskin, 286
Budnitskii, D. Z., 156, 211
Bukharin, N. I., 144, 149, 169, 232; on cultural revolution, 188–189; on planning of R and D, 151
Bulgakov, N. A., 35
Bunin, Ivan, 42

413

Recent and Selected Studies of the Harriman Institute

1905 in St. Petersburg: Labor, Society, and Revolution by Gerald D. Surh, Stanford University Press, 1989.

The Menshevik Leaders in the Russian Revolution: Social Realities and Political Strategies by Ziva Galili, Princeton University Press, 1989.

Soldiers in the Proletarian Dictatorship: The Red Army and the Soviet Socialist State, 1917–1930 by Mark von Hagen, Cornell University Press, 1990.

Moscow, Germany, and the West from Khrushchev to Gorbachev by Michael Sodaro, Cornell University Press, 1991.

Folklore for Stalin by Frank Miller, M. E. Sharpe, 1991.

Thinking Theoretically About Soviet Nationalities, edited by Alexander J. Motyl, Columbia University Press, 1991.

Geographic Perspective on Central Asia, edited by Robert Lewis, Unwin Hyman, 1991.

Odessa 1905: Workers in Revolt, by Robert Weinberg, Indiana University Press, 1991.

Designer: U.C. Press Staff
Compositor: Asco Trade Typesetting Ltd., Hong Kong
Text: 10/13 Palatino
Display: Palatino
Printer: Braun-Brumfield, Inc.
Binder: Braun-Brumfield, Inc.